Microbial Biofilms

Current Research and Applications

Edited by

Gavin Lear

Lincoln University
Christchurch
New Zealand

and

Gillian D. Lewis

University of Auckland
Auckland
New Zealand

Caister Academic Press

Caister Academic Press
Norfolk, UK

www.caister.com

British Library Cataloguing-in-Publication Data
A catalogue record for this book is available from the British Library

ISBN: 978-1-904455-96-7

Cover image: A computer-generated visualization of bacterial microcolony formation. Designed by Andrew Dopheide, 2011.

Printed and bound in Great Britain

Contents

Contributors

Pierre-Yves Ancion
School of Biological Sciences
The University of Auckland
Auckland
New Zealand

panc002@aucklanduni.ac.nz

Joan Artigas
Ecologie Microbienne des Hydrosystemes Anthropises (UR MALY)
CEMAGREF
Lyon
France

joan.artigas-alejo@cemagref.fr

Harsh P. Bais
Department of Plant and Soil Sciences
University of Delaware;
Delaware Biotechnology Institute
Newark, DE
USA

bais@dbi.udel.edu

James D. Bryers
Department of Bioengineering
University of Washington
Seattle, WA
USA

jbryers@u.washington.edu

Katja Buehler
Department of Biochemical and Chemical Engineering
Laboratory of Chemical Biotechnology
TU Dortmund University
Dortmund
Germany

katja.buehler@bci.tu-dortmund.de

Mette Burmølle
Department of Biology
University of Copenhagen
Copenhagen
Denmark

burmolle@bio.ku.dk

Shanika A. Crusz
School of Molecular Medical Sciences
University Park
University of Nottingham
Nottingham
UK

shanika.crusz@nottingham.ac.uk

Stephen P. Diggle
School of Molecular Medical Sciences
University Park
University of Nottingham
Nottingham
UK

steve.diggle@nottingham.ac.uk

Andrew Dopheide
School of Biological Sciences
The University of Auckland
Auckland
New Zealand

a.dopheide@auckland.ac.nz

G.A. Clark Ehlers
School of Biological Sciences
The University of Auckland
Auckland
New Zealand

c.ehlers@auckland.ac.nz

Matthew P. Fletcher
School of Molecular Medical Sciences
University Park
University of Nottingham
Nottingham
UK

matthew.fletcher@nottingham.ac.uk

Steve Flint
Institute of Food Nutrition and Human Health
Massey University
Palmerston North
New Zealand

s.h.flint@massey.ac.nz

Giulio Gazzola
Microbial Electrochemistry Group
Dublin City University
Dublin
Ireland

giulio.gazzola@gmail.com

Robert J. Goldstone
School of Molecular Medical Sciences
University Park
University of Nottingham
Nottingham
UK

r.j.goldstone@gmail.com

Rainer Gross
Department of Biochemical and Chemical Engineering
Laboratory of Chemical Biotechnology
TU Dortmund University
Dortmund
Germany

ragr@gmx.ch

Seratna Guadarrama
Microbial Electrochemistry Group
Dublin City University
Dublin
Ireland

seratna.guadarrama@gmail.com

Souichiro Kato
Hashimoto Light Energy Conversion Project
ERATO, JST
Bunkyo-ku, Tokyo
Japan

kato@light.t.u-tokyo.ac.jp

Annelise Kjøller
Department of Biology
University of Copenhagen
Copenhagen
Denmark

ak@bio.ku.dk

Atsushi Kouzuma
Hashimoto Light Energy Conversion Project
ERATO, JST
Bunkyo-ku, Tokyo
Japan

kouzuma@light.t.u-tokyo.ac.jp

Amutha Sampath Kumar
Department of Plant and Soil Sciences
University of Delaware;
Delaware Biotechnology Institute
Newark, DE
USA

amutha@udel.edu

Venkatachalam Lakshmanan
Department of Plant and Soil Sciences
University of Delaware;
Delaware Biotechnology Institute
Newark, DE
USA

venki@udel.edu

Gavin Lear
Department of Soil and Physical Sciences
Faculty of Agriculture and Life Sciences
Lincoln University
Christchurch
New Zealand

gavin.lear@lincoln.ac.nz

Gillian D. Lewis
School of Biological Sciences
The University of Auckland
Auckland
New Zealand

gd.lewis@auckland.ac.nz

Enrico Marsili
Microbial Electrochemistry Group
Dublin City University
Dublin
Ireland

enrico.marsili@dcu.ie

Koichi Nishio
Department of Applied Chemistry
The University of Tokyo
Bunkyo-ku, Tokyo
Japan

nishi@light.t.u-tokyo.ac.jp

Gabriele Pastorella
Microbial Electrochemistry Group
Dublin City University
Dublin
Ireland

gabriele.pastorella@dcu.ie

Roman Popat
School of Molecular Medical Sciences
University Park
University of Nottingham
Nottingham
UK

romanpopat@gmail.com

Kelly Roberts
School of Biological Sciences
The University of Auckland
Auckland
New Zealand

k.roberts@auckland.ac.nz

Anna M. Romaní
Institute of Aquatic Ecology
University of Girona
Girona
Spain

anna.romani@udg.edu

Andreas Schmid
Department of Biochemical and Chemical Engineering
Laboratory of Chemical Biotechnology
TU Dortmund University
Dortmund
Germany

andreas.schmid@bci.tu-dortmund.de

Jo Smith
School of Biological Sciences
The University of Auckland
Auckland
New Zealand

jsmi067@aucklanduni.ac.nz

Søren J. Sørensen
Department of Biology
University of Copenhagen
Copenhagen
Denmark

sjs@bio.ku.dk

Susan J. Turner
School of Biological Sciences
The University of Auckland
Auckland
New Zealand

s.turner@auckland.ac.nz

Vidya Washington
School of Biological Sciences
The University of Auckland
Auckland
New Zealand

v.washington@auckland.ac.nz

Kazuya Watanabe
Hashimoto Light Energy Conversion Project
ERATO, JST
Bunkyo-ku, Tokyo
Japan;
Research Center for Advanced Science and Technology
The University of Tokyo
Meguro-ku, Tokyo
Japan

kazuyaw@toyaku.ac.jp

Gideon Wolfaardt
Department of Chemistry and Biology
Ryerson University
Toronto
Canada

gwolfaar@ryerson.ca

Irene Ylla
Institute of Aquatic Ecology
University of Girona
Girona
Spain

irene.ylla@gmail.com

Preface

Biofilms are the default mode of life for many, if not most, bacterial species. Despite this, most microbial investigations to date have focused on planktonic cultures of bacterial cells. Reasons for this include the relative ease with which certain planktonic forms of bacteria can be grown in liquid media, and the better level of experimental control and replication provided by these relatively homogeneous mixes of microbial cells. However, the physical, chemical and biological complexity of biofilm-associated communities has major implications for the functioning of these highly diverse communities. The three-dimensional structure of biofilm provides the associated microbial communities with additional protection from predation, toxic substances and physical perturbation. The variety of microniches provided by the biofilm also promotes a huge diversity of microbial life and metabolic potential. These complex and highly structured communities provide essential ecosystem services that help to maintain the health of soils and waters, where they are significant and essential basal components of intricate food webs. The unique features of biofilm-associated microbial communities are now being capitalized on at commercial scales. For example, the metabolic potential of biofilm communities is increasingly used to degrade and transform toxic substances or excessive nutrient loads within soil and water, for the commercial production of a wide variety of chemical substances, and even for the generation of electricity. It is well recognized that the complex structures and processes provided by microbial biofilms are by no means accidental, but the result of a complicated set of interactions which include communication and cooperation among the many members of most biofilm-associated populations. We are improving our knowledge of how these complex communities work together to develop and maintain their biofilm mode of life. As a consequence we are becoming better equipped to manipulate the colonization and development of biofilm-based infections that currently cause harm to millions of humans annually, and reduce the quality and yield of vital crop plants and animals. Better knowledge of biofilm formation is also being used to reduce the enormous economic impact of both biofouling and microbially induced corrosion, which effect vital industrial processes around the globe, including those related to food manufacture and the treatment and disposal of wastewaters.

In this book we provide an up-to-date review of the latest scientific research that is revolutionizing our understanding of these fascinating, surface-associated microbial communities, and predict future trends and growth areas in biofilm-related research.

Gavin Lear and Gillian D. Lewis

Current Books of Interest

Small DNA Tumour Viruses	2012
Extremophiles: Microbiology and Biotechnology	2012
Bacillus: Cellular and Molecular Biology	2012
Microbial Biofilms: Current Research and Applications	2012
Bacterial Glycomics: Current Research, Technology and Applications	2012
Non-coding RNAs and Epigenetic Regulation of Gene Expression	2012
Brucella: Molecular Microbiology and Genomics	2012
Molecular Virology and Control of Flaviviruses	2012
Bacterial Pathogenesis: Molecular and Cellular Mechanisms	2012
Bunyaviridae: Molecular and Cellular Biology	2011
Emerging Trends in Antibacterial Discovery: Answering the Call to Arms	2011
Epigenetics: A Reference Manual	2011
Metagenomics: Current Innovations and Future Trends	2011
Nitrogen Cycling in Bacteria: Molecular Analysis	2011
The Biology of Paramyxoviruses	2011
Helicobacter pylori	2011
Microbial Bioremediation of Non-metals: Current Research	2011
Lactic Acid Bacteria and Bifidobacteria: Current Progress in Advanced Research	2011
Viruses and Interferon: Current Research	2011
Essentials of Veterinary Parasitology	2011
Hepatitis C: Antiviral Drug Discovery and Development	2011
Streptomyces: Molecular Biology and Biotechnology	2011
Alphaherpesviruses: Molecular Virology	2011
Recent Advances in Plant Virology	2011
Vaccine Design: Innovative Approaches and Novel Strategies	2011
Salmonella: From Genome to Function	2011
PCR Troubleshooting and Optimization: The Essential Guide	2011
Insect Virology	2010
Environmental Microbiology: Current Technology and Water Applications	2011
Sensory Mechanisms in Bacteria: Molecular Aspects of Signal Recognition	2010
Bifidobacteria: Genomics and Molecular Aspects	2010
Molecular Phylogeny of Microorganisms	2010
Nanotechnology in Water Treatment Applications	2010
Iron Uptake and Homeostasis in Microorganisms	2010
Caliciviruses: Molecular and Cellular Virology	2010
Epstein-Barr Virus: Latency and Transformation	2010
Anaerobic Parasitic Protozoa: Genomics and Molecular Biology	2010
Lentiviruses and Macrophages: Molecular and Cellular Interactions	2010
Microbial Population Genetics	2010
Borrelia: Molecular and Cellular Biology	2010
Influenza: Molecular Virology	2010

Quorum Sensing and Social Interactions in Microbial Biofilms

1

Robert J. Goldstone, Roman Popat, Matthew P. Fletcher, Shanika A. Crusz and Stephen P. Diggle

Abstract

It is now well recognized that population of bacteria from many Gram-positive and Gram-negative species cooperate and communicate to perform diverse social behaviours including swarming, toxin production and biofilm formation. Communication between bacterial cells involves the production and detection of diffusible signal molecules and has become commonly known as quorum sensing (QS). In addition, an evolutionary perspective on QS illuminates important phenomena which help in understanding the prevalence and diversity of QS phenotypes and strategies under various conditions. The research fields of QS and biofilm formation often overlap with a number of studies demonstrating that QS is an important regulatory mechanism of biofilm formation in a variety of bacterial species. However, in contrast, there are conflicting reports, demonstrating that QS appears to play a minimal role in the development of biofilms. Our aim in this review is to highlight the key findings with respect to QS and the subsequent impact on biofilm formation. We also discuss QS and cooperation in the context of social evolution and how this may impact on the development and maintenance of microbial biofilms.

What is quorum sensing?

Quorum sensing (QS) is a widely used term to describe cell-to-cell signalling between bacterial cells in many Gram-negative and Gram-positive species. QS involves the accumulation of signalling molecules (Fig. 1.1) in the surrounding environment which enables a single cell to sense the number of bacteria (density) and therefore the population as a whole can make a coordinated response. The signal produced regulates its own production (autoinduction) and so this leads to a positive feedback response and greatly increased signal production. At critical densities, the binding of a regulator protein to the signal leads to the switching on of genes controlled by QS and a coordinated population response (Williams *et al.*, 2007; Ng and Bassler, 2009).

Work on cell-to-cell signalling in bacteria primarily began in the 1960s and an early classic QS experiment was described in 1970 by Nealson and colleagues, who demonstrated that addition of spent culture supernatants of the marine luminescent bacterium *Vibrio fischeri* to low density cultures of the same organism, induced bioluminescence production owing to the presence of an autoinducing substance (Nealson *et al.*, 1970).

The autoinducer responsible for the regulation of bioluminescence was subsequently identified as *N*-(3-oxohexanoyl) homoserine lactone (3-oxo-C6-HSL) which is an *N*-acyl homoserine lactone (AHL) molecule (Eberhard *et al.*, 1981). The structural and regulatory genes necessary for bioluminescence and 3-oxo-C6-HSL production were later identified and termed the *lux* regulon (Engebrecht *et al.*, 1983). This regulon is organized into two divergently transcribed operons. The leftward operon comprises the *luxR* gene which encodes the transcriptional regulator protein LuxR. The rightward operon consists of six genes arranged as *luxICDABE*. The *luxI* gene

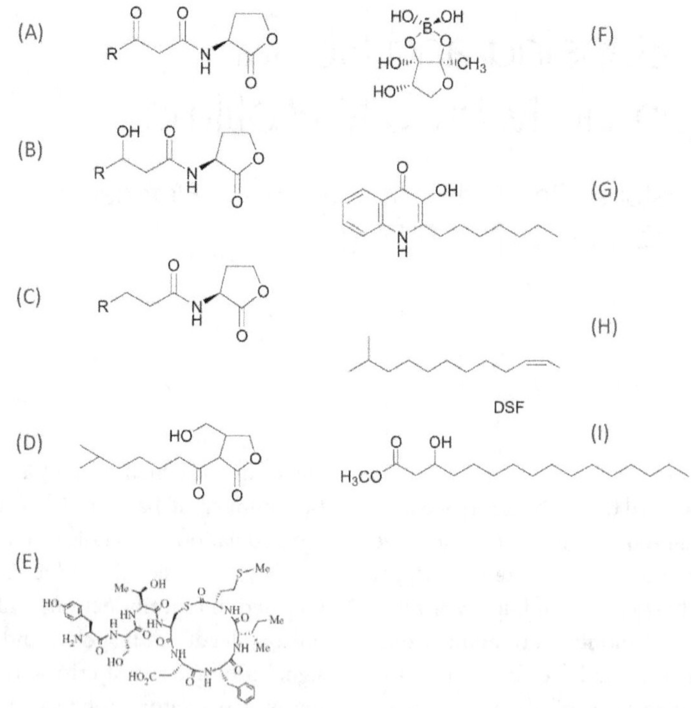

Figure 1.1 Quorum sensing signalling molecules described in Gram-negative and Gram-positive species. (A) 3-Oxo-AHL, *N*-(3-oxoacyl)homoserine lactone. (B) 3-Hydroxy-AHL, *N*-(3-hydoxyacyl)homoserine lactone. (C) AHL, *N*-acylhomoserine lactone, where R ranges from C1 to C15. (D) A-factor, 2-isocapryloyl-3-hydroxy-methyl-g-butyrolactone. (E) Autoinducing peptide-1 (AIP-1) from *Staphylococcus aureus*. (F) AI-2, autoinducer-2, furanosyl borate ester form. (G) PQS, *Pseudomonas* quinolone signal, 2-heptyl-3-hydroxy-4(1H)-quinolone. (H) DSF, 'diffusible factor', methyl dodecenoic acid. (I) PAME, hydroxyl-palmitic acid methyl ester.

encodes an autoinducer synthase responsible for the synthesis of 3-oxo-C6-HSL. The *luxCDABE* genes are involved in generating the products required for the luciferase reaction and the induction of bioluminescence.

It is now known that similar QS systems are widespread in Gram-negative bacteria. Early work in this area revealed that the production of the β-lactam antibiotic, 1-carbapen-2-em-3-carboxylic acid (carbapenem) by the terrestrial plant pathogen *Erwinia carotovora* was also regulated by 3-oxo-C6-HSL (Bainton *et al.*, 1992b), suggesting that AHL-based signalling may be common in Gram-negative bacteria and not restricted to marine *Vibrio* spp. The generation of plasmid-based AHL-biosensors to detect AHL molecules from spent culture supernatants confirmed this and revealed that *Pseudomonas aeruginosa, Serratia marcescens, Erwinia herbicola, Citrobacter freundii, Enterobacter agglomerans* and *Proteus mirabilis* all produced AHLs (Bainton *et al.*, 1992a). Since this early work, many other Gram-negative species have been shown to produce AHL molecules and have homologues of the LuxI and LuxR proteins from *V. fischeri* (Williams *et al.*, 2007; Ng and Bassler, 2009).

QS signalling is not restricted to Gram-negative bacteria, a number of Gram-positive bacteria have been shown to employ small, modified oligopeptides as extracellular signalling molecules. These peptides activate gene expression by interacting with two-component histidine protein kinase signal transduction systems (Kleerebezem *et al.*, 1997). For example, in *Staphylococcus aureus,* the expression of a number of density-dependent virulence factors are controlled by the global regulatory locus *agr* (accessory gene regulator) (Novick and Geisinger, 2008).

A third class of quorum sensing signal molecule was first identified in the marine bacterium *Vibrio harveyi.* Bioluminescence in this organism is regulated by AHLs and a molecule termed autoinducer-2 (AI-2) which is a furanosyl borate diester produced by the enzyme LuxS (Chen *et al.*, 2002). The identification of the *luxS* gene required for the production of AI-2 production (Surette *et al.*, 1999) sparked an exponential increase into AI-2 signalling research. The reason being that the *luxS* gene can be found in a wide variety of bacterial genera (Winzer *et al.*, 2002). Importantly, representatives of both Gram-negative and Gram-positive bacteria carry this particular gene and consequently, AI-2 production has been demonstrated in many species. This has led to the hypothesis that AI-2 is employed as an interspecies signalling molecule, although its role as a signal between species has been questioned (Diggle *et al.*, 2007).

There are other molecules that have been described in individual species as QS signals. For example, *P. aeruginosa* possesses a non-AHL based QS signalling system based upon 2-alkyl-4-quinolone (AQ) signalling (Pesci *et al.*, 1999). The primary signal molecules of this system are 2-heptyl-3-hydroxy-4(1*H*)-quinolone, known as the Pseudomonas quinolone signal (PQS) and 2-heptyl-4(1*H*)-quinolone (HHQ) (Pesci *et al.*, 1999; Déziel *et al.*, 2004). AQ production is intricately linked into the AHL QS system of *P. aeruginosa* (Dubern and Diggle, 2008).

QS has been shown to regulate a number of diverse behaviours and phenotypes in bacteria including antibiotic production, exotoxin production, virulence, competence and motility. The importance of QS to a bacterium can be seen when studying *P. aeruginosa.* In this organism, the QS system has been estimated to regulate at least 6% of the genome (Hentzer *et al.*, 2003; Schuster *et al.*, 2003; Wagner *et al.*, 2003), which is a possible reason why *P. aeruginosa* is so highly adaptable and able to inhabit a diverse range of environmental niches. One of the most complex social behaviours performed by microbes is the development of biofilms and research examining the role of QS on biofilm development has been performed in a number of bacterial species.

Evidence for quorum sensing in biofilms

There is an increasing body of evidence linking QS with biofilm formation in a number of species of bacteria. Detecting QS activity in biofilm communities can be achieved by either looking for the production of QS signals within biofilms, or by determining whether genes involved in QS regulatory circuits are expressed in biofilms.

AHLs have been detected in a variety of natural and clinical settings in association with bacterial biofilms. For example, AHLs were isolated from natural biofilms formed on limestone rocks from the San Marcos River in Texas (McLean *et al.*, 1997). Based on colony morphology, these rocks were colonized by at least 30 bacterial species, including the AHL producer *Pseudomonas putida.* Since then, AHLs have been detected in marine biofilm communities, such as those produced by *Roseobacter* spp. in association with marine-snow (Gram *et al.*, 2002) and *Roseobacter* and *Vibrio* spp. associated with Australian coastal sponges (Taylor *et al.*, 2004), microbial mats (Bachofen and Schenk, 1998) and also subtidal marine biofilms grown in Petri dishes off the Hong Kong shore (Huang *et al.*, 2007). Additionally, *V. fischeri* activates the production of bioluminescence in the light organs of a squid via the LuxR/I QS system; and recently the community of bacteria in this light organ has been suggested to be a biofilm (Karatan and Watnick, 2009).

Evidence for QS activity in clinically relevant biofilms has also been obtained. AHLs have been detected *in situ* and *in vitro* in biofilms growing on urinary catheters, where 14 isolated *P. aeruginosa* strains were found to produce AHLs (Stickler *et al.*, 1998). *P. aeruginosa* strains isolated from the cystic fibrosis (CF) lung, where biofilms are thought to play an important role in pathogenesis, were shown to produce AHLs *in vitro* (Singh *et al.*, 2000). In addition, AHLs have also been isolated directly from CF sputum (Erickson *et al.*, 2002; Middleton *et al.*, 2002). However, whilst some studies make efforts to characterize the range of AHLs present by reliable, quantitative methods such as gas or liquid chromatography/mass spectrometry, caution must be taken when interpreting results from studies relying on AHL responsive biosensor assays since other compounds present in a very heterogeneous sample may result in false positives, such has been seen with cyclic dipeptides in *P. aeruginosa* (Holden *et al.*, 1999). Using gas chromatography and mass spectrometry, the presence of AHLs in biofilms has been measured *in vitro* and shows more than 600 μM 3-oxo-C12-HSL to be present in flow-cell grown *P. aeruginosa* biofilms (Charlton *et al.*, 2000). This study, using a method to detect 3-oxo-AHL species, also found significant levels of 3-oxo-C14-HSL in biofilms, which is not typically identified in flask grown cultures of *P. aeruginosa*.

The activity of QS genes has also been directly detected during the biofilm mode of growth. For *P. aeruginosa*, the expression of the AHL synthase genes *lasI* and *rhlI* have been investigated using a Green Fluorescent (*gfp*) reporter and it was shown that expression is maximal in cells at the biofilm substratum, decreasing with biofilm height (De Kievit *et al.*, 2001). The authors also found that whilst *lasI* expression is reduced over time, *rhlI* expression remains constant (De Kievit *et al.*, 2001). RNA transcripts for *lasI* and *lasR* have also been identified in CF sputum infected with *P. aeruginosa* (Erickson *et al.*, 2002) and a study by Sauer *et al.* used a *lacZ* reporter to measure the expression of *lasA* and *rhlA* as a measure of *las* and *rhl* QS activity. They found the *las* system to be more active during the early stages of biofilm attachment, while the *rhl* system appears to be more active during the maturation stage (Sauer *et al.*, 2002).

Finally, the importance of QS in biofilms is revealed by the use of compounds which can inhibit QS and also have an impact on the ability of bacteria to form biofilms. Halogenated furanone compounds similar to those produced by the red algae *Delisea pulchra* can interfere with AHL QS in *P. aeruginosa*, which affects maturation of the biofilm architecture, enhancing dispersal and causing the formation of substantially thinner biofilms (Hentzer *et al.*, 2002). These compounds appear to function by displacing the AHL from the receptor protein, or by increasing the turnover rate of the receptor (Manefield *et al.*, 1999, 2002). However, such furanone compounds have also been shown to affect biofilm formation in *Salmonella enterica* serovar Typhimurium (Janssens *et al.*, 2008) and *Escherichia coli* (Ren *et al.*, 2001), bacteria which are not thought to utilize AHL-based QS. In these species, the furanone compounds are thought to be interfering with the LuxS/AI-2 signalling system and in this way affecting biofilm formation. In both cases, the furanone may suppress QS by displacing the signal ligand from its cognate receptor (Manefield *et al.*, 1999; Ren *et al.*, 2001).

Effect of quorum sensing on biofilms

Whilst a number of studies have demonstrated the presence of AHLs in natural *in situ* or *in vivo* biofilms, the precise role of QS in regulating biofilm formation and maturation is not always clear. For example, *in vitro* biofilm studies have suggested a role for QS in biofilm formation in a range of bacterial species. However, some studies have found QS not to play a critical role in the formation of biofilms in natural and clinical settings. For instance, Schaber *et al.* found that biofilms formed by *P. aeruginosa* in acute burn wound infections in mice to be unaffected by mutation of the *lasI* and *rhlI* QS genes, despite the fact that the mutant strains were significantly less virulent than their wild-type counterpart (Schaber *et al.*, 2007). The authors speculated that the role of the biofilm may be to accumulate the high cell densities necessary for the activation of QS-regulated virulence products leading to systemic dissemination of infection (Schaber *et al.*, 2007).

However in many studies, QS does appear to play a role in regulating the formation of biofilms (Table 1.1). In several cases, including *P. aeruginosa* and a common co-habitant of CF patient lung infections, *Burkholderia cenocepacia*, QS does not appear to affect the initial attachment of cells to a surface, but rather the maturation of biofilm structure (Fig. 1.2A). Mutations in the QS systems of these bacteria result in the formation of flat, undifferentiated biofilms (Davies *et al.*, 1998; Huber *et al.*, 2001).

An AQ-based QS system in *P. aeruginosa* also affects biofilms. Addition of high concentrations of exogenous PQS to a culture of *P. aeruginosa* PAO1 enhances biofilm development (Diggle *et al.*, 2003). This effect may be due to an increased positive effect on the expression of the *lecA* gene,

Table 1.1 Overview of the use of QS in the regulation of biofilm formation by Gram-negative and Gram-positive bacteria

Species	Type of QS signalling	Effect on biofilm formation	Reference
Gram negative			
P. aeruginosa	PQS	Positive effect on architecture	Diggle (2006b)
P. aeruginosa	AHL	Positive effect on maturation	Davies *et al.* (1998)
P. aureofaciens	AHL	Negative effect on matrix production	Zhang and Pierson (2001)
B. cenocepacia	AHL	Positive effect on maturation	Huber (2001)
S. marcescens	AHL	Positive effect on maturation	Rice *et al.* (2005)
S. liquefaciens	AHL	Positive effect on maturation	Labbate *et al.* (2004)
A. hydrophila	AHL	Positive effect on maturation	Lynch *et al.* (2002)
Y. pseudotuberculosis	AHL	Negative effect on floc formation	Atkinson *et al.* (1999)
R. sphaeroides	AHL	Negative effect on floc formation	Puskas *et al.* (1997)
P. stewartii	AHL	Positive effect on attachment to surfaces, negative effect on floc formation, negative effect on matrix production	von Bodman *et al.* (1998); Koutsoudis *et al.* (2006)
V. anguillarium	AHL	Positive effect on attachment	Croxatto *et al.* (2002)
V. cholerae	AHL	Negative effect on matrix production; positive effect on dispersal	Karatan and Watnick (2009)
V. cholerae	AI-2	Negative effect on matrix production; positive effect on dispersal	Karatan and Watnick (2009)
S. enterica	AI-2	Positive effect on biofilm formation	Prouty *et al.* (2002)
E. coli	AI-2	Positive effect on biofilm maturation	Li *et al.* (2007)
H. influenzae	AI-2	Negative effect on biofilm formation	Daines *et al.* (2005)
Gram positive			
B. subtilis	Peptide	Positive effect on matrix production	Lopez *et al.* (2009)
L. monocytogenes	Peptide	Positive effect on attachment	Riedel *et al.* (2009)
E. faecalis	Peptide	Positive effect on biofilm growth	Nakayama *et al.* (2006)
S. mutans; S. gordonii	Peptid, peptide	Positive effect on matrix production, positive effect on biofilm formation	Loo *et al.* (2000), Zhang *et al.* (2009)
S. aureus; S. epidermidis	Peptide	Positive effect on dispersal	Yarwood *et al.* (2004)
S. epidermidis	AI-2	Negative effect on matrix production	Xu *et al.* (2006)
S. mutans	AI-2	Affects biofilm morphology	Yoshida *et al.* (2005)
B. cereus	AI-2	Positive effect on dispersal	Auger *et al.* (2006)

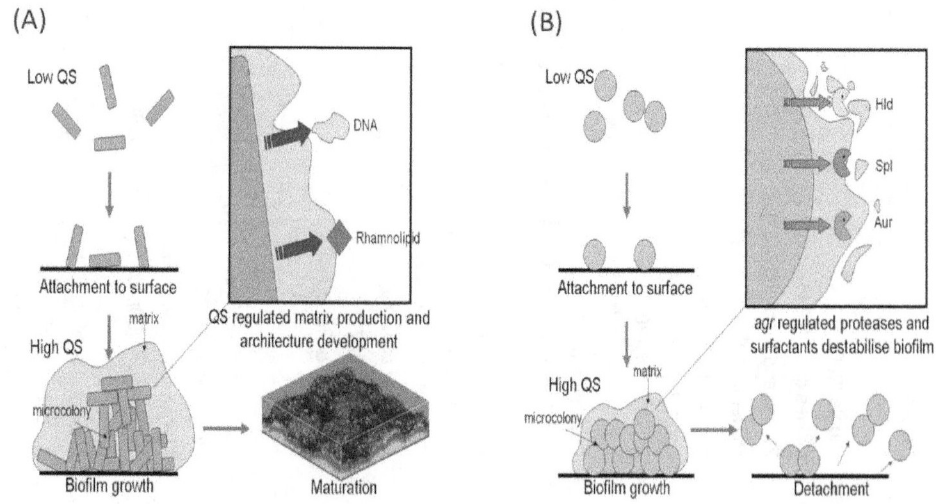

Figure 1.2 QS can be used to generate biofilm mass and architecture as in *P. aeruginosa* (A), and can also be used as a signal for the dispersal of cells from the biofilm as in *S. aureus* (B).

responsible for directing synthesis of a cytotoxic lectin. LecA binds galactose monomers in extracellular matrix (ECM) polysaccharides and has been shown to play an important role in maintaining biofilm architecture (Diggle *et al.*, 2006). In addition, biofilms contain extracellular DNA (eDNA) that can function as a cell-to-cell interconnecting matrix component in the biofilm ECM (Whitchurch *et al.*, 2002). *P. aeruginosa* biofilms also contain this eDNA, which presumably originates from the lysis of a subpopulation of cells. AQs have been linked to the release of eDNA into the extracellular environment and this may be another method by which they exert their positive effect on biofilm formation. The maximal release of eDNA from cells correlates with maximal PQS release in late log phase. An AQ-negative *pqsA* mutant culture releases low amounts of eDNA and correspondingly forms flat, thin, unstructured biofilms that contain little eDNA and a lower biomass than a wild-type strain (Allesen-Holm *et al.*, 2006).

For bacteria such as *Aeromonas hydrophila*, the deletion of the AHL synthase gene *ahyI*, responsible for the synthesis of both C4 and C6-HSL, results in the reduction of biofilm differentiation and the failure of the bacteria to form microcolonies (Lynch *et al.*, 2002). However, in this study, a mutant constructed in the AHL receptor gene *ahyR* does not share the same phenotype as the AHL synthase mutant. Indeed, biofilms formed by the *ahyR* mutant were shown to exhibit increased surface coverage (Lynch *et al.*, 2002). The effect of QS regulating the maturation stages of biofilm formation appears to be shared by *Serratia liquefaciens* and *Serratia marcescens*. In *S. liquefaciens*, a mutation in the AHL synthase gene *swrI* results in biofilm development which is apparently halted after 24 hours and does not reach the level of cellular differentiation typical of wild-type cells. The mutant biofilm has also been shown to be thinner and more homogeneous with less cellular aggregation (Labbate *et al.*, 2004). In *S. marcescens*, AHL-dependent QS affects biofilms grown in rich media similar to *S. liquefaciens*, with a signal synthase mutant forming a flat, undifferentiated biofilm which exhibits poorly attached cells. This is in contrast to the wild-type, which forms a highly filamentous-chain and aggregate cluster biofilms (Rice *et al.*, 2005).

Not all Gram-negative bacteria positively regulate biofilm formation using QS. High cell densities of *Vibrio cholerae* are known to inhibit biofilm formation (Karatan and Watnick, 2009), whilst studies have shown that *V. cholerae* 'locked' in a low-cell density mode, produce more pronounced biofilms (Hammer and Bassler, 2003). *Pseudomonas aureofaciens,* which possesses two AHL signalling systems, uses the CsaR/I QS system to repress the synthesis of biofilm matrix components and also induces a state of cellular aggregation or clumping (Zhang and Pierson, 2001). This process is shared by the plant pathogen *Pantoea stewartii.* In this organism, the formation of a biofilm in the plant xylem blocks the mass-transit of water, causing vascular wilt and the plant to die. The production of the matrix is repressed by the EsaR/I AHL QS system (von Bodman *et al.,* 1998), where a mutation in the AHL synthase gene *esaI* results in the formation of dense mats of bacteria whilst the wild-type adheres poorly to surfaces (Koutsoudis *et al.,* 2006). However, whilst the production of the matrix is uncontrolled, an *esaRI* double mutant is unable to attach to a surface (Koutsoudis *et al.,* 2006). It was also observed that bacterial clumping was induced in an *esaRI* double mutant (Koutsoudis *et al.,* 2006) and the development of a clumping phenotype has also been observed in response to the deletion of the *ypsR* AHL receptor gene in *Y. pseudotuberculosis,* however, a mutant in the AHL synthase gene *ypsI* does not exhibit the same effect (Atkinson *et al.,* 1999). Floc formation is also controlled by QS in the marine bacterium *Rhodobacter sphaeroides.* In this species, the deletion of the AHL synthase *cerI* gene results in the formation of bacterial aggregates in a way that can be reversed by the addition of an exogenous source of the CerI generated AHL 7,8-*cis*-*N*-(tetradecanoyl) homoserine lactone. In addition, this mutant produces more exopolysaccharide matrix than wild-type cells (Puskas *et al.,* 1997).

Taken together, this suggests a complex and species-specific role for AHL-dependent QS in the formation of biofilms. However, in spite of this, other studies have demonstrated that this conclusion may not be so clear-cut. In *P. aeruginosa,* the effect of AHL QS on biofilms appears to depend on other environmental factors such as the type of carbon source available (De Kievit *et al.,* 2001). Interestingly, it was shown that QS affected *P. aeruginosa* biofilms only in the presence of glucose and not citrate (De Kievit *et al.,* 2001). It has also been shown that biofilms formed by *P. aureofaciens* are affected by citrate concentration where mean biofilm thickness increases with increasing citrate but biofilm roughness increases with decreasing citrate (Heydorn *et al.,* 2000). In this way, the effect of QS on the formation of biofilms may be itself tightly regulated by other environmental cues such as nutrient availability.

In the Gram-positive bacteria, much research points to a critical role for QS in the regulation of the biofilm mode of growth. Bacteria in the genera *Staphylococcus* use the *agr* (accessory gene regulation) system to regulate biofilm formation, in a way that appears to repress biofilms or to induce the dispersal of bacteria (Yarwood *et al.,* 2004). This view of QS repressing biofilms is supported by a study of 105 *S. aureus* isolates in which 78% of *agr*-negative strains formed biofilms on polystyrene, compared with only 6% of *agr*-positive strains (Vuong *et al.,* 2000). The effect of the *agr* system repressing biofilms has been reinforced by the use of a *gfp* reporter, which showed activation of the *agr* system to occur in temporal waves in a subset of cells in the biofilm in a way that correlates with cellular detachment from the biofilm (Yarwood *et al.,* 2004) (Fig. 1.2B).

Another Gram-positive species in which QS and biofilm formation has been extensively studied is *Streptococcus mutans.* The *com* system is typically associated with the development of genetic competence in *S. mutans,* and appears to function optimally in biofilms (Li *et al.,* 2002). This QS system is well characterized and is regulated by the *comCDE* and *comAB* loci. These encode a peptide based QS system, controlled by a heptadecapeptide signal molecule (*comC*), also termed competence stimulating peptide (CSP), a mechanism for export (*comAB*) and an associated membrane receptor (*comD*) and response regulator (*comE*). Evidence suggests a *comC* signal mutant forms biofilms with altered architecture, being aggregated in mesh-like microcolonies, with large intercolony voids (Li *et al.,* 2002). These voids are thought to be due to detachment of cells from the microcolonies in the *comC* mutant and this effect could be rescued with exogenous supply of CSP. Mutants in *comD, comE* and *comX* all form biofilms with reduced biomass (Li *et al.,* 2002). The authors suggest the effect of *comC* could be related to an effect on cell division or septation, with the possibility that a second receptor for CSP acts to control this phenomenon while the *com* system positively affects biofilm growth. In support of this effect of CSP on biofilms, the *com* system of the related species *S. gordonii* affects biofilms, where a *comD* transposon mutant is defective in biofilm formation (Loo *et al.*, 2000). Additionally, the application of exogenous CSP to a *comC* mutant appears to increase the maturation of *S. mutans* biofilms and results in increased biomass, with cells forming large aggregates with abundant extracellular matrix (Zhang *et al.*, 2009). CSP also appears to affect cell viability (Qi *et al.*, 2005; Zhang *et al.*, 2009), inducing cell death in up to 50% of the population with CSP addition (Qi *et al.*, 2005; Zhang *et al.*, 2009). The release of DNA from these dead cells probably contributes to the enhancing effect of CSP on biofilm maturation, however CSP does not affect early stage biofilm development (Qi *et al.*, 2005; Zhang *et al.*, 2009). A final line of evidence regarding the influence of QS in biofilms in *S. mutans* comes from the observation that the *com* system also positively affects the acid-tolerance response in *S. mutans*. This acid-tolerance may be especially important in biofilms grown in sucrose-containing media, where the formation of thick biofilms is correlated with increased acid-resistance over planktonic cells (Li *et al.*, 2001).

While the Staphylococci appear to mainly use QS to repress biofilm formation and induce detachment of cells from a biofilm, many other Gram-positive species use QS to induce or promote the biofilm mode of growth. In *Listeria monocytogenes*, a mutation in the *agrD* gene, encoding the peptide signal is reduced in its ability to attach to PVC relative to a wild type, in a way that can be genetically complemented or by supplying wild-type supernatant (Riedel *et al.*, 2009). This function of *agrD* has also been shown to be similar to *agrA* (Rieu *et al.*, 2008). An increasing body of evidence suggests that for *L. monocytogenes*, the effect of the *agr* system is to induce the cells to bind to a surface, rather than being involved in the later stages of biofilm development and maturation. Rieu *et al.* (2008) found that whilst initial attachment was reduced in an *agrA* mutant, the propensity for *L. monocytogenes* to grow in a network of knitted chains of cells under flowing media conditions was unchanged. Expression of *agrB*, *agrC* and *agrD* do not seem to be controlled by growth phase in planktonic growth (Rieu *et al.*, 2008), though studies are conflicting over *agrA*, with the finding that *agrA* is expressed two-fold more in stationary phase than exponential (Autret *et al.*, 2003). However, during sessile growth, *agrB*, *agrC* and *agrD* appear to be expressed increasingly over the course of growth between 2 and 24 hours, but not after this point, indicating a plateau, while there was no difference in the expression of *agrA* (Rieu *et al.*, 2008). Similarly in *E. facealis*, the *fsr* peptide based QS system has also been shown to positively influence biofilm formation (Hancock and Perego, 2004; Nakayama *et al.*, 2006).

Quorum sensing regulation of biofilm genes

Underlying the conclusion that AHL-dependent QS affects biofilm formation is the characterization of some of the genetic pathways by which this regulation is accomplished. Probably the best example of this comes from *P. aeruginosa*. The rhl QS system was identified because of the effect that the *rhlI* and *rhlR* genes exert on the production of the biosurfacant rhamnolipid (Brint and Ohman, 1995; Ochsner and Reiser, 1995). Rhamnolipid production is well known to have consequences for biofilm architecture (Davey *et al.*, 2003), probably by mediating detachment and dispersal of cells from the biofilm (Boles *et al.*, 2005). Synthesis of rhamnolipid appears to be spatially and temporally localized, apparently occurring preferentially in the stalks rather than the caps of the biofilm 'mushroom' structures (Lequette and Greenberg, 2004). In addition to the RhlRI system and its effect on rhamnolipid synthesis, the *pel* operon is responsible for the synthesis of a glucose-rich polymeric component of the biofilm matrix. The expression of the *pel* genes has been shown to be reduced by mutation of *lasI* and *rhlI*, an effect which can be ameliorated by the addition of 3-oxo-C12-HSL (Sakuragi and Kolter, 2007). This could explain the apparent defect in biofilm development that some authors have reported as a consequence of mutation in QS genes.

In *S. liquefaciens*, QS has also been

heavily studied owing to the regulation by the SwrRI system on the synthesis of a lipopeptide biosurfactant, via an effect on the peptide synthetase gene *swrA*. This biosurfactant is involved in swarming motility (Lindum *et al.*, 1998) and it is possible that this regulatory relationship underpins the consequence of QS mutation on biofilm development. In addition to this, QS also regulates the genes *bsmA* and *bsmB*, which are involved in the development of the highly differentiated, filamentous biofilms characteristic of this species (Labbate *et al.*, 2004). However, these genes are also involved in biofilm formation independent of QS, when the surface colonized is biotic rather than abiotic (Labbate *et al.*, 2007).

For *V. cholerae*, the effect of QS in repressing biofilm formation appears to be dependent on the gene *hapR*, a transcriptional regulator which is activated by QS in this species (Karatan and Watnick, 2009). In certain strains of *V. cholerae*, high cell densities lead to the repression of matrix polysaccharide production via the activity of HapR. In addition, *hapR* activation may also lead to enhanced dispersal from the biofilm via the activation of the haemagglutinin/protease gene *hapA*. This protein also encourages detachment of *V. cholerae* from epithelial cells (Karatan and Watnick, 2009).

Interestingly, there may also be a degree of feedback from the biofilm mode of growth towards QS. In *P. aeruginosa*, the gene *algR*, which regulates the production of the biofilm matrix component alginate, has been shown to also bind to the promoter of *rhlI* and is thought to repress the expression of this AHL synthase, since a mutant in *algR* produces more C4-HSL and rhamnolipid and this is thought to occur in a biofilm-specific manner (Morici 2007). While the importance of AHL QS in biofilm formation has been relatively well studied, there is also an increasing body of research concerning the involvement of a second quorum sensing network controlled by the synthesis of a signal derived from 4,5-dihydroxy-2,3-pentanedione by the enzyme LuxS (Miller *et al.*, 2004). Mutational analysis has shown that this type of QS, termed autoinducer-2 (AI-2) signalling, activated biofilm formation in *Salmonella enterica* (Prouty *et al.*, 2002), but represses biofilm formation in *Haemophilus influenzae* (Daines *et al.*, 2005). It has also been shown that biofilm formation can be increased in *Escherichia coli* by the addition of exogenous AI-2 (González Barrios *et al.*, 2006). This study identified a key regulator in this process, termed MqsR, which mediates the effect of AI-2 via a positive effect on motility, controlled by the QscBC two-component system (González Barrios *et al.*, 2006). However, there appears to be conflict regarding the involvement of AI-2, as some studies have shown no effect of AI-2 signalling on the formation of biofilms by the *E. coli* strain K-12 W3110 (Wang *et al.*, 2005). In contrast, others have seen consequences on the architecture of the developing biofilm, owing to an effect of AI-2 signalling on the expression of the genes *wza*, which is responsible for the production of the important *E. coli* biofilm matrix component colanic acid, and fimbrial genes (Li *et al.*, 2007).

In the Gram-positive bacteria, for staphylococci, the effect of the *agr* system in inducing detachment from the biofilm may be related to the activation of the gene *hld*, encoded within the *agr* locus. The product of this gene, delta-haemolysin, has surfactant properties, and the exogenous addition of this molecule decreases the ability of *S. aureus* to attach to polystyrene (Vuong *et al.*, 2000). Another haemolysin produced by *S. aureus*, alpha-haemolysin encoded by the gene *hlaA*, has also been shown to affect biofilm formation (Caiazza and O'Toole, 2003) and is also under the direct control of the *agr* system. In addition to the haemolysins, *agr* in *S. aureus* also leads to the activation of the proteases aureolysin, Aur, and Spl, which are also known to promote dispersal from biofilms (Boles and Horswill, 2008). The *S. epidermidis* autolysin gene *atlE* is also repressed by *agr* (Kong *et al.*, 2006) and evidence suggests a major role for autolysin in facilitating the release of DNA. This eDNA may be a major component required for the early stages of biofilm attachment to both PVC and glass (Qin *et al.*, 2007). In contrast, the *atlE* homologue in *S. aureus* is not under the control of *agr* (Vuong *et al.*, 2000). In *S. aureus*, *atlE* is regulated by SarA, a virulence regulator which joins *agr* in regulating a range of virulence factors.

In addition to the staphylococci, some of the genetic mechanisms underlying the role of the *agr* system in biofilm formation in *L. monocytogenes* have recently been elucidated. The virulence regulator *prfA* has been shown to be expressed in a growth phase-dependent manner (Schwab *et al.*, 2005), and is also regulated by the *agr* system (Riedel *et al.*, 2009). This virulence regulator *prfA* has also been shown to promote biofilm formation (Lemon *et al.*, 2010). Similarly, in *Enterococcus faecalis*, an *agr*-type system termed the *fsr* system operates on a pheromone peptide encoded by the *fsrD* gene that has been termed the gelatinase biosynthesis activating pheromone (Nakayama *et al.*, 2006). This system is known to regulate the production of proteases including gelatinase *gelE*, while it has also been shown that mutants in *fsr* genes and *gelE* are reduced in their ability to form biofilms (Hancock and Perego, 2004; Nakayama *et al.*, 2006).

For *Bacillus subtilis*, the lipopeptide surfactin has recently been identified as a quorum sensing molecule and appears to influence biofilm formation by causing potassium leakage in the cell. This leakage leads to the activation of the kinase *kinC* and the expression of biofilm matrix component genes encoded by the *espA-O* and *yqxM-sipW-tsaA* operons (Chu *et al.*, 2008) (López *et al.*, 2009). In this species, the matrix genes are repressed by the regulator SinR, which is antagonized by SinI. The gene *sinI* is only expressed when KinC-mediated phosphorylation of the transcription factor SpoOA~*p* rises above a threshold level (Branda *et al.*, 2004; Fujita *et al.*, 2005; Kearns *et al.*, 2005). A bistable switch is known to govern the expression of *spoOA* so that this gene is only expressed in a subset of the population (López *et al.*, 2009). Correlated with this, *sinI* has also been shown to behave in this way, probably a result of *spoOA* expression (Chai *et al.*, 2008). Thus, matrix production appears to be a task allocated to a subset of cells. The nature of surfactin causing potassium leakage from the cell is an interesting type of QS, since this effect could be achieved by a variety of natural products, including surfactin. This may be an elegant way for the bacteria to respond to a variety of signals in the environment through biofilm formation (López *et al.*, 2009).

Like in some Gram-negative species, there is an emerging body of research regarding the role of AI-2 in the biofilms formed by some Gram-positive species. In *S. epidermidis*, a *luxS* mutant reveals that AI-2 suppresses biofilm formation by suppressing the production of the PIA polysaccharide and this effect can be complemented by adding culture filtrate containing AI-2 (Xu *et al.*, 2006). AI-2 also has a role in *S. mutans* biofilm formation. Studies have shown that a *luxS* mutant is disrupted in the appearance of the biofilm, which is smooth and confluent in the wild-type (WT), but granular with large gaps between aggregates of cells in the mutant (Merritt *et al.*, 2003). The *luxS* mutant expresses the glucosyltransferase genes *gtfB* and *gtfC* 5.5 and 3-fold higher than wild-type in mid-log phase of growth, but not in early or late growth. These genes are responsible for the production of the insoluble glucans in the biofilm matrix. In contrast, the soluble glucan glucosyltransferase gene *gtfD* is slightly repressed in the *luxS* mutant in mid-log phase (Yoshida *et al.*, 2005). The authors suggest that this effect on glucan biosynthesis may explain the differences in biofilm morphology (Yoshida *et al.*, 2005). However, recent evidence suggests that any role for AI-2 in the formation of biofilms by *S. mutans* is dependent on both the carbon source available, since biofilm formation in a *luxS* mutant has also been seen to be attenuated in sucrose containing media but not in glucose (Yoshida *et al.*, 2005) and on the substratum for biofilm formation. It has been shown that AI-2 production is increased following growth on hydroxyapatite versus PVC (Shemesh *et al.*, 2010). *Bacillus cereus*, unlike *B. subtilis*, contains an *lsrACDB*-like operon. In other species, these genes are used for the uptake and processing of AI-2. It has been shown that exogenously supplied AI-2 appears to suppress attachment and can induce the detachment of cells from a pre-formed biofilm (Auger *et al.*, 2006).

Despite this body of evidence, some authors still speculate that AI-2 may not operate as a true signal, but as an environmental cue (Diggle *et al.*, 2007a). This has been investigated in *Mycobacterium avium*, where AI-2 induces biofilm formation via the oxidative-stress response (Geier *et al.*, 2008). The surfactin peptide of *B. subtilis* could also be viewed in this way. These are important cautionary examples against the quick conclusion that any molecule deriving a cellular response in terms of biofilm formation is acting as a true measure of population density. The next section will clarify what is meant by a signal, and why it is important to distinguish between signals, cues and coercion.

Quorum sensing and cooperation in biofilms: a social evolution perspective

In a microbial biofilm, innumerable cells form complex structures in which each individual cell plays only a small role. Much of the rhetoric traditionally used to describe biofilms alludes to a coordinated and cooperative unit where the activity of each cell benefits the group. Upon closer inspection however, this is not always the case. First, the architectural features such as mushroom like structures, water channels and stratifications may be the result of chemical gradients, simple growth parameters and physiological adaptation to the micro-niche. These can arise with no coordination or communication between participant cells (Stewart and Franklin, 2008; Nadell *et al.*, 2008a). Second, when a behaviour is performed that confers a cost to the individual performing the behaviour and a benefit to the surrounding population, that behaviour is vulnerable to cheating strategies which do not contribute a benefit to the group (West *et al.*, 2006; Nadell *et al.*, 2008a). Here we define cooperation and communication in microbes. The conditions under which these could evolve will be explored and QS will be examined in this light. We will also discuss whether biofilm formation and QS are cooperative ventures.

Sociality in biofilms

A social trait is one that has fitness implications for both the actor performing the behaviour and the recipient of such a behaviour. If a trait can positively or negatively affect the fitness of both actor and recipient then four categories of social traits emerge (Fig. 1.3) (West *et al.*, 2006). Explaining altruistic behaviours, those which incur a cost to the actor but benefit the recipient remains one of the greatest challenges to evolutionary biologists. This problem can be observed in simple microcosm experiments. Using experimental populations of the soil bacterium *Pseudomonas fluorescens*, which forms liquid–air interface biofilms by production of a cellulosic polymer, it is possible to observe that polymer 'cheating' mutants arise which do not contribute to the integrity of the biofilm but nonetheless experience the advantage gained by higher oxygen levels at the air liquid interface (Rainey and Rainey, 2003). This is damaging to the biofilm as a whole, because when the

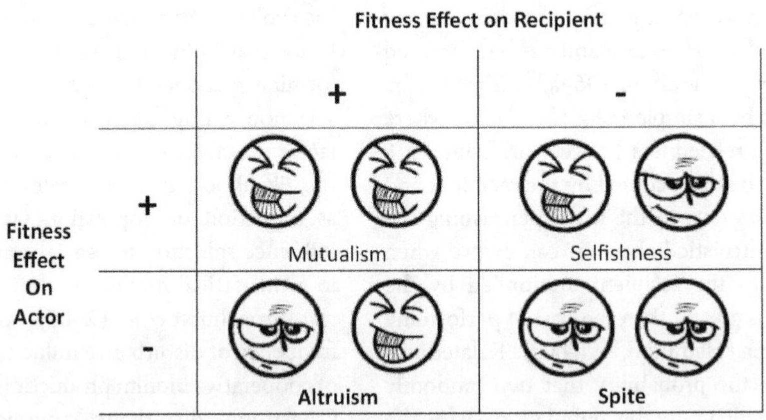

Figure 1.3 There are four types of social behaviour classified with respect to their fitness consequences for both actor and recipient. Mutualism and selfishness pose few evolutionary dilemmas, but those which incur a fitness cost to the actor require evolutionary explanations.

cheat load becomes too great, the entire biofilm sinks into the anoxic liquid phase.

Microorganisms engage in a plethora of cooperative behaviours such as the fruiting body formation of *Dictyostelium discoideum* (Strassmann *et al.*, 2000) and *Myxococcus xanthus* (Velicer *et al.*, 2000; Fiegna and Velicer, 2003), iron scavenging (siderophore production) and cooperative protease production in *Pseudomonas aeruginosa* (Griffin *et al.*, 2004; Diggle *et al.*, 2007b) all of which are vulnerable to exploitation by non-cooperative cheats. Many of the products of microbial social behaviours can be considered 'public goods'. These are products which are individually costly to produce but benefit the local population. All exoproducts such as enzymes which unlock nutrients in the environment or toxins which provide access to the host or even the biofilm matrix polymers can be considered public goods.

A central question in this body of work is: 'What mechanisms can maintain cooperative traits in the face of vulnerability to cheating?' Many ideas have been suggested and empirically examined (West *et al.*, 2007), but perhaps most worthy of discussion regarding microorganisms and biofilms is 'inclusive fitness'. This is the idea that an individual maximizes its inclusive fitness not just by maximizing its own reproduction, but the reproduction of its genes even if they happen to be in other individuals. For example, in the absence of an opportunity to reproduce, helping a relative reproduce still increases inclusive fitness. This theory was developed by the evolutionary biologist William Hamilton and has been termed 'kin selection' (Hamilton, 1964a,b). This can be summarized by a simple inequality, $rb > c$, where r defines the relatedness between individuals, b is the fitness benefit accrued by the recipient and c is the fitness cost to the actor performing the behaviour. Altruistic behaviour can evolve when the benefit to the recipient, multiplied by the relatedness, is greater than the cost of performing the behaviour (Hamilton, 1964a,b). Relatedness (r) refers to the probability that two randomly chosen individuals in the population share the allele responsible for cooperation over and above the average of the population (West *et al.*, 2006). This theory is extensive and a full discussion is beyond the remit of this chapter but the message becomes clearer with an example. Many eusocial insects have sterile worker castes that completely forgo their own reproduction in favour of that of the queen. How is this altruism inherited if the altruists never reproduce? The behaviour can be inherited if the benefits accrue to individuals which share the genes for altruism. This is the case in insects but how does this manifest in bacterial populations? Selection for cooperative iron-scavenging siderophores in *P. aeruginosa* is maximized when relatedness is kept high (Griffin, 2004). When closely related genotypes are kept together, cooperators avoid cheats. Since cheats do not grow well in the absence of cooperators selection for cheating is reduced at high relatedness.

There are two more noteworthy implications of this theory for prokaryotic clonal microorganisms. Both arise from the fact that relatedness is measured with respect to a single locus responsible for cooperation. First, this means that a single null mutation in this locus generates a relatedness of 0 between two individuals. Second, it means that two individuals with no genetic similarities apart from sharing the cooperative locus have a relatedness of 1. In populations of microorganisms where generation times are short, mutations occur frequently and lateral gene transfer is possible, the patterns of relatedness in populations can take forms unrecognisable in the eukaryotic world from which much of this work originates.

Hamilton's inequality also helps to delineate the ecological conditions that favour or disfavour the evolution of cooperative traits. Any ecological factor that influences the cost or benefit of performing a cooperative behaviour, influences the selection acting on that trait. In addition, any factor which structures the population and alters the likelihood of clone-mates interacting, such as migration or population viscosity, will also influence selection for social traits. For example in a theoretical model followed by an empirical test, Brockhurst *et al.* (2007) examined how the frequency of disturbance influences the evolution of cooperative biofilm production by *P. fluorescens*. Catastrophic disturbance is mimicked by sampling a small proportion (0.1%) of the population and transferring it into fresh media at varying intervals. The authors conclude from both theory and empirical data that cooperative biofilm formation in experimental populations of *P. fluorescens* peaks at an intermediate level of disturbance (Brockhurst *et al.*, 2007). The explanation is that at a low frequency of disturbance, the biofilms are invaded by non-cooperative cheats. This is prevented with increasing disturbance frequency because the disturbances alter the population structure by inflating relatedness. At very high disturbance frequencies, the population never reaches sufficient density for cooperation (biofilm polymer production) to be beneficial thus cooperation peaks at intermediate disturbance frequency.

Though it is possible to observe cheating mutants invading cooperative biofilms, biofilm formation need not always be cooperative. In fact, the scale at which a biofilm is observed and the biological details of its formation may cause us to speculate whether biofilm formation is cooperative. Inspired by spectacular modern microscopic observations of biofilms such as the mushroom-like structures of *P. aeruginosa*, many researchers have modelled *in silico*, the growth and spatial patterns of cells growing on a surface with interesting results. First, it is possible to observe the types of vertical structures typical of *in vitro* biofilms using only basic growth limitation criteria in the environment and simple growth parameters of the organism (Kreft *et al.*, 1999, 2001; Van Loosdrecht *et al.*, 2002). Second, in competitive growth simulations between a wild type bacterium and a mutant that does not produce a 'scaffold' polymer, the mutant is outgrown and covered by the cooperative wild type (Xavier and Foster, 2007). Where oxygen gradients are present, an overproducer of the scaffold polymer pushes its cell lineage vertically into favourable oxygen conditions. This generates an opposite prediction of the experiments described earlier with *P. fluorescens*. In the microcosms, mixing genotypes with respect to biofilm polymer production reduces relatedness and selects for cheating (Rainey and Rainey, 2003). In the *in silico* model of growth on a surface, mixing genotypes does not select for cheating owing to the local benefit of polymer secretion (Xavier and Foster, 2007). However, if you introduce a new parameter to the model, the possibility that one genotype can mutate into the other, the result is different. The cheating mutant now gains an advantage if it occurs in an oxygen rich region where it can quickly produce lots of cells (Foster and Xavier, 2007).

The social and environmental conditions and the biological details of a social behaviour all determine where we expect to observe biofilm formation and how it will appear. We will now discuss another type of social behaviour, communication. The types of biological communication and how we might expect these to evolve will be outlined and QS will be examined in light of this.

Quorum sensing as biological communication

Communication is an act or structure in one individual that influences the behaviour of another. As such, communication is inherently social and we can categorize types of communication based on the conditions that are required for them to evolve. There is a large body of literature dedicated to definitions of biological communication but those proposed by Maynard-Smith and Harper (2003) are used in this chapter (Smith and Harper, 2003). Based on two criteria, their classification reveals how different types of communication evolve. First, 'did the act or structure evolve due to a particular response?' And secondly, 'is the act or structure effective because the response has evolved accordingly?' Answering these questions leads us to determining that an act or structure may be a true signal, an environmental cue that guides an action or a coercion (Fig. 1.4) (Smith and Harper, 2003). Is this important? Yes, because understanding the precise biological nature of an interaction between individuals allows us to make predictions about how such interactions evolve and are maintained in natural populations.

It is useful to explore a few examples to illuminate this. Consider a cow grazing on a field. It respires releasing carbon dioxide which is detected by a mosquito. The mosquito follows the carbon dioxide gradient to find the cow and feeds on it. The carbon dioxide is effective owing to an evolved response on the part of the mosquito but its production by the cow did not evolve for that reason so in this scenario we should consider carbon dioxide to be a cue and not a signal. Now consider two rhinoceros beetles. They fight on a

	Evolved to Elicit that response	Effective due to evolved response	Example
Cue	-	+	
Coercion	+	-	
Signal	+	+	

Figure 1.4 The different types of biological communication (Maynard Smith and Harper, 2003). This type of classification enables us to distinguish scenarios which evolve very differently.

branch and one pushes the other over the side. This act evolved owing to its effect on the receiver, but the receiver does not benefit from an accordingly evolved response. This scenario is a coercion. Finally, consider an alarm call from a putty nosed monkey. When a predator is detected, the alarm call is sounded and all in earshot rush to safety. The alarm has evolved owing to the response from its receivers and the receivers benefit from an evolved response and so in this case the alarm call can be termed a signal. It is interesting that in the QS literature, there is a tendency to term all the molecules and interactions described as 'signalling' even if the true biological nature of the interactions might not support this. This has been discussed in detail elsewhere (Keller and Surette, 2006; Diggle *et al.*, 2007a).

Let us turn our attention to a QS bacterial population. When a QS-regulated exo-enzyme is important for growth, many deviations from the normal strategy would confer a benefit (Diggle *et al.*, 2007b; Sandoz *et al.*, 2007; Czárán and Hoekstra, 2009). A mutant that does not respond to QS by producing the enzyme would benefit nonetheless from the enzyme production of others. A mutant that does not signal benefits from sensing population density as signalled by others. A mutant that overproduces the signal, yet does not respond to it, benefits from overproduction of the costly enzyme by its neighbours. Experimentally it has been demonstrated that QS is subject to invasion by cheats and that QS cheating can be detrimental to the population (Diggle *et al.*, 2007b; Sandoz *et al.*, 2007).

Since signalling is vulnerable to various types of social exploitation we may ask what forces maintain it in nature? Three broad mechanisms can maintain reliable signalling. First, a signal can be a reliable index of quality. For example, the pitch of a deer roar is strongly correlated with the size of its larynx and linked to the body size of the male. Therefore, a roar provides reliable information about the male. Second, if a signal is very costly, this reduces the likelihood of cheating. For instance, a nightingale sings for most of the night and loses up to 10% of its body weight in doing so. An inferior male could also sing but is less likely to do so if the metabolic cost is unaffordable. Third, a reliable signal can be maintained by common interest. Common interests in bacterial populations are likely to arise through Hamiltonian relatedness, as described previously. For instance, if, on average, a cooperative cell interacts with other cells that share the gene for cooperation then the shared interest in reproduction removes the advantage of cheating.

With an evolutionary model of QS, Brown and Johnstone (2001) explore how signalling and response to the signal evolves with varying relatedness. Their model assumes that QS signals and response are costly to the individual cells producing them, beneficial to the group of interacting cells (cooperative) and that QS is more beneficial at high population density. First they find that selection for public goods production in response to the signal is reduced with decreasing relatedness. When, on average, a cooperative individual encounters individuals that do not share the genes for cooperation, it does not pay to cooperate. Second, the authors find that selection for signalling to elicit cooperation peaks at intermediate relatedness. At high relatedness all individuals are cooperative and only the minimum signal required to elicit cooperation is needed. At low relatedness there are no cooperative individuals so signalling is wasteful. At intermediate relatedness however, as individuals decrease investment in public goods, selection favours a higher signal level to compensate. Brown and Johnstone's model suggests that QS may be a mechanism to enforce cooperation under conditions of intermediate relatedness.

In a spatially explicit simulation model Czaran and Hoekstra (2009) explore the social evolution of different QS strategies. They allow for variation in three hypothesized loci, C, S and R, standing for cooperation, signalling and response, which can be either functional or inactive. This creates eight possible genotypes. When set into motion the authors observed a stable coexistence of two of these, 'Honest', denoted CSR, which cooperates, signals and responds, and 'Ignorant' denoted, csr ,which does none of these. The equilibrium shifts in the favour of 'Ignorant' when diffusion rates are increased as this mixes interactions between non-related individuals (i.e. decreases relatedness). As the threshold number of cells resulting in a quorum increases, the system becomes highly sensitive to changes in relatedness. For example, when a large number of cells is required for a quorum, even the smallest diffusion will disrupt interactions between related cells and prevent a quorum from being reached. Most interestingly, across much of the parameter space and particularly at intermediate levels of effective relatedness, 'Liar' genotypes which do no cooperate or respond but nonetheless produce signal are present. For example when diffusion is sufficient to allow exploitation of cooperating genotypes, selection for signal production allows 'Liar' genotypes to prosper more than silent exploitative genotypes.

The two theoretical studies, though very different in their approach, produce similar conclusions. First, exploitation of cooperative signalling is commonplace and can be stabilized by high relatedness. Second, when QS regulates cooperation, a plethora of social interactions are possible and can coexist, in particular, cooperative, exploitative and coercive strategies should be relatively common in nature.

Biofilms, quorum sensing and cooperation

A consideration of social evolution provides useful insights into the biology of both biofilms and communication but of course these do not occur independently of each other. When the benefits of biofilm formation and QS can be shared between cells and therefore exploited, a double Darwinian dilemma occurs. To explore how biofilm formation and QS evolve in concert, three aspects of this interaction will be examined in turn.

Let us consider how a cooperation scenario would unfold in a biofilm. It is theoretically and empirically well established that selection for cheating on secreted public goods is moderated by the spatial structure of a population, how far the public good is shared and the rate at which cells migrate (all things which alter relatedness within a population). For instance, a sessile biofilm community reproducing clonally, would be expected to have a higher relatedness than a well mixed planktonic population (Fig. 1.5) (Kreft, 2004; Nadell *et al.*, 2010). Although this would increase cooperative activity between clone-mates and therefore increase selection for cooperation, it would also increase local competition between clone-mates. These two could cancel each other out and so have no effect on selection for cooperation (Queller, 1994; Frank, 1998). The effect of local competition between relatives can be overcome if the production of the public good in question increases the patch size or local availability of nutrients, i.e. a hydrolytic enzyme (Pfeiffer and Bonhoeffer, 2003; Kreft, 2004; Grafen, 2007; Kümmerli *et al.*, 2009). Therefore, the effect of a population's spatial structure on selection for cooperation depends on the nature of the cooperative behaviour. It could be argued,

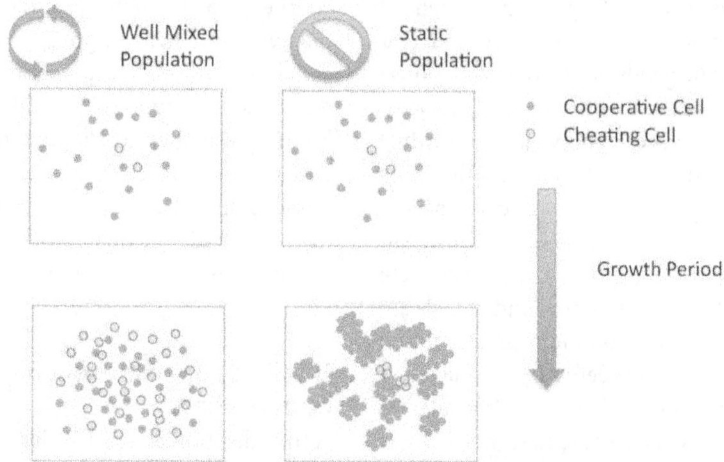

Figure 1.5 The effect of population mixing (spatial structure) on the selection for cheating. In a well mixed population, cooperators and cheats are likely to interact so cheats can exploit cooperators. In a static or viscous population such as those encountered in biofilms, neighbours are more likely to be clonemates and like interacts with like. In this scenario, cheats cannot exploit cooperators to the same degree as is found in a well mixed culture.

however, that in many cases public goods increase local capacity and therefore selection for those public goods increases with spatial structuring in spite of local competition.

Since it is known that QS is involved in biofilm formation in many species, we may ask the question: what role do we expect QS to play in biofilm formation? Nadell *et al.* (2008b) do exactly this using individual based biofilm simulations. They compete four hypothetical strains in which QS and biofilm polymer secretion are linked or not. A non-polymer producer, a constitutive polymer producer, a strain which produces polymer only at high cell density and a strain which ceases polymer production at high cell density. Given that polymer production is beneficial by pushing the cell lineage into favourable nutrient conditions (Xavier and Foster, 2007), the authors find that polymer secretion at low cell density is beneficial in the short term as it allows immediate access to favourable nutrient conditions. Conversely, polymer production at high cell density is beneficial in the long term (Nadell *et al.*, 2008b). The subtle implication of these results is that selection for either positive or negative QS-dependent biofilm regulation should depend on the duration of biofilm growth. For instance, in an acute infection where the window of opportunity is small and the QS induced reduction in biofilm formation coincides with a dispersal event, a negative relationship between QS and biofilm formation is adaptive (e.g. *V. cholerae*). However, in chronic infections where local competition is sustained and dispersal is rare, a positive relationship between QS and biofilm formation enhances competitive success (e.g. *P. aeruginosa*).

It is not only the spatial structure of the population that changes in the transition from planktonic to biofilm modes of growth, this usually also entails profound changes in the physiology of the organism. Some of these physiological changes may influence competition and evolution of social behaviours. First, a biofilm population grows slower than its planktonic counterpart, because of the additional polymeric matrix that must be produced and secreted. This means that it may be outcompeted by faster growing planktonic cells where there is the opportunity for them to grow such as in the absence of flow or in the absence of antibiotics that target faster growing cells. Since the growth rate of polymer-secreting biofilm cells is slowed, any cells defecting from polymer production gain an advantage. Second, conjugative transfer of DNA occurs more readily in biofilms than in planktonic cells owing to their proximity to each other. This inflates relatedness and thus may select for a greater level of cooperation (Smith, 2001; Nogueira *et al.*, 2009).

So far it is evident that many behaviours in the microbial world are social and therefore open to conflict between participants and non-participants. The severity and outcome of such conflicts will in part be determined by ecology (i.e. dispersal, population structuring, catastrophic disturbance) and the nature of the behaviour in question such as the diffusibility of the exoproduct. All of these factors can be explained by their effect on either the cost/benefit of the behaviour or the relatedness of cells within the population (Hamilton's rule). Communication is a special case of such social conflict and increases the complexity of social interactions that can occur, specifically when it is linked to other social behaviours, such as in the case of QS. Many of the studies discussed here either implicitly or explicitly reveal that diffusion of extracellular public goods or signal molecules is an important part of the ecology and influences the outcomes of social conflicts. Some authors have suggested that the origins of QS may have been to sense the diffusion rate in the environment providing a direct benefit to cells that need not waste energy producing expensive enzymes only for them to diffuse away into the environment (diffusion sensing) (Redfield, 2002). This need not be an alternative to QS and whilst small populations or even single cells may become quorate, there is an inevitable social element to QS (Diggle *et al.*, 2007b; Hense *et al.*, 2007; Köhler *et al.*, 2009; Rumbaugh *et al.*, 2009) and its maintenance in nature requires the resolution of social conflicts.

The role of quorum sensing on *in vivo* biofilms

It is well established that, rather than being free-swimming and planktonic, most bacteria in nature exist as biofilms. Within the human body, these structured bacterial communities may occur as part of the normal commensal flora of non-sterile sites such as the intestine and oral cavity. In the human gut, biofilms of mixed bacterial species attach to intestinal epithelial cells and form an important barrier against invading foodborne pathogens (Lee *et al.*, 2000). Indeed, dental plaque is known to be a biofilm of diverse species and is of historical significance as it represents the first biofilm to be formally described when Antoine van Leeuwenhoek visualized the 'animalcules' in the plaque on his own teeth in the seventeenth century (Donlan and Costerton, 2002). Since the 1970s it has become increasingly apparent that microbial biofilms are of huge importance in the pathogenesis of a wide range of infections. Probably the best studied example is that of *P. aeruginosa* infections in the cystic fibrosis lung (Bjarnsholt *et al.*, 2009). Other examples include middle ear infections, infections of prosthetic material including prosthetic joints, intravenous lines and urinary catheters, infective endocarditis and chronic wounds. These involve a range of microbial pathogens such as *S. aureus*, *S. epidermidis*, *E. coli* and eukaryotic organisms such as *Candida albicans*.

Bacterial biofilms represent a strategy for causing chronic infections that can persist despite the use of antibiotic treatment deemed effective when tested *in vitro* and the actions of the host organism's innate and adaptive immune response (Høiby *et al.*, 2010). They therefore pose a serious therapeutic challenge and are frequently extremely hard or impossible to manage without physical or surgical removal of the infected prosthesis or tissue. However, given that QS may influence the development of the biofilm and the evidence that QS signalling occurs *in vivo* (Storey *et al.*, 1998; Middleton *et al.*, 2002; Guina *et al.*, 2003), strategies aimed at disrupting this bacterial communication system offer a novel and alternative approach to tackling biofilm-associated infections. Indeed, the demonstration that QS has been shown to determine the tolerance of *P. aeruginosa* biofilms to aminoglycoside antibiotic therapy and the innate inflammatory response of the host including the action of hydrogen peroxide and phagocytosis by polymorphonuclear (PMN) white blood cells (Bjarnsholt *et al.*, 2005a) is further evidence that measures aimed at tackling QS will be of use in biofilm control.

Approaches to disrupting QS include removal of the signal in a process termed 'quorum quenching' (Dong *et al.*, 2007) and the use of QS antagonists or inhibitors (Rasmussen and Givskov, 2006). These latter agents have been particularly studied with respect to biofilm control, starting with the identification of naturally occurring brominated furanones produced as secondary metabolites by the Australian macroalga *D. pulchra,* which have structural similarities to AHLs (Givskov *et al.*, 1996). Biofilms grown in the presence of synthetic furanone derivatives were rendered susceptible to antimicrobial killing with the antibiotic tobramycin and detergent sodium dodecyl sulfate in contrast to the untreated biofilms. In a mouse pulmonary infection model, the synthetic furanone was able to inhibit QS in the infecting bacteria and enable their clearance by the host's immune response (Hentzer *et al.*, 2003).

Screening has been undertaken for QS inhibitors that may occur in nature and extracts from different *Penicillium* species, the phytotoxins patulin and penicillic acid, have been identified as being biologically active QS inhibitory compounds. In a similar fashion to the previously described work, patulin was found to enhance biofilm susceptibility to tobramycin treatment and in a mouse pulmonary infection model, *P. aeruginosa* was more rapidly cleared from the mice that were treated with patulin compared with the placebo group. Furthermore, the *P. aeruginosa* QS-dependent mechanisms that block development of the oxidative burst in PMN neutrophils was overcome when bacteria were treated with either patulin or penicillic acid, enabling the neutrophils to become activated (Rasmussen *et al.*, 2005).

Garlic extract is a further example of a naturally occurring QS inhibitor detected following the screening of natural and synthetic compound libraries. Garlic has been shown to have anti-fungal, anti-cancer and anti-microbial properties (Ankri and Mirelman, 1999), and has been shown to specifically inhibit QS-regulated gene expression in *P. aeruginosa,* as demonstrated by DNA microarray-based transcriptomic analysis. It also increased *P. aeruginosa* biofilm susceptibility to tobramycin treatment and the respiratory burst and phagocytosis by PMNs as well as leading to an improved outcome in a mouse pulmonary infection model and reduced virulence in a eukaryotic pathogenesis model using *C. elegans* (Bjarnsholt *et al.*, 2005b; Rasmussen *et al.*, 2005).

The possible mechanisms as to how disruption of QS may render biofilms more susceptible to antibiotic or host immune clearance has not been fully explored or explained. Potential pathways in *P. aeruginosa* include two bacterially produced carbohydrate-binding lectin molecules, termed LecA and LecB (Gilboa-Garber, 1972). These are QS-controlled virulence determinants which additionally may contribute to biofilm formation by promoting binding to or cross-linking of biofilm carbohydrate-containing structures (Tielker *et al.*, 2005; Diggle *et al.*, 2006).

Overall, the rise of resistant organisms challenges the therapeutic use of conventional bacteriocidal and bacteriostatic antibiotics. This is compounded by the tolerance of bacteria within a biofilm to the actions of antibiotics and the immune system. Given the widespread occurrence and clinical impact of biofilm-associated infections in the healthcare setting, alternative strategies for control such as QS inhibition represent promising and novel approaches. Furthermore, based on current knowledge, resistance to QS inhibitors can only arise by mutations which render the bacteria unable to produce any virulence factors, i.e. they become avirulent (Lee *et al.*, 2005). If this is indeed the case, resistance problems as encountered currently would not be a clinical issue with the use of QS inhibitor therapy (Høiby *et al.*, 2010).

Acknowledgements

We thank the Royal Society (S.P.D.), BBSRC (R.J.G., R.P., M.P.F.) and Mologic (R.J.G.) for funding.

References

Allesen-Holm, M., Barken, K.B., Yang, L., Klausen, M., Webb, J.S., Kjelleberg, S., Molin, S., Givskov, M., and Tolker-Nielsen, T. (2006). A characterization of DNA release in *Pseudomonas aeruginosa* cultures and biofilms. Mol. Microbiol. 59, 1114–1128.

Ankri, S., and Mirelman, D. (1999). Antimicrobial properties of allicin from garlic. Microbes Infect. *1*, 125–129.

Atkinson, S., Throup, J.P., Stewart, G.S., and Williams, P. (1999). A hierarchical quorum-sensing system in *Yersinia pseudotuberculosis* is involved in the regulation of motility and clumping. Mol. Microbiol. 33, 1267–1277.

Auger, S., Krin, E., Aymerich, S., and Gohar, M. (2006). Autoinducer 2 affects biofilm formation by *Bacillus cereus*. Appl. Environ. Microbiol. 72, 937–941.

Autret, N., Raynaud, C., Dubail, I., Berche, P., and Charbit, A. (2003). Identification of the *agr* locus of *Listeria monocytogenes*: role in bacterial virulence. Infect. Immun. *71*, 4463–4471.

Bachofen, R., and Schenk, A. (1998). Quorum sensing autoinducers: Do they play a role in natural microbial habitats? Microbiol. Res. *153*, 61–63.

Bainton, N.J., Bycroft, B.W., Chhabra, S.R., Stead, P., Gledhill, L., Hill, P.J., Rees, C.E., Winson, M.K., Salmond, G.P., and Stewart, G.S. (1992a). A general role for the lux autoinducer in bacterial cell signalling: control of antibiotic biosynthesis in *Erwinia*. Gene. *116*, 87–91.

Bainton, N.J., Stead, P., Chhabra, S.R., Bycroft, B.W., Salmond, G.P., Stewart, G.S., and Williams, P. (1992b). N-(3-oxohexanoyl)-L-homoserine lactone regulates carbapenem antibiotic production in *Erwinia carotovora*. Biochem. J. *288*, 997–1004.

Bjarnsholt, T., Jensen, P., Burmølle, M., Hentzer, M., Haagensen, J.A.J., Hougen, H.P., Calum, H., Madsen, K.G., Moser, C., *et al.* (2005a). *Pseudomonas aeruginosa* tolerance to tobramycin, hydrogen peroxide and polymorphonuclear leukocytes is quorum-sensing dependent. Microbiology *151*, 373–383.

Bjarnsholt, T., Jensen, P., Rasmussen, T.B., Christophersen, L., Calum, H., Hentzer, M., Hougen, H.-P., Rygaard, J., Moser, C., *et al.* (2005b). Garlic blocks quorum sensing and promotes rapid clearing of pulmonary *Pseudomonas aeruginosa* infections. Microbiology *151*, 3873–880.

Bjarnsholt, T., Jensen, P.Ø., Fiandaca, M.J., Pedersen, J., Hansen, C.R., Andersen, C.B., Pressler, T., Givskov, M., and Høiby, N. (2009). *Pseudomonas aeruginosa* biofilms in the respiratory tract of cystic fibrosis patients. Pediatr. Pulmonol. *44*, 547–558.

von Bodman, S.B., Majerczak, D.R., and Coplin, D.L. (1998). A negative regulator mediates quorum-sensing control of exopolysaccharide production in *Pantoea stewartii* subsp. *stewartii*. Proc. Natl. Acad. Sci. U.S.A. *95*, 7687–7692.

Boles, B.R., Thoendel, M., and Singh, P.K. (2005). Rhamnolipids mediate detachment of *Pseudomonas aeruginosa* from biofilms. Mol. Microbiol. *57*, 1210–1223.

Branda, S.S., González-Pastor, J.E., Dervyn, E., Ehrlich, S.D., Losick, R., and Kolter, R. (2004). Genes involved in formation of structured multicellular communities by *Bacillus subtilis*. J. Bacteriol. *186*, 3970–3979.

Brint, J.M., and Ohman, D.E. (1995). Synthesis of multiple exoproducts in *Pseudomonas aeruginosa* is under the control of RhlR-RhlI, another set of regulators in strain PAO1 with homology to the autoinducer-responsive LuxR-LuxI family. J. Bacteriol. *177*, 7155–7163.

Brockhurst, M.A., Buckling, A., and Gardner, A. (2007). Cooperation peaks at intermediate disturbance. Curr. Biol. *17*, 761–765.

Brown, S.P., and Johnstone, R.A. (2001). Cooperation in the dark: signalling and collective action in quorum-sensing bacteria. Proc. Roy. Soc. B. *268*, 961–965.

Caiazza, N.C., and O'Toole, G.A. (2003). Alpha-toxin is required for biofilm formation by *Staphylococcus aureus*. J. Bacteriol. *185*, 3214–3217.

Chai, Y., Chu, F., Kolter, R., and Losick, R. (2008). Bistability and biofilm formation in *Bacillus subtilis*. Mol. Microbiol. *67*, 254–263.

Charlton, T.S., de Nys, R., Netting, A., Kumar, N., Hentzer, M., Givskov, M., and Kjelleberg, S. (2000). A novel and sensitive method for the quantification of N-3-oxoacyl homoserine lactones using gas chromatography–mass spectrometry: application to a model bacterial biofilm. Environ. Microbiol. *2*, 530–541.

Chen, X., Schauder, S., Potier, N., Van Dorsselaer, A., Pelczer, I., Bassler, B.L., and Hughson, F.M. (2002). Structural identification of a bacterial quorum-sensing signal containing boron. Nature *415*, 545–549.

Chu, F., Kearns, D.B., McLoon, A., Chai, Y., Kolter, R., and Losick, R. (2008). A novel regulatory protein governing biofilm formation in *Bacillus subtilis*. Mol. Microbiol. *68*, 1117–1127.

Croxatto A., Chalker V.J., Lauritz J., Jass J., Hardman A., Williams P., Camara M., and Milton D.L. (2002). VanT, a homologue of *Vibrio harveyi* LuxR, regulates serine, metalloprotease, pigment, and biofilm production in *Vibrio anguillarum*. J. Bacteriol. *184*, 1617–1629.

Czárán, T., and Hoekstra, R.F. (2009). Microbial communication, cooperation and cheating: quorum sensing drives the evolution of cooperation in bacteria. PLoS ONE *4*, e6655.

Daines, D.A., Bothwell, M., Furrer, J., Unrath, W., Nelson, K., Jarisch, J., Melrose, N., Greiner, L., Apicella, M., and Smith, A.L. (2005). *Haemophilus influenzae luxS* mutants form a biofilm and have increased virulence. Microb. Pathog. *39*, 87–96.

Davey, M.E., Caiazza, N.C., and O'Toole, G.A. (2003). Rhamnolipid surfactant production affects biofilm architecture in *Pseudomonas aeruginosa* PA01. J. Bacteriol. *185*, 1027–1036.

Davies, D.G., Parsek, M.R., Pearson, J.P., Iglewski, B.H., Costerton, J.W., and Greenberg, E.P. (1998). The involvement of cell-to-cell signals in the development of a bacterial biofilm. Science *280*, 295–298.

De Kievit, T.R., Gillis, R., Marx, S., Brown, C., and Iglewski, B.H. (2001). Quorum-sensing genes in *Pseudomonas aeruginosa* biofilms: their role and expression patterns. Appl. Environ. Microbiol. *67*, 1865–1873.

Déziel, E., Lépine, F., Milot, S., He, J., Mindrinos, M.N., Tompkins, R.G., and Rahme, L.G. (2004). Analysis of *Pseudomonas aeruginosa* 4-hydroxy-2-alkylquinolines (HAQs) reveals a role for 4-hydroxy-2-heptylquinoline in cell-to-cell communication. Proc. Natl. Acad. Sci. U.S.A. *101*, 1339–1344.

Diggle, S.P., Winzer, K., Chhabra, S.R., Worrall, K.E., Cámara, M., and Williams, P. (2003). The *Pseudomonas aeruginosa* quinolone signal molecule overcomes the cell density-dependency of the quorum sensing hierarchy, regulates rhl-dependent genes at the onset of stationary phase and can be produced in the absence of LasR. Mol. Microbiol. *50*, 29–43.

Diggle, S.P., Stacey, R.E., Dodd, C., Cámara, M., Williams, P., and Winzer, K. (2006). The galactophilic lectin, LecA, contributes to biofilm development in *Pseudomonas aeruginosa*. Environ. Microbiol. *8*, 1095–1104.

Diggle, S.P., Gardner, A., West, S.A., and Griffin, A.S. (2007a). Evolutionary theory of bacterial quorum sensing: when is a signal not a signal? Phil. Trans. Roy. Soc. B. *362*, 1241–1249.

Diggle, S.P., Griffin, A.S., Campbell, G.S., and West, S.A. (2007b). Cooperation and conflict in quorum-sensing bacterial populations. Nature. *450*, 411–414.

Dong, Y.H., Wang, L.H., and Zhang, L.H. (2007). Quorum-quenching microbial infections: mechanisms and implications. Phil. Trans. Roy. Soc. B. *362*, 1201–1211.

Donlan, R.M., and Costerton, J.W. (2002). Biofilms: survival mechanisms of clinically relevant microorganisms. Clin. Microbiol. Rev. *15*, 167–193.

Dubern, J.F., and Diggle, S.P. (2008). Quorum sensing by 2-alkyl-4-quinolones in *Pseudomonas aeruginosa* and other bacterial species. Mol. Biosyst. *4*, 882–888.

Eberhard, A., Burlingame, A.L., Eberhard, C., Kenyon, G.L., Nealson, K.H., and Oppenheimer, N.J. (1981). Structural identification of autoinducer of *Photobacterium fischeri* luciferase. Biochemistry *20*, 2444–2449.

Engebrecht, J., Nealson, K., and Silverman, M. (1983). Bacterial bioluminescence: isolation and genetic analysis of functions from *Vibrio fischeri*. Cell. 32, 773–781.

Erickson, D.L., Endersby, R., Kirkham, A., Stuber, K., Vollman, D.D., Rabin, H.R., Mitchell, I., and Storey, D.G. (2002). *Pseudomonas aeruginosa* quorum-sensing systems may control virulence factor expression in the lungs of patients with cystic fibrosis. Infect. Immun. *70*, 1783–1790.

Fiegna, F., and Velicer, G.J. (2003). Competitive fates of bacterial social parasites: persistence and self-induced extinction of *Myxococcus xanthus* cheaters. Proc. Biol. Sci. *270*, 1527–1534.

Foster, K.R., and Xavier, J.B. (2007). Cooperation: bridging ecology and sociobiology. Curr. Biol. *17*, R319–321.

Frank, S.A. (1998). Foundations of Social Evolution. (Princeton University Press, Princeton, NJ).

Fujita, M., Gonzalez-Pastor, J.E., and Losick, R. (2005). High-and low-threshold genes in the Spo0A regulon of *Bacillus subtilis*. J. Bacteriol. *187*, 1357–1368.

Geier, H., Mostowy, S., Cangelosi, G.A., Behr, M.A., and Ford, T.E. (2008). Autoinducer-2 triggers the oxidative stress response in *Mycobacterium avium*, leading to biofilm formation. Appl. Environ. Microbiol. *74*, 1798–1804.

Gilboa-Garber, N. (1972). Inhibition of broad spectrum hemagglutinin from *Pseudomonas aeruginosa* by D-galactose and its derivatives. FEBS Lett. *20*, 242–244.

Givskov, M., de Nys, R., Manefield, M., Gram, L., Maximilien, R., Eberl, L., Molin, S., Steinberg, P.D., and Kjelleberg, S. (1996). Eukaryotic interference with homoserine lactone-mediated prokaryotic signalling. J. Bacteriol. *178*, 6618–6622.

González Barrios, A.F., Zuo, R., Hashimoto, Y., Yang, L., Bentley, W.E., and Wood, T.K. (2006). Autoinducer 2 controls biofilm formation in *Escherichia coli* through a novel motility quorum-sensing regulator (MqsR, B3022). J. Bacteriol. *188*, 305–316.

Grafen, A. (2007). Detecting kin selection at work using inclusive fitness. Proc. Roy. Soc. B *274*, 713–719.

Gram, L., Grossart, H.-P., Schlingloff, A., and Kiørboe, T. (2002). Possible quorum sensing in marine snow bacteria: production of acylated homoserine lactones by Roseobacter strains isolated from marine snow. Appl. Environ. Microbiol. *68*, 4111–4116.

Griffin, A.S., West, S.A., and Buckling, A. (2004). Cooperation and competition in pathogenic bacteria. Nature *430*, 1024–1027.

Guina, T., Purvine, S.O., Yi, E.C., Eng, J., Goodlett, D.R., Aebersold, R., and Miller, S.I. (2003). Quantitative proteomic analysis indicates increased synthesis of a quinolone by *Pseudomonas aeruginosa* isolates from cystic fibrosis airways. Proc. Natl. Acad. Sci. U.S.A. *100*, 2771–2776.

Hamilton, W.D. (1964a). The genetical evolution of social behaviour. I. J. Theor. Biol. *7*, 1–16.

Hamilton, W.D. (1964b). The genetical evolution of social behaviour. II. J. Theor. Biol. *7*, 17–52.

Hancock, L.E., and Perego, M. (2004). The *Enterococcus faecalis fsr* two-component system controls biofilm development through production of gelatinase. J. Bacteriol. *186*, 5629–5639.

Hense, B.A., Kuttler, C., Müller, J., Rothballer, M., Hartmann, A., and Kreft, J.U. (2007). Does efficiency sensing unify diffusion and quorum sensing? Nat. Rev. Microbiol. *5*, 230–239.

Hentzer, M., Riedel, K., Rasmussen, T.B., Heydorn, A., Andersen, J.B., Parsek, M.R., Rice, S.A., Eberl, L., Molin, S., *et al.* (2002). Inhibition of quorum sensing in *Pseudomonas aeruginosa* biofilm bacteria by a halogenated furanone compound. Microbiology *148*, 87–102.

Hentzer, M., Wu, H., Andersen, J.B., Riedel, K., Rasmussen, T.B., Bagge, N., Kumar, N., Schembri, M.A., Song, Z., *et al.* (2003). Attenuation of *Pseudomonas aeruginosa* virulence by quorum sensing inhibitors. EMBO J. *22*, 3803–3815.

Heydorn, A., Ersbøll, B.K., Hentzer, M., Parsek, M.R., Givskov, M., and Molin, S. (2000). Experimental reproducibility in flow-chamber biofilms. Microbiology *146*, 2409–2415.

Høiby, N., Bjarnsholt, T., Givskov, M., Molin, S., and Ciofu, O. (2010). Antibiotic resistance of bacterial biofilms. Int. J. Antimicrobial Agents. *35*, 322–332.

Holden, M.T., Chhabra, S.R., de Nys, R., Stead, P., Bainton, N.J., Hill, P.J., Manefield, M., Kumar, N., Labatte, M., *et al.* (1999). Quorum–sensing cross talk: isolation and chemical characterization of cyclic dipeptides from *Pseudomonas aeruginosa* and other Gram-negative bacteria. Mol. Microbiol. 33, 1254–1266.

Huang, Y.L., Dobretsov, S., Ki, J.S., Yang, L.H., and Qian, P.Y. (2007). Presence of acyl-homoserine lactone in subtidal biofilm and the implication in larval behavioral response in the polychaete *Hydroides elegans*. Microbial Ecol. *54*, 384–392.

Huber, B., Riedel, K., Hentzer, M., Heydorn, A., Gotschlich, A., Givskov, M., Molin, S., and Eberl, L. (2001). The *cep* quorum-sensing system of *Burkholderia cepacia* H111 controls biofilm formation and swarming motility. Microbiology *147*, 2517–2528.

Janssens, J.C., Steenackers, H., Robijns, S., Gellens, E., Levin, J., Zhao, H., Hermans, K., De Coster, D., Verhoeven, T.L., *et al.* (2008). Brominated furanones inhibit biofilm formation by *Salmonella enterica* serovar Typhimurium. Appl. Environ. Microbiol. *74*, 6639–6648.

Karatan, E., and Watnick, P. (2009). Signals, regulatory networks, and materials that build and break bacterial biofilms. Microbiol. Mol. Biol. Rev. 73, 310–347.

Kearns, D.B., Chu, F., Branda, S.S., Kolter, R., and Losick, R. (2005). A master regulator for biofilm formation by *Bacillus subtilis*. Mol. Microbiol. *55*, 739–749.

Keller, L., and Surette, M.G. (2006). Communication in bacteria: an ecological and evolutionary perspective. Nat. Rev. Microbiol. *4*, 249–258.

Kleerebezem, M., Quadri, L.E.N., Kuipers, O.P., and de Vos, W.M. (1997). Quorum sensing by peptide pheromones and two-component signal-transduction systems in Gram-positive bacteria. Mol. Microbiol. *24*, 895–904.

Kong, K.F., Vuong, C., and Otto, M. (2006). Staphylococcus quorum sensing in biofilm formation and infection. Int. J. Med. Microbiol. *296*, 133–39.

Koutsoudis, M.D., Tsaltas, D., Minogue, T.D., and von Bodman, S.B. (2006). Quorum-sensing regulation governs bacterial adhesion, biofilm development, and host colonization in *Pantoea stewartii* subspecies *stewartii*. Proc. Natl. Acad. Sci. U.S.A. *103*, 5983–5988.

Köhler, T., Buckling, A., and van Delden, C. (2009). Cooperation and virulence of clinical *Pseudomonas aeruginosa* populations. Proc. Natl. Acad. Sci. U.S.A. *106*, 6339–6344.

Kreft, J.-U. (2004). Biofilms promote altruism. Microbiology *150*, 2751–2760.

Kreft, J.U., Booth, G., and Wimpenny, J.W. (1999). BacSim, a simulator for individual-based modelling of bacterial colony growth. Microbiology *144*, 3275–3287.

Kreft, J.U., Picioreanu, C., Wimpenny, J.W., and Loosdrecht, M.C.V. (2001). Individual-based modelling of biofilms. Microbiology *147*, 2897–2912.

Kümmerli, R., Griffin, A.S., West, S.A., Buckling, A., and Harrison, F. (2009). Viscous medium promotes cooperation in the pathogenic bacterium *Pseudomonas aeruginosa*. Proc. Biol. Sci. *276*, 3531–3538.

Labbate, M., Queck, S.Y., Koh, K.S., Rice, S.A., Givskov, M., and Kjelleberg, S. (2004). Quorum sensing-controlled biofilm development in *Serratia liquefaciens* MG1. J. Bacteriol. *186*, 692–698.

Labbate, M., Zhu, H., Thung, L., Bandara, R., Larsen, M.R., Willcox, M.D., Givskov, M., Rice, S.A., and Kjelleberg, S. (2007). Quorum-sensing regulation of adhesion in *Serratia marcescens* MG1 is surface dependent. J. Bacteriol. *189*, 2702–2711.

Lee, B., Haagensen, J.A., Ciofu, O., Andersen, J.B., Høiby, N., and Molin, S. (2005). Heterogeneity of biofilms formed by nonmucoid *Pseudomonas aeruginosa* isolates from patients with cystic fibrosis. J. Clin. Microbiol. *43*, 5247–5255.

Lee, Y.K., Lim, C.Y., Teng, W.L., Ouwehand, A.C., Tuomola, E.M., and Salminen, S. (2000). Quantitative approach in the study of adhesion of lactic acid bacteria to intestinal cells and their competition with enterobacteria. Appl. Environ. Microbiol. *66*, 3692–3697.

Lemon, K.P., Freitag, N.E., and Kolter, R. (2010). The virulence regulator PrfA promotes biofilm formation by *Listeria monocytogenes*. J. Bacteriol. *192*, 3969–3976.

Lequette, Y., and Greenberg, E.P. (2004). Timing and localization of rhamnolipid synthesis gene expression in *Pseudomonas aeruginosa* biofilms. J. Bacteriol. *187*, 37–44.

Li, J., Attila, C., Wang, L., Wood, T.K., Valdes, J.J., and Bentley, W.E. (2007). Quorum sensing in *Escherichia coli* is signaled by AI-2/LsrR: Effects on sRNA and biofilm architecture. J. Bacteriol. *189*, 6011–6020.

Li, Y.H., Hanna, M.N., Svensäter, G., Ellen, R.P., and Cvitkovitch, D.G. (2001). Cell density modulates acid adaptation in *Streptococcus mutans*: implications for survival in biofilms. J. Bacteriol. *183*, 6875–6884.

Li, Y.H., Tang, N., Aspiras, M.B., Lau, P.C.Y., Lee, J.H., Ellen, R.P., and Cvitkovitch, D.G. (2002). A quorum-sensing signaling system essential for genetic competence in *Streptococcus mutans* is involved in biofilm formation. J. Bacteriol. *184*, 2699–2708.

Lindum, P.W., Anthoni, U., Christophersen, C., Eberl, L., Molin, S., and Givskov, M. (1998). *N*-Acyl-l-homoserine lactone autoinducers control production of an extracellular lipopeptide biosurfactant required for swarming motility of *Serratia liquefaciens* MG1. J. Bacteriol. *180*, 6384–6388.

Loo, C.Y., Corliss, D.A., and Ganeshkumar, N. (2000). *Streptococcus gordonii* biofilm formation: identification of genes that code for biofilm phenotypes. J. Bacteriol. *182*, 1374–1382.

López, D., Fischbach, M.A., Chu, F., Losick, R., and Kolter, R. (2009). Structurally diverse natural products that cause potassium leakage trigger multicellularity in *Bacillus subtilis*. Proc. Natl. Acad. Sci. U.S.A. *106*, 280–285.

Lopez, D., Vlamakis, H., Kolter, R. (2009). Generation of multiple cell types in *Bacillus subtilis*. FEMS Microbiol. Rev. 33, 152–163.

Lynch, M.J., Swift, S., Kirke, D.F., Keevil, C.W., Dodd, C.E.R., and Williams, P. (2002). The regulation of biofilm development by quorum sensing in *Aeromonas hydrophila*. Environ. Microbiol. 4, 18–28.

McLean, R.J., Whiteley, M., Stickler, D.J., and Fuqua, W.C. (1997). Evidence of autoinducer activity in naturally occurring biofilms. FEMS Microbiol. Lett. *154*, 259–263.

Manefield, M., de Nys, R., Kumar, N., Read, R., Givskov, M., Steinberg, P., and Kjelleberg, S. (1999). Evidence that halogenated furanones from *Delisea pulchra* inhibit acylated homoserine lactone (AHL)-mediated gene expression by displacing the AHL signal from its receptor protein. Microbiology *145*, 283–291.

Manefield, M., Rasmussen, T.B., Henzter, M., Andersen, J.B., Steinberg, P., Kjelleberg, S., and Givskov, M. (2002). Halogenated furanones inhibit quorum sensing through accelerated LuxR turnover. Microbiology *148*, 1119–1127.

Merritt, J., Qi, F., Goodman, S.D., Anderson, M.H., and Shi, W. (2003). Mutation of luxS affects biofilm formation in *Streptococcus mutans*. Infect. Immun. *71*, 1972–1979.

Morici, L.A., Carterson, A.J., Wagner, V.E., Frisk, A., Schurr, J.R., Höner zu Bentrup, K,. Hassett, D.J., Iglewski, B.H., Sauer, K., and Schurr, M.J. (2007). *Pseudomonas aeruginosa* AlgR represses the Rhl quorum-sensing system in a biofilm-specific manner. J. Bacteriol. *189*, 7752–7764.

Middleton, B., Rodgers, H.C., Cámara, M., Knox, A.J., Williams, P., and Hardman, A. (2002). Direct detection of *N*-acylhomoserine lactones in cystic fibrosis sputum. FEMS Microbiol. Lett. *207*, 1–7.

Miller, S.T., Xavier, K.B., Campagna, S.R., Taga, M.E., Semmelhack, M.F., Bassler, B.L., and Hughson, F.M. (2004). *Salmonella typhimurium* recognizes a chemically distinct form of the bacterial quorum-sensing signal AI-2. Mol. Cell. *15*, 677–687.

Nadell, C., Xavier, J., and Foster, K. (2008a). The sociobiology of biofilms. FEMS Microbiol. Rev. *33*, 206–224.

Nadell, C.D., Xavier, J.B., Levin, S.A., and Foster, K.R. (2008b). The evolution of quorum sensing in bacterial biofilms. PLoS Biol. *6*, e14.

Nadell, C.D., Foster, K.R., and Xavier, J.B. (2010). Emergence of spatial structure in cell groups and the evolution of cooperation. PLoS Comput. Biol. *6*, e1000716.

Nakayama, J., Chen, S., Oyama, N., Nishiguchi, K., Azab, E.A., Tanaka, E., Kariyama, R., and Sonomoto, K. (2006). Revised model for *Enterococcus faecalis fsr* quorum-sensing system: the small open reading frame *fsrD* encodes the gelatinase biosynthesis-activating pheromone propeptide corresponding to staphylococcal *agrD*. J. Bacteriol. *188*, 8321–8326.

Nealson, K.H., Platt, T., and Hastings, J.W. (1970). Cellular control of the synthesis and activity of the bacterial luminescent system. J. Bacteriol. *104*, 313–322.

Ng, W.L., and Bassler, B.L. (2009). Bacterial quorum-sensing network architectures. Annu. Rev. Genet. *43*, 197–222.

Nogueira, T., Rankin, D.J., Touchon, M., Taddei, F., Brown, S.P., and Rocha, E.P. (2009). Horizontal gene transfer of the secretome drives the evolution of bacterial cooperation and virulence. Curr. Biol. *19*, 1683–1691.

Novick, R.P., and Geisinger, E. (2008). Quorum sensing in staphylococci. Ann. Rev. Genet. *42*, 541–564.

Ochsner, U.A., and Reiser, J. (1995). Autoinducer-mediated regulation of rhamnolipid biosurfactant synthesis in *Pseudomonas aeruginosa*. Proc. Natl. Acad. Sci. U.S.A. *92*, 6424–6428.

Pesci, E.C., Milbank, J.B., Pearson, J.P., McKnight, S., Kende, A.S., Greenberg, E.P., and Iglewski, B.H. (1999). Quinolone signaling in the cell-to-cell communication system of *Pseudomonas aeruginosa*. Proc. Natl. Acad. Sci. U.S.A. *96*, 11229–11234.

Pfeiffer, T., and Bonhoeffer, S. (2003). An evolutionary scenario for the transition to undifferentiated multicellularity. Proc. Natl. Acad. Sci. U.S.A. *100*, 1095–1098.

Prouty, A.M., Schwesinger, W.H., and Gunn, J.S. (2002). Biofilm formation and interaction with the surfaces of gallstones by Salmonella spp. Infect. Immun. *70*, 2640–2649.

Puskas, A., Greenberg, E.P., Kaplan, S., and Schaefer, A.L. (1997). A quorum-sensing system in the free-living photosynthetic bacterium *Rhodobacter sphaeroides*. J. Bacteriol. *179*, 7530–7537.

Qi, F., Kreth, J., Lévesque, C.M., Kay, O., Mair, R.W., Shi, W., Cvitkovitch, D.G., and Goodman, S.D. (2005). Peptide pheromone induced cell death of *Streptococcus mutans*. FEMS Microbiol. Lett. *251*, 321–26.

Qin, Z., Ou, Y., Yang, L., Zhu, Y., Tolker-Nielsen, T., Molin, S., and Qu, D. (2007). Role of autolysin-mediated DNA release in biofilm formation of *Staphylococcus epidermidis*. Microbiology *153*, 2083–2092.

Queller, D.C. (1994). Genetic relatedness in viscous populations. Evol. Ecol. *8*, 70–73.

Rainey, P.B., and Rainey, K. (2003). Evolution of cooperation and conflict in experimental bacterial populations. Nature *425*, 72–74.

Rasmussen, T.B., and Givskov, M. (2006). Quorum sensing inhibitors: a bargain of effects. Microbiology *152*, 895–904.

Rasmussen, T.B., Bjarnsholt, T., Skindersoe, M.E., Hentzer, M., Kristoffersen, P., Köte, M., Nielsen, J., Eberl, L., and Givskov, M. (2005). Screening for quorum-sensing inhibitors (QSI) by use of a novel genetic system, the QSI selector. J. Bacteriol. *187*, 1799–1814.

Rasmussen, T.B., Skindersoe, M.E., Bjarnsholt, T., Phipps, R.K., Christensen, K.B., Jensen, P.O., Andersen, J.B., Koch, B., Larsen, T.O., *et al.* (2005). Identity and effects of quorum-sensing inhibitors produced by Penicillium species. Microbiology *151*, 1325–1340.

Redfield, R.J. (2002). Is quorum sensing a side effect of diffusion sensing? Trends Microbiol. *10*, 365–370.

Ren, D., Sims, J.J., and Wood, T.K. (2001). Inhibition of biofilm formation and swarming of *Escherichia coli* by (5Z)-4-bromo-5-(bromomethylene)-3-butyl-2 (5H)-furanone. Environ. Microbiol. *3*, 731–736.

Rice, S.A., Koh, K.S., Queck, S.Y., Labbate, M., Lam, K.W., and Kjelleberg, S. (2005). Biofilm formation and sloughing in *Serratia marcescens* are controlled by quorum sensing and nutrient cues. J. Bacteriol. *187*, 3477–3485.

Riedel, C.U., Monk, I.R., Casey, P.G., Waidmann, M.S., Gahan, C.G., and Hill, C. (2009). AgrD-dependent quorum sensing affects biofilm formation, invasion, virulence and global gene expression profiles in *Listeria monocytogenes*. Mol. Microbiol. *71*, 1177–1189.

Rieu, A., Briandet, R., Habimana, O., Garmyn, D., Guzzo, J., and Piveteau, P. (2008). *Listeria monocytogenes* EGD-e biofilms: no mushrooms but a network of knitted chains. Appl. Environ. Microbiol. *74*, 4491–4497.

Rumbaugh, K.P., Diggle, S.P., Watters, C.M., Ross-Gillespie, A., Griffin, A.S., and West, S.A. (2009). Quorum sensing and the social evolution of bacterial virulence. Curr. Biol. *19*, 341–345.

Sakuragi, Y., and Kolter, R. (2007). Quorum-sensing regulation of the biofilm matrix genes (*pel*) of *Pseudomonas aeruginosa*. J. Bacteriol. *189*, 5383–5386.

Sandoz, K.M., Mitzimberg, S.M., and Schuster, M. (2007). Social cheating in *Pseudomonas aeruginosa* quorum sensing. Proc. Natl. Acad. Sci. USA. *104*, 15876–15881.

Sauer, K., Camper, A.K., Ehrlich, G.D., Costerton, J.W., and Davies, D.G. (2002). *Pseudomonas aeruginosa* displays multiple phenotypes during development as a biofilm. J. Bacteriol. *184*, 1140–1154.

Schaber, J.A., Triffo, W.J., Suh, S.J., Oliver, J.W., Hastert, M.C., Griswold, J.A., Auer, M., Hamood, A.N., and Rumbaugh, K.P. (2007). *Pseudomonas aeruginosa* forms biofilms in acute infection independent of cell-to-cell signaling. Infect. Immun. *75*, 3715–3721.

Schuster, M., Lostroh, C.P., Ogi, T., and Greenberg, E.P. (2003). Identification, timing, and signal specificity of *Pseudomonas aeruginosa* quorum-controlled genes: a transcriptome analysis. J. Bacteriol. *185*, 2066–2079.

Schwab, U., Bowen, B., Nadon, C., Wiedmann, M., and Boor, K.J. (2005). The *Listeria monocytogenes prfAP2* promoter is regulated by sigma B in a growth phase dependent manner. FEMS Microbiol. Lett. *245*, 329–336.

Shemesh, M., Tam, A., Aharoni, R., and Steinberg, D. (2010). Genetic adaptation of *Streptococcus mutans* during biofilm formation on different types of surfaces. BMC Microbiol. *10*, 51.

Singh, P.K., Schaefer, A.L., Parsek, M.R., Moninger, T.O., Welsh, M.J., and Greenberg, E.P. (2000). Quorum-sensing signals indicate that cystic fibrosis lungs are infected with bacterial biofilms. Nature. *407*, 762–764.

Smith, J. (2001). The social evolution of bacterial pathogenesis. Proc. Roy. Soc. B. *268*, 61–69.

Smith, J.M., and Harper, D. (2003). Animal Signals (Oxford University Press, Oxford, UK).

Stewart, P.S., and Franklin, M.J. (2008). Physiological heterogeneity in biofilms. Nat. Rev. Microbiol. *6*, 199–210.

Stickler, D.J., Morris, N.S., McLean, R.J.C., and Fuqua, C. (1998). Biofilms on indwelling urethral catheters produce quorum-sensing signal molecules *in situ* and *in vitro*. Appl. Environ. Microbiol. *64*, 3486–3490.

Storey, D.G., Ujack, E.E., Rabin, H.R., and Mitchell, I. (1998). *Pseudomonas aeruginosa lasR* transcription correlates with the transcription of *lasA*, *lasB*, and *toxA* in chronic lung infections associated with cystic fibrosis. Infect. Immun. *66*, 2521–2528.

Strassmann, J.E., Zhu, Y., and Queller, D.C. (2000). Altruism and social cheating in the social amoeba *Dictyostelium discoideum*. Nature *408*, 965–967.

Surette, M.G., Miller, M.B., and Bassler, B.L. (1999). Quorum sensing in *Escherichia coli*, *Salmonella typhimurium*, and *Vibrio harveyi*: a new family of genes responsible for autoinducer production. Proc. Natl. Acad. Sci. U.S.A. *96*, 1639–1644.

Taylor, M.W., Schupp, P.J., Baillie, H.J., Charlton, T.S., De Nys, R., Kjelleberg, S., and Steinberg, P.D. (2004). Evidence for acyl homoserine lactone signal production in bacteria associated with marine sponges. Appl. Environ. Microbiol. *70*, 4387–4389.

Tielker, D., Hacker, S., Loris, R., Strathmann, M., Wingender, J., Wilhelm, S., Rosenau, F., and Jaeger, K.-E. (2005). *Pseudomonas aeruginosa* lectin LecB is located in the outer membrane and is involved in biofilm formation. Microbiology *151*, 1313–1323.

Van Loosdrecht, M.C.M., Heijnen, J.J., Eberl, H., Kreft, J., and Picioreanu, C. (2002). Mathematical modelling of biofilm structures. Antonie Van Leeuwenhoek *81*, 245–256.

Velicer, G.J., Kroos, L., and Lenski, R.E. (2000). Developmental cheating in the social bacterium *Myxococcus xanthus*. Nature *404*, 598–601.

Vuong, C., Saenz, H.L., Götz, F., and Otto, M. (2000). Impact of the *agr* quorum-sensing system on adherence to polystyrene in *Staphylococcus aureus*. J. Infect. Dis. *182*, 1688–1693.

Wagner, V.E., Bushnell, D., Passador, L., Brooks, A.I., and Iglewski, B.H. (2003). Microarray analysis of *Pseudomonas aeruginosa* quorum-sensing regulons: effects of growth phase and environment. J. Bacteriol. *185*, 2080–2095.

Wang, L., Li, J., March, J.C., Valdes, J.J., and Bentley, W.E. (2005). *luxS*-dependent gene regulation in *Escherichia coli* K-12 revealed by genomic expression profiling. J. Bacteriol. *187*, 8350–8360.

West, S.A., Griffin, A.S., Gardner, A., and Diggle, S.P. (2006). Social evolution theory for microorganisms. Nat. Rev. Microbiol. *4*, 597–607.

West, S.A., Griffin, A.S., and Gardner, A. (2007). Evolutionary explanations for cooperation. Curr. Biol. *17*, R661–672.

Whitchurch, C.B., Tolker-Nielsen, T., Ragas, P.C., and Mattick, J.S. (2002). Extracellular DNA required for bacterial biofilm formation. Science *295*, 1487.

Williams, P., Winzer, K., Chan, W.C., and Cámara, M. (2007). Look who's talking: communication and quorum sensing in the bacterial world. Phil. Trans. Roy. Soc. B. *362*, 1119–1134.

Winzer, K., Hardie, K.R., and Williams, P. (2002). Bacterial cell-to-cell communication: sorry, cannot talk now – gone to lunch! Curr. Opin. Microbiol. *5*, 216–222.

Xavier, J.B., and Foster, K.R. (2007). Cooperation and conflict in microbial biofilms. Proc. Natl. Acad. Sci. U.S.A. *104*, 876–881.

Xu, L., Li, H., Vuong, C., Vadyvaloo, V., Wang, J., Yao, Y., Otto, M., and Gao, Q. (2006). Role of the luxS quorum-sensing system in biofilm formation and virulence of *Staphylococcus epidermidis*. Infect. Immun. *74*, 488–496.

Yarwood, J.M., Bartels, D.J., Volper, E.M., and Greenberg, E.P. (2004). Quorum sensing in *Staphylococcus aureus* biofilms. J. Bacteriol. *186*, 1838–1850.

Yoshida, A., Ansai, T., Takehara, T., and Kuramitsu, H.K. (2005). LuxS-based signaling affects *Streptococcus mutans* biofilm formation. Appl. Environ. Microbiol. *71*, 2372–2380.

Zhang, K., Ou, M., Wang, W., and Ling, J. (2009). Effects of quorum sensing on cell viability in *Streptococcus mutans* biofilm formation. Biochem. Biophys. Res. Commun. *379*, 933–938.

Zhang, Z., and Pierson, L.S. (2001). A second quorum-sensing system regulates cell surface properties but not phenazine antibiotic production in *Pseudomonas aureofaciens*. Appl. Environ. Microbiol. *67*, 4305–4315.

Biofilms in Disease

2

James D. Bryers

Abstract

Clinically related research on biofilms has expanded exponentially in the past 10 years owing to the pandemic of nosocomial (hospital-related) infections. Biofilms are thought to cause a significant amount of all human microbial infections, according to the Centers for Disease Control and Prevention. Nosocomial infections are the fifth leading cause of death in the USA, with >2 million cases annually (or ~10% of American hospital patients). The difficulty of eradicating biofilm bacteria with classic systemic antibiotic treatments is a prime concern of medicine. Biofilm bacteria can be up to a thousand times less susceptible to antimicrobial stress than their freely suspended counterparts. This chapter discusses the pathogenesis of a number of biofilm-mediated infections, including: oral infections, biomedical device based infections, osteomyelitis, otitis media, and others. Emerging research in biofilm control and prevention is also reviewed.

Introduction

Bacteria–host interactions range from mutual symbiosis that benefits both parties to rampant infections that could kill the host. Lying somewhere in between, chronic infections result from a stalemate between reduced bacterial virulence and partial success of host defences. In chronic infections, bacteria adapt to the specific environment to be more persistent rather than outright invasive. The 'biofilm mode of persistence' plays a critical role in this adaptation (Singh *et al.*, 2000; Høiby *et al.*, 2001).

In this chapter I first discuss how bacteria, especially within biofilms are able to successfully ignore their host's most aggressive defence mechanism in order to create acute or chronic infections. Then I review biofilms in disease within two contexts: (1) biofilm infections related to host tissues versus (2) biomedical implant-related infections.

Ignoring their hosts

Immune evasion by bacteria

The primary defence against infection is the innate protection provided by neutrophils, macrophages,and dendritic immune cells. However, *Staphylococcus aureus*, for example, has the ability to thwart neutrophils and macrophages by (a) inhibiting chemotaxis (by secreting molecules that block receptors that recognize formylated peptides, complement C5a, and/or LFA1-ICAM1 mediated extravasation; and by secreting leukotoxins); (b) negating 'opsonization', a process by which pathogens are marked (by opsonin) for destruction by a phagocyte, via protein G reverse IgG binding, plasmin degradation of bound IgG and C3b, and blocking C3b binding; and (c) thwarting phagocytosis (via cell wall modifications to resist low endosomal pH, or enzymatic degradation of endosomes) (Foster, 2005). *Staphylococcus epidermidis*, the main species isolated in the majority of nosocomial infections, avoids the immune system through adhesion and biofilm formation (Vuong *et al.*, 2004; Sun *et al.*, 2005). Li *et al.* (2007) documented that Gram-positive bacteria (such as staphylococci) can actually sense antimicrobial peptides released by neutrophils and macrophage and then coordinate a directed defensive response. The authors discovered an antimicrobial peptide sensor system that controls major specific resistance mechanisms; the sensor contains a classical two-component signal transducer and an unusual third protein, all of which are indispensable for signal transduction and antimicrobial peptide resistance.

Complement evasion by many Gram-positive bacteria involves incorrect binding of complement recognition factors by bacterial cell wall proteins (proteins A, G, M) (Rooijakkers and van Strijp, 2007). Gram-negative bacteria have also evolved secretory proteins that can degrade complement factors or their binding components, or that prove anti-chemotactic or toxic to immune cells.

Entry of the oral pathogen, *Porphyromonas gingivalis*, into monocyte-derived dendritic cells (DC) *in vitro*, leading to suboptimal DC maturation, requires *P. gingivalis* to express the major fimbriae, FimA (Evans *et al.*, 1992; Jotwani and Cutler, 2004). *P. gingivalis* also expresses unique immunosuppressive lipopolysaccharides (LPS) (Pulendran *et al.*, 2001; Jotwani *et al.*, 2003; Cohen *et al.*, 2004) and proteolytic gingipains (cytokine-degrading proteases; Potempa *et al.*, 2003). LPS of *P. gingivalis*, related to those of *E. coli*, stimulate dendritic cells to secrete the cytokine interleukin 10 (IL-10), but not IL-12, *in vitro* (Jotwani *et al.*, 2003) and *in vivo* (Pulendran *et al.*, 2001). These two factors (suboptimal DC maturation and truncated cytokine expression) leads to the induction of a type 2 helper T cell (Th2)-effector response, which suggests that *P. gingivalis* may target dendritic cell C-type lectin receptors (e.g. dendritic cell-specific ICAM-3-grabbing non-integrins) for entry, and for blunting of dendritic cell maturation. Bacteria that target C-type lectin receptors are unique in their ability to survive in the host DCs and induce chronic life-long infections. C-type lectin receptors function primarily to interact with conserved carbohydrates shared by a large group of microbes, and internalize these pathogens for processing and antigen-presentation. While Toll-like receptors rapidly transmit pathogen-associated molecular patterns through Toll-Interleukin 1 receptor activation domains, C-type lectin receptors appear to lack these activation domains, and thus do not generally lead directly to dendritic cell maturation and cytokine secretion (van Kooyk and Geijtenbeck, 2003; van Kooyk *et al.*, 2004). Thus, some pathogens obviate the immune response by specifically targeting C-type lectin receptors, particularly dendritic cell-specific ICAM-3-grabbing non-integrins and mannose receptors, which benefits the pathogen by down-regulating intracellular signalling, inhibiting maturation, and cytokine secretion. Pathogens, like *P. gingivalis*, that interact with dendritic cell-specific ICAM-3-grabbing non-integrins cause chronic infections that can last a lifetime by manipulating the balance of Type 1 (Th1) and Type 2 (Th2) helper T cells (van Kooyk and Geijtenbeck, 2003).

Biofilm–immune cell interactions

As indicated above, freely suspended microorganisms have evolved a multitude of ways to avoid detection by the immune system. One of these mechanisms is to form biofilms. Historically, there have been few studies that focused on biofilm-bound bacterial cell interaction with cells of the immune system, but this is changing. As I discuss here, there appears to be a dynamic feedback between immune cells and bacteria that may be a causative mechanism in chronic inflammation and reduced wound healing. Biofilm bacteria may subtly manipulate immune responses while reactions of immune cells may exacerbate biofilm formation and reinforce virulence.

For years, one explanation behind a biofilm's resistance to the immune system was that immune cells could not penetrate into the biofilm. However, Leid *et al.* (2002) documented such penetration in an *in vitro* study of freshly isolated human leucocytes contacting 2-day-old (early, maturing) and 7-day-old (fully developed) *S. aureus* biofilms, under static- and flowing-fluid conditions that mimic physiological shear. Leucocytes penetrated 7-day-old *S. aureus* biofilms under laminar-shear conditions but did not penetrate 2-day-old *S. aureus* biofilms grown under static conditions. Leucocytes produced a Th1-type response to both the 7-day-old shear and 2-day-old static *S. aureus* biofilms. That leucocytes were able to penetrate biofilms but unable to phagocytose bacteria suggests that other mechanisms may inhibit normal leucocyte function. Leid *et al.* (2005) continued this investigation to report that the exopolysaccharide alginate matrix protects *P. aeruginosa* biofilm bacteria from interferon-γ (IFN-γ, a soluble cytokine)-mediated leucocyte killing.

Jesaitis *et al.* (2003) exposed *P. aeruginosa* biofilms to neutrophils then measured *in vitro* oxygen consumption, cell morphology, and host defensive processes. They report that the consumption of oxygen increases in both the bacteria and neutrophils, but with little accumulation of H_2O_2, when leucocytes settled on established biofilm. Neutrophils did not exhibit significant motility or the characteristic polarized morphology of motile cells, but were still able to engage in phagocytosis while retaining a roughly spherical shape and modest membrane ruffling. Bacteria, on the other hand, exhibited increased oxygen consumption and a subsequent detachment response that released a cloud of bacteria from the biofilm that eventually enveloped and obscured the settled neutrophils.

Walker *et al.* (2005) report that human neutrophils serve to enhance the initial development of *P. aeruginosa* biofilms. The mechanism of biofilm enhancement by neutrophils was attributed to neutrophil-generated polymers comprising actin and DNA. The bacteria bind to F-actin while the free DNA promotes biofilm matrix stability. Disruption of the polymers with DNase results in dispersion of the bacteria and a reduction in biofilm development. The presence of these actin-DNA polymers, with co-localization of *P. aeruginosa,* was confirmed in both neutrophil lysates and sputum from cystic fibrosis (CF) patients. Introduction of additional neutrophils after 24 and 48 hours further enhanced *P. aeruginosa* biofilm development, while exposure to fewer neutrophils resulted in a lesser degree of biofilm enhancement.

Similarly, Chandra *et al.* (2007) showed that peripheral blood mononuclear cells (PBMCs) enhanced the ability of the yeast, *Candida albicans,* to form biofilms and that the majority of PBMCs were localized to the basal and middle layers of the biofilm. They also found that only viable PBMCs enhanced *Candida* biofilm formation and that PBMCs cell surface components did not contribute to this biofilm enhancement; the biofilm-enhancing effect is mediated by a soluble factor released into the medium, and the supernatant collected from co-cultures contained differential levels of pro- and anti-inflammatory cytokines.

Zimmermann *et al.* (2006) reports that the quorum-sensing (QS) molecule *N*-(3-oxododecanoyl)-l-homoserine lactone (3OC12-HSL) synthesized by *P. aeruginosa* induces directed migration (chemotaxis) of neutrophils in an *in vitro* assay. They proposed that this early attraction of polymorphonuclear leucocytes (PMN) could contribute to prevention of biofilm formation, which may be an inaccurate conclusion. It could be possible that these bacterial signals are actually used by adherent bacteria to stimulate an immune cell response, which in turns enhances biofilm formation and virulence. In a mini-review, Wagner *et al.* (2006) discusses the topic of communication between adherent bacteria and host immune cells. Wu *et al.* (2005) demonstrate that IFN- binds to an outer membrane protein in *P. aeruginosa,* OprF, resulting in the expression of a quorum-sensing dependent virulence determinant, the PA-I lectin. To determine whether IFN-γ induced PA-I by activation of the QS signalling system, they measured *rhlI* gene expression in response to IFN-γ. *RhlI* is the gene required for the synthesis of C4-HSL (C4-homoserine lactone), a core QS signalling molecule that plays a central role in the expression of PA-I. IFN-γ induced *rhlI* transcription in *P. aeruginosa* PAO1 and C4-HSL synthesis increased significantly after exposure to IFN-γ. Activation of the QS system by IFN-γ also led to the increase of pyocyanin production, another quorum sensing-dependent virulence product. The results by Wu *et al.* (2005) and others (Smith and Iglewski, 2003) suggest that adherent bacteria not only actively sense but can also manipulate host immune response (Smith and Iglewski, 2003; Tateda *et al.*, 2003).

Mittal *et al.* (2006) report that strains of *Pseudomonas aeruginosa* vary in their ability to induce macrophage response, which effects the composition of secreted products (cytokines, chemokines). Concomitantly, these different compositions of macrophage-secreted products influenced overall biofilm formation and up-regulated expression of a litany of virulence factors in a cytokine composition-dependent manner, again indicating the need to study macrophage response to biofilm, and not planktonic bacteria.

Tissue-related biofilm diseases

For tissue-related biofilm infections, I will highlight four diseases: cystic fibrosis (CF) airway infections, chronic otitis media, oral biofilms and periodontitis, and osteomyolytis. Owing to page constraints, there are other tissue-related biofilm diseases I will not address here. The reader is directed to a number of specific review articles that address these omissions; they include infectious kidney stones (Abrahams and Stoller, 2003; Healy and Ogan, 2007; Marcus *et al.*, 2008), bacterial endocarditis (Chopra and Kaatz, 2010; Fitzsimmons *et al.*, 2010; Timóteo *et al.*, 2010), vaginosis (Fichorova, 2009; Reid *et al.*, 2009; Verstraelen and Verhelst, 2009) and urinary tract infections (Hatt and Rather, 2008).

Cystic fibrosis infections

Of all tissue-related biofilm infections, CF infections are by far the most extensively investigated and yet CF remains the most mysterious of biofilm diseases. CF, an inherited disease, is caused by mutations in the cystic fibrosis transmembrane regulator (CFTR) chloride channel found primarily in apical membranes of epithelial cells. Exactly how the mutated CFTR protein causes subsequent infection remains a controversy.

There are two clinically distinct stages of CF infections (Govan and Deretic, 1996; Rajan and Saiman, 2002). First, patients develop intermittent 'bronchitis-like' respiratory infections comprising a variety of organisms, including *Haemophilus influenzae*, *Staphylococcus aureus* and *Pseudomonas aeruginosa* (Rajan and Saiman, 2002). The second stage of CF is a permanent infection with only *P. aeruginosa* that will persist for the rest of the patient's life (Pedersen, 1992; Govan and Deretic, 1996). DNA fingerprinting of clinical isolates show that in the latter stage the same *P. aeruginosa* strain can persist in the airway for decades (Kersulyte *et al.*, 1995).

While intensive treatment may postpone and reduce the damage caused by CF chronic infection, it cannot eliminate it. During chronic *P. aeruginosa* infection, CF patients experience a continuous degradation of lung tissue caused equally by the infection and by inflammatory processes. Bjarnsholt *et al.* (2009a,b) recently evaluated the orientation and distribution of *P. aeruginosa* in the various zones of airways of chronic *P. aeruginosa*-infected CF patients that received *intensive* treatment with antibiotic regimens. Despite the use of a universal eubacterial peptide nucleic acid fluorescence *in situ* hybridization (PNA FISH) probe (designed to detect all bacteria) in combination with a specific *P. aeruginosa* PNA FISH probe, *P. aeruginosa* was the only bacterium detected in their study. Post-mortem examinations show *P. aeruginosa* mainly localized in the upper part of the lung; all of the recovered bacteria were embedded in mucus plugs that varied considerably in size and spatial orientation. Very few planktonic bacteria were detected. A large number of host polymorphonuclear leucocytes (PMNs) surrounded the mucoid aggregates. In the lower, respiratory part of the lungs (the alveoli), Burmølle *et al.* (2010) observed both dead alveoli and air-filled healthy alveoli. Relatively few bacteria were detected in the respiratory part of the lungs as compared with the conductive part. Planktonic bacteria always appeared phagocytosed by PMNs and the aggregated bacteria were closely surrounded by PMNs. When examining lungs from *non-intensively* treated chronically infected CF patients, Burmølle *et al.* (2010) observed a different pattern of bacteria distribution; here, the alveoli of the respiratory zone were also filled with aggregating bacteria and PMNs. Most of the lung tissue was extensively destroyed, caused by the heavy infection and inflammation.

More in-depth review of CF biofilms can be found in recent articles by Moreau-Marquis *et al.* (2008), Tart and Wozniak (2008), Wagner and Iglweski (2008), Hassett *et al.* (2009), Hall-Stoodley and Stoodley (2009), Davies and Bilton (2009), George *et al.* (2009), Hassett *et al.* (2010) and Høiby *et al.* (2010).

Chronic otitis media

Chronic suppurative otitis media (middle ear infection) without any tympanostomy tube insertion (a procedure used to keep the middle ear aerated) is characterized by recurrent chronic suppuration (the discharge of pus) succeeded by varying lengths of dry periods. Such suppurative infections occur in patients first afflicted with chronic otitis media with dry perforations or following acute otitis media (AOM), where treatment fails or is not instigated. Once established, chronic OM can often be recalcitrant and difficult to treat. Chronic OM is characterized by polymicrobial aerobic and anaerobic bacteria, such as: *P. aeruginosa, Escherichia coli, Staphylococcus aureus,* pneumococci and *Haemophilus influenzae,* all known biofilm formers.

Numerous articles over the last decade have documented biofilm presence experimentally and clinically in different chronic middle ear infectious diseases. The first observation was carried out experimentally in the middle ear of chinchillas with chronic otitis media with effusion (Swords *et al.,* 2004; Jurcisek *et al.,* 2005; Reid *et al.,* 2009) and later directly with clinical samples of the mucosal surface lining from children with recurrent acute otitis media (Rayner *et al.,* 1998; Post, 2001; Hall-Stoodley *et al.,* 2006). Biofilms have been found in human cholesteatoma, which is another chronic middle ear disease, as well as in experimentally induced cholesteatomas in gerbils (Chole and Faddis, 2002). Biofilms are also frequently found on tympanostomy tubes (small tubes inserted into the eardrum in order to keep the middle ear aerated for a prolonged period of time, and to prevent the accumulation of mucus in the middle ear) and most recently on a human cochlear implants (Saidi *et al.,* 1999; Bothwell *et al.,* 2003; Pawlowski *et al.,* 2005; see also 'Medical device-based infections', below).

Thus, biofilms have been found in several chronic infectious middle ear diseases (Bakaletz, 2007). However, the potential pathogenic role of biofilm in these diseases and the cause and relationship still remain to be elucidated. Recent review articles addressing the latest advances in biofilm mediated middle ear infections include Kilty and Desrosiers (2008), Macassey *et al.* (2008), Marchisio *et al.* (2010) and Pagella *et al.* (2010).

Oral biofilms, dental caries and periodontal complications

Dental caries is an infectious, communicable disease resulting in destruction of tooth structure by acid-forming bacteria found in dental plaque, an intraoral biofilm. The disease results in tooth loss if left untreated. Although significant progress has been made in reducing and controlling dental caries, the disease still remains a problem for many children and adults in the USA The problem is much worse for children from poor families: more than one-third of all children aged 2 to nine years old living in poverty in the USA have one or more untreated primary teeth (Edelstein, 2002). The USA also suffer from significant and persistent racial disparities in dental health and treatment.

During the past few decades, changes have been observed not only in the prevalence but also in the distribution and pattern of dental caries in the population. Most notably, the relative distribution of dental caries on different tooth surfaces has changed, and the rate of lesion progression through teeth has slowed. These changes have important implications for diagnosing and treating incipient lesions, predicting caries risk, and conducting effective disease prevention programs for individuals or populations. Scientific research continues to make progress in the identification of best practices for diagnosing, treating, and preventing dental caries. Traditional approaches for treating carious lesions in a surgical manner are being replaced by newer strategies that emphasize disease prevention and conservation of tooth structure. However, the mainstays of caries prevention, topical and systemic fluorides and pit and fissure sealants, are technologies developed in the 1950s and 1960s. The US and state governments bear the majority of the cost of trying to control this pandemic through Medicaid, the Public Health Service, Indian Health Service and other similar programs. These costs continue to escalate, while continued applications of existing technology are unlikely to markedly reduce disparities.

Periodontitis is a bacterial-associated inflammatory disease characterized by the destruction of supporting tissues of the teeth and resulting, if untreated, in the exfoliation of teeth. Periodontitis is a major public health problem affecting 30% of the adult dentate population, with severe forms affecting 5–10% (Oliver *et al.*, 1998). The annual economic burden of periodontitis was estimated at $14 billion in the USA in 1999 (Brown *et al.*, 2002). More insidiously, several epidemiological and animal studies indicate a link between chronic periodontitis and an increased risk of cardiovascular diseases (Wu *et al.*, 2000; Jain *et al.*, 2003; Lin *et al.*, 2003; Miyakawa *et al.*, 2004), spontaneous preterm birth and preterm low birth weight (Offenbacher *et al.*, 1998; Pucher and Stewart, 2004), diabetes (Mercado *et al.*, 2001) and rheumatoid arthritis (Socransky and Haffajee, 2000). Colonization of the subgingival crevice by a consortium of Gram-negative bacteria within a biofilm is strongly associated with the onset and progression of periodontitis (Socransky *et al.*, 1998). Three bacterial species within such biofilms have been strongly associated with periodontitis: *Porphyromonas gingivalis*, *Bacteroides forsythus*, and *Treponema denticola* (Slots, 1982), with *P. gingivalis* presence correlating strongly with disease severity. *P. gingivalis* represents ≤50% of the anaerobically cultivatable subgingival bacteria from active disease sites (van Winkelhoff *et al.*, 2002) but *P. gingivalis* is rarely cultured from subjects without periodontitis (Mayrand and Holt, 1988). *P. gingivalis* possesses a number of virulence factors including LPS, fimbriae, haemagglutinin, haemolysin and extracellular hydrolytic enzymes (especially the Arg- and Lys-specific proteinases) (O'Brien-Simpson *et al.*, 2003). A virulence factor common to all strains of *P. gingivalis* is the cell surface Arg-specific (RgpA) and Lys-specific (Kgp) proteinases and their associated adhesins, designated the RgpA–Kgp complex (O'Brien-Simpson *et al.*, 2003). Both outer membrane-associated and released forms of the proteinases are considered to be major virulence determinants in the onset and progression of chronic periodontitis (Kadowaki *et al.*, 2000). Interestingly, Beikler *et al.* (2005) report that the *rgpA* sequence and thus assumingly the encoded RgpA protein of clinical *P. gingivalis* strains is highly homogeneous, thus indicating that the RgpA protein may be an ideal candidate for vaccination studies against *P. gingivalis*.

Osteomyelitis

Osteomyelitis is defined as an infection of the bone (Mader *et al.*, 1997). There are two types of osteomyelitis: *haematogenous* (long bone and vertebral) osteomyelitis and *contiguous focus* osteomyelitis. Haematogenous osteomyelitis is caused either by the direct seeding of bone from a bacterial species in the blood or by secondary infection, where the bacteria gain access to the bloodstream and seed distal bone and marrow. Contiguous focus osteomyelitis implies that the infection stems from an adjacent soft tissue infection. Chronic contiguous focus osteomyelitis can also begin as an acute infection, with the microorganisms being directly inoculated into the bone at the time of trauma (e.g. warfare, accidents, medical device implantation). Osteomyelitis infections commonly occur in the long bones of the legs, but may also be found in nearly every bone of the body. *Staphylococcus aureus* is, by far, the most commonly isolated organism in monomicrobiotic osteomyelitis cases. The prevalence of *Staphylococcus aureus* in osteomyelitis is due to its wide array of virulence factors, including secreted products for host damage and immune avoidance, as well as adherence factors.

During early stage infections, the infection will usually clear after 2–4 weeks of continuous antibiotic treatment (Berendt and Byren, 2004) if started early. However, diagnosing early stage infections is problematic. Acute osteomyelitis usually does not feature any radiographic changes until 1–1.5 weeks after inception of the disease. Magnetic resonance imaging (MRI) is effective in diagnosing bone infections in the absence of metal implants, but it suffers from a lag time after previous surgery or prior infections (Berendt and Byren, 2004). In chronic osteomyelitis, there are large areas of devitalized cortical and cancellous bone within the wound. Since antibiotics do not penetrate well into devitalized bone (Healy and Freedman, 2006), dead areas must be completely debrided (surgically removed), including scar tissue, marrow, and cortex. Diagnosis of chronic bone infection is made by radiography, while other methods such as radionuclide scans, lack specificity (Berendt and Byren, 2004).

The absence of general vascular sufficiency makes suitable therapy and management of chronic contiguous osteomyelitis complicated. Most of the patients with contiguous osteomyelitis have diabetes mellitus (Calhoun *et al.*, 1988) and range from 35–70 years of age. The small bones of the feet, as well as the talus, calcaneus, distal fibula, and tibia, are commonly involved in contiguous osteomyelitis. Often, the infection begins with a minor trauma to the feet, such as infected nail beds, cellulitis, or trophic skin ulceration. Neuropathy in these patients impairs the proper functioning of the foot, leading to progression of soft tissue infections into the underlying bone (Berendt and Byren, 2004). Osteomyelitis in patients with compromised vasculature can be difficult to diagnose. While the patient may present with an ingrown toenail, a perforating foot ulcer, cellulitis, or a deep space infection, fever and systemic toxicity are often absent. The recalcitrant nature of this type of infection often leads to recurrent bone infections, even after suitable therapy. Debridement and ablation of the infected tissue is often necessary. Because staphylococci are the most common infectious agent in osteomyelitis, it is advisable to begin therapy upon patient presentation with a regimen that includes an anti-staphylococcal agent (Berendt and Byren, 2004). In adults, *S. aureus* is generally treated with penicillin [for methicillin-sensitive *S. aureus* (MSSA)] or nafcillin [for methicillin-resistant *S. aureus* (MRSA)], with cefazolin, clindamycin, vancomycin, ciprofloxacin, or levofloxacin being given as alternatives (Lew and Waldvogel, 2004).

Summary: tissue-related biofilm diseases

Burmølle *et al.* (2010) suggest that chronic biofilm infections differ ecologically from natural symbiotic biofilms (such as resident in the oral cavity, nasal or vaginal mucosa, or the intestinal tract). Based on their observations, chronic infections most often are technically 'multispecies', although the individual aggregates themselves are predominantly monocultures per aggregate. Different bacterial species can form individual aggregates in the same infection, but at different locations. Thus these biofilms are not truly multispecies; only the infection can be characterized as multispecies. This is in contrast to natural symbiotic biofilms, where several bacterial species are spatially intermixed throughout the matrix. Reasons for this difference in bacterial organization might be (a) selection pressure, (b) availability of nutrients and (c) commensalism. Co-habitation in natural benign biofilms could be justified by beneficial catabolism and anabolism of compounds shared among the different inhabitants. The situation is very different in biofilm infections. Here, the main obstacle for colonizing bacteria is the host's defence system. The initial phase of biofilm formation in infections is the crucial point; by chance, the right (opportunistic) pathogenic bacteria needs to be at the right place at the right time in order to establish an infection in spite of the activities of the host defence system. Chances are low for another species being able to out-compete the first species when the first is within an established biofilm surrounded by a heightened immune system. Further, dead cells (both eukaryotic and prokaryotic) and a constant blood supply result in high nutrient availability in these infected environments, imposing a lower selection pressure for coexistence and niche differentiation (i.e. symbiosis is not a crucial requisite for survival).

Biomedical implant-related infections

As already stated, biofilms cause a significant amount of all human microbial infections. Nosocomial (hospital acquired) infections are the fifth leading cause of death in the USA with 2 million cases annually (or ~10% of American hospital patients) leading to more than \$5 billion in added medical cost per annum (Wenzel, 2007).

About 60–70% of nosocomial infections are associated with some type of implanted medical device. It is estimated that over 5 million medical devices or implants are used per annum in the USA alone (Ratner *et al.*, 2004). Microbial infections have been observed on most, if not all, such devices, including: prosthetic heart valves, orthopaedic implants, intravascular catheters, artificial hearts, left ventricular assist devices, cardiac pacemakers, vascular prostheses, cerebrospinal fluid shunts, urinary catheters, ocular prostheses and contact lenses, and intrauterine contraceptive devices. Worldwide production of biomedical devices and tissue engineering-related materials is a $180 billion per year industry and expanding rapidly (Ratner *et al.*, 2004). Regardless of the sophistication of the biomedical implant (catheters vs. a three-dimensional stem cell-containing polymer scaffold), all medical devices or tissue engineering constructs are susceptible to microbial colonization and infection (Bryers and Ratner, 2004; Castelli *et al.*, 2006).

Biomaterials infection, inflammation and host immune response

The host response to implanted medical devices follows a cascade of events initially similar to, but ultimately diverging from, normal wound healing. The initial injury (a wound or the implantation of a medical device) causes a perturbation of the homeostatic mechanisms, setting off the process of healing (Anderson, 2001). The classic view of the host response to biomaterials divides the response into several overlapping phases: blood–material interactions, acute inflammation, chronic inflammation, foreign body reaction, and fibrous encapsulation.

Inflammatory cell recruitment requires enhanced leucocyte chemotaxis to the implant site, as well as leucocyte transmigration through the endothelium. Various molecular signals at the implant site can act as chemoattractants for the initial recruitment of inflammatory cells. Initial injury from the implantation procedure induces injured cells to release a variety of distress signals such as, ATP, uric acid, and bioactive lipids. Activated platelets release several growth factors and activated mast cells likewise release various degranulation products (e.g. histamine, proteoglycans, proteases) and cytokines [e.g. tumour necrosis factor (TNF)-α, and interleukin 16 (IL-16)] (Mekori and Metcalf, 2000). Mast cells can also initiate the production of various other cytokines and chemokines, such as granulocyte–macrophage colony-stimulating factor (GM-CSF), IL-4, IL-6, IL-10, IL-13, macrophage inflammatory protein (MIP)-α, and monocyte chemotactic protein (MCP) 1. As leucocytes arrive at the site of injury, they, too, begin to produce various growth factors and cytokines that attract more inflammatory cells. MIP-1α and MCP-1 have been implicated in phagocyte chemotaxis towards the implant (Tang *et al.*, 1998).

Following their chemotaxis towards the medical implant and transmigration through the endothelium, leucocytes adhere to the implant surface and become activated to carry out their functions. Since neutrophils arriving first at the implant site are 'spent' within hours to less than a day (Savill, 1996), the focus of most cell–biomaterial interaction studies has been the next incoming cell line, macrophages. Macrophages are widely considered the central cell type in directing and determining the outcome of the host response against implanted materials, because they are the predominant cell type found at the tissue–material interface in the later stages of the host response (Schwartz *et al.*, 1999). It is commonly hypothesized that host cells interact with surface-adsorbed proteins, rather than with the biomaterial surface itself (Hu *et al.*, 2001; Wilson *et al.*, 2005). Macrophage function is a dynamic process, which is dependent on the modulation of macrophage activation states by either the changing local microenvironment (Stout and Suttles; 2004; Stout *et al.*, 2005) or by the macrophages themselves, as they release a wide variety of cytokines, extracellular matrix (ECM) proteins, and growth factors. The presence of adherent infecting bacteria can lead to a chronic inflammatory situation at the device interface. Chronic inflammation is associated with macrophages, lymphocytes, proliferation of connective tissues, deposition of matrix proteins, and capillary neogenesis (or tissue regeneration; Shankar and Greisler, 1996). Ultimately, macrophages may form foreign body giant cells (FBGCs) in an attempt to 'wall off' the implanted device, leading to fibrous encapsulation.

Good intentions gone bad

Since native immunity can be circumvented or compromised (by drugs or disease), the medical profession has attempted to eradicate biofilm-based infections by resorting to disinfectants and antibiotics. These are largely synthetic compounds evaluated for the most part on their ability to inactivate or kill *suspended* bacteria but exhibit little efficacy when applied to biofilm infections (Stewart and Costerton, 2001; Stewart, 2002). Apparently, biofilm bacteria are significantly less responsive to antibiotics and antimicrobial stressors than planktonic organisms of the same species (Gilbert *et al.*, 2002; Stewart, 2002). Further, recent studies have shown that sublethal doses of antibiotics can actually enhance biofilm formation. Bagge *et al.* (2004) reports that genes coding for alginate biosynthesis were induced by exposure to the β-lactam antibiotic, imipenem. Exposure to subinhibitory concentrations of imipenem caused structural changes in the biofilm, e.g. an increased biofilm volume and alginate polymer matrix. Increased levels of alginate matrix production may be an unintended adverse consequence of imipenem treatment in cystic fibrosis patients. Similarly, Hoffman *et al.* (2005) report that subinhibitory concentrations of aminoglycoside antibiotics (e.g. tobramycin) induced biofilm formation in *Pseudomonas aeruginosa* and *Escherichia coli*. We have also observed this enhanced biofilm formation response for *Pseudomonas putida* biofilms exposed to sublethal doses of kanamycin, ciprofloxacin and norfloxacin (Ma and Bryers, unpublished data). These results suggest enhanced biofilm formation in the presence of antibiotics may be one universal defence mechanism of bacteria in avoiding the lethal effects of antibiotics.

The biomaterials community over the past 25 years has attempted to produce anti-infective devices or implants by either (1) mechanical design alternatives (liquid–air breaks, skin cuffs, and antibiotic fills; all for indwelling catheters), (2) tethered anti-infective agents, bound directly to the surface of the material (silver coatings, tethered quaternary ammonium, synthetic antibiotics) or (3) the release of soluble toxic agents (chlorhexidine, antibiotics) into the adjacent surroundings. Mechanical design alternatives have had only marginal success and are only applicable for short-term indwelling catheters. Tethered anti-infective agents are only toxic to the initial wave of incoming bacteria and provide little residual effects once layers of dead cells accumulate, which are also inflammatory. Finally, regardless of the type of 'drug release' method used (passive vs. sustained vs. responsive), release of a toxic agent from a biomaterial of a soluble anti-infective agent will inevitably stop once the entrapped agent is depleted. Further, as shown above, delivery of sublethal dosages of antibiotics can lead to accelerated biofilm formation and induced virulence factor expression.

New directions in medical biofilm control

Traditional treatment of microbial infections is based on compounds that kill or inhibit growth of the microbe. One major concern with this approach is the frequent development of resistance to antibiotics (Geddes, 2000). As stated above, biofilm communities tend to be significantly less responsive to antibiotics and antimicrobial stressors than planktonic organisms of the same species. A further complication is that the spread of antibiotic resistance genes borne on plasmid DNA, within and between species, is greatly exacerbated in biofilm communities (Beaudoin *et al.*, 1998a,b). As a consequence of this increase in resistance, researchers have turned to a number of alternatives to synthetic antibiotics, including: bacteriophage (Sulakvelidze *et al.*, 2001) and bacteriophage lytic enzymes (Fischetti, 2005), probiotics (Persson, 2005; Hong *et al.*, 2005), and human antimicrobial peptides (defensins, cathelicidins and histatins) (De Smet and Contreras, 2005). The success of these alternatives awaits much development and optimization. Unfortunately, most of these alternatives are still based upon some mechanism of killing or terminating the target bacteria; an approach some feel preordains the development of resistance in bacteria.

It has been recently proposed to develop substances that specifically inhibit bacterial virulence. Such 'antipathogenic' drugs, in contrast to antibacterial drugs, do not kill bacteria or stop their growth and are assumed not to lead to the development of resistant strains. A very elegant approach comprises the inhibition of regulatory systems that govern the expression of a series of bacterial virulence factors, e.g. anti-adhesion therapy, passive antibody therapy (Casadevall *et al.*, 2004; Martinez and Casadevall, 2005) and synthetic peptide vaccine and antibody therapy (Cachia and Hodges, 2003), inhibiting or negating cell-cell signalling (Otto, 2004), negating biofilm formation by disrupting iron metabolism (Kaneko *et al.*, 2007), and up-regulation of biofilm detachment promoters (rhamnolipids) (Boles *et al.*, 2005).

Here, I present examples of three such 'anti-biofilm' strategies that could be deployed from a biomedical device coating or implant as a defence aimed at negating biofilm formation, defences based on (1) disrupting bacterial iron metabolism, (2) negating quorum sensing, and (3) 'self-vaccinating' biomaterials.

Iron metabolism interference

Iron is critical for bacterial growth and the function of key metabolic enzymes (Finkelstein *et al.*, 1983; Ankenbauer, *et al.*, 1985; Barclay and Ratledge, 1986) and sequestration of iron is an early evolutionary strategy of host defence (Jurado, 1997). Gallium and zinc have many features similar to Fe^{3+}, including a nearly identical ionic radius, and biological systems are often unable to distinguish Ga from Fe^{3+}. Unlike Fe^{3+}, Ga cannot be reduced to the divalent state and sequential enzymatic reduction/oxidation is critical for Fe to function in many enzymes. Thus, placing Ga, rather than Fe, in such enzymes renders them non-functional (Narasimhan *et al.*, 1992; Singh, 2004). Further, Ga readily binds to the siderophores (bacterially derived iron chelators) of *Pseudomonas sp.* (Chitamber and Narasimhan, 1991; Narasimhan *et al.*, 1992), *S. aureus* and *S. epidermidis*, *E. coli*, *E. faecalis*, and *S. typhimurium* (Atkinson *et al.*, 1998; Hubbard *et al.*, 1986).

Kaneko *et al.* (2007) recently reported that concentrations 1–2 µM $Ga(NO_3)_3$ did not impact the growth rate or extent of suspended cultures of *P. aeruginosa*. Concentrations of Ga >2 µM decreased *P. aeruginosa* growth rate in a dose-dependent manner. For their biofilm studies, they used a Ga concentration (0.5 µM) that did not impair the growth of suspended *P. aeruginosa*, since they were interested specifically in an *anti-biofilm* effect of Ga (in a therapeutic application, both growth inhibitory and anti-biofilm actions would be desirable). Gallium effects were studied using a green fluorescent protein (gfp)-expressing *P. aeruginosa* strain. At subinhibitory concentrations of gallium, *P. aeruginosa* attached to the growth surface, but biofilm formation was completely inhibited. Biofilm formation by mucoid and non-mucoid *P. aeruginosa* isolates from cystic fibrosis patients was also blocked by 1 µM $Ga(NO_3)_3$ (Kaneko *et al.*, 2007).

To determine if Ga applications could kill *P. aeruginosa* biofilms, Kaneko *et al.* (2007) also reported growing biofilms for 3 days (with no Ga present) and then switching to medium containing Ga for 48 h. Bacterial viability was assayed using a live–dead stain. While most commercial antibiotic agents show markedly less activity against biofilms than against planktonic organisms, *P. aeruginosa* bacteria within mature biofilms were killed by concentrations of Ga similar to those that killed planktonic cells.

Negating quorum sensing

Salicylic acid or salicylate is the major *in vivo* metabolite of acetylsalicylic acid, i.e. aspirin. Salicylate has a significant antibacterial activity, including the ability to enhance the activities of certain antibiotics. Salicylic acid (SA) enhances the *in vitro* and *in vivo* activities of amikacin against *Klebsiella pneumoniae* and increases the synergistic activity of imipenem and amikacin when they are used to treat *K. pneumoniae* infections in animals (Domenico *et al.*, 1989, 1990, 1993). Farber and Wolff (1992) reports that salicylic acid inhibited adherence (55%), growth, and biofilm production of *S. epidermidis*. Farber *et al.* (1995) further illustrated that non-steroidal anti-inflammatory drugs (NSAIDs), including sodium salicylate, inhibit bacterial biofilm production. The authors studied the effect of sodium salicylate on bacterial adherence and biofilm formation on contact lenses, lens cases, and commonly used medical polymers and observed that biofilm formation was inhibited for *S. epidermidis* and *P. aeruginosa* with sodium salicylate in a dose-related manner. Related studies (Bandara *et al.*, 2004) demonstrated that salicylic acid and other NSAIDs decreased bacterial colonization on contact lenses in a dose-dependent manner.

A study by Stepanovic *et al.* (2004) explored the possible effect of aspirin on *Candida* spp. biofilm-producing capacity. Two strains of *C. guilliermondii* and one strain per species of *C. kefyr*, *C. glabrata*, *C. albicans* and *C. parapsilosis* were included. The antifungal property of aspirin was tested by a broth microdilution method, while the effect of aspirin on biofilm formation was determined by a microtitre plate test. The concentrations of aspirin, which induced statistically significant decreases in biofilm formation, ranged from 0.43 mM to 1.73 mM and depended on the tested yeast strain. However, the significant effects of aspirin on growth and biofilm formation of *Candida* spp. were achieved only with excessively high concentrations of the drug that exceed pharmaceutical levels.

We reported on the development of poly(anhydride-ester) materials composed of non-steroidal anti-inflammatory drugs that biodegrade at 'tunable' rates into salicylic acid (SA) and adipic acid that could prevent colonization by *P. aeruginosa in vitro* while reducing the effects of the foreign-body response *in vivo* (Bryers *et al.*, 2006). Soluble SA in bacterial medium at concentrations up to 300 mg/l did not affect the growth rate or viability of *P. aeruginosa*, indicating that SA does not exhibit a direct toxicity effect on the bacterium. Short-term (3 h) bacterial adhesion studies in agitated batch systems indicated a 47% reduction in the rate of *P. aeruginosa* adhesion relative to a control polymer that does not release SA upon biodegradation. Long-term (3-day) biofilm accumulation studies indicated a dramatic reduction in biofilm formation on salicylate-based polymer versus controls. A recombinant *P. aeruginosa* pMHLAS strain, containing a fluorescent reporter gene (GFP) prior to the QS *las* regulon, was employed to determine whether the salicylate-based polymer prevents biofilm formation by the released SA inhibiting QS pathways. Long-term biofilm accumulation studies with *P. aeruginosa* pMHLAS insinuate that salicylate-based polymer did prevent biofilm accumulation by inhibiting the *las* QS system. Furthermore, unlike the control polymer, salicylate-based polymer implanted subcutaneously for a period of 4 weeks, resisted cell-mediated degradation and remained intact. Histological and immunohistochemical analysis indicated a reduction in overall encapsulation and paucity of macrophages in the area of the salicylate-based polymer implant.

Self-vaccinating biomaterials

Imagine a biomedical device that would be capable of generating a life-long protection for itself against infection. Our research group is developing a biomaterial's platform that will engineer infection immunity by releasing vaccine complexes that have as antigen targets- those bacterial surface structures used for bacterial specific adhesion.

The goal of any vaccine is to produce a long-term protective immune response against a pathogen. For most bacteria, initial attachment to a eukaryotic cell surface or ligand-coated biomedical device leads to biofilm formation, then up-regulation of virulence factors leading to infection. Both an innate and an induced antibody response could prevent attachment and abrogate colonization. The ideal antigens to promote both levels of immune response would be the very surface proteins (bacterial 'adhesins') that mediate specific bacterial cell adhesion to ligands present on host tissue or device surfaces (Wizeman *et al.*, 1999). Given that bacterial specific adhesion can trigger expression of many virulence factors leading to acute and chronic inflammation, a vaccine approach that also blocks bacterial adhesion may have multiple advantages. Recent research has greatly expanded the molecular details of the specific adhesin:ligand couples employed by Gram-positive bacteria (*S. epidermidis*, *S. aureus*, group A and B *Streptococcus*, *Enterococcus faecalis*), Gram-negative bacteria (*P. aeruginosa*, *Klebsiella pneumoniae*, *Porphyromonas gingivalis*, *E. coli*), and yeast (*Candida* spp.) to colonize and infect living tissue and implanted biomedical devices (Wizeman *et al.*, 1999). Thus, the concept of biomaterials designed to engineer infection immunity, targeting specific adhesins, could be applied to numerous situations.

There are essentially three ways to activate a dendritic cell to present antigen with the subsequent immune response being dependent upon the form of the antigen: direct antigen presentation (e.g. whole attenuated pathogen, isolated antigen molecule, recombinant protein), DNA vaccine (i.e. plasmid DNA (pDNA) encoding for the antigen protein), and mRNA vaccine (i.e. RNA encoding for the antigen protein). *Exogenously* acquired protein antigens are processed by late endosomes into major histocompatibility complex class-II (MHC-II) molecules that initiate $CD4^+$ T cell activation, and then activate B cell antibody secretion. *Endogenously* acquired antigen (e.g. viral infection; DNA/mRNA vaccine) is tagged with ubiquitin, then partially degraded in the proteosome, transported next to the endoplasmic recticulum, and presented on the DC's surface as a MHC-I complex, thus activating an immediate (but short-lived) cytotoxic $CD8^{+}$ T cell response.

In contrast to protein vaccines, DNA- or RNA-based vaccines can provide the ability to *potentially* generate both a strong cytotoxic T cell response and humoral response. Upon internalization of a pDNA vaccine carrier by dendritic cells, the carrier must escape endosomal entrapment or be degraded, the carrier must release the pDNA into the cytoplasm, and then the pDNA must be incorporated into the dendritic nucleus for expression. Subsequently expressed antigen can be processed as above into MHC-I complexes to initiate priming of CD8+ T-cell or the antigen can be secreted. The secreted antigen can be taken up exogenously by the same or other DCs and presented by the MHC-II pathway to $CD4^{+}$ T cells, which can secrete soluble cytokine signals to T- or to B-cells to induce antibody secretion. In dendritic cells, in a process known as 'cross-priming', endogenously expressed antigen could also be routed to the MHC-II complex.

DNA vaccines have certain advantages over protein antigens including (a) DNA can serve as a natural adjuvant by including unmethylated CpG motifs; (b) by coding for multiple gene expression, DNA vaccine can also induce co-stimulatory molecules; (c) pDNA can target expression to certain cellular locations, thus fine-tuning the immune response; and (d) pDNAs allow the possibility of multiple antigen expression. Disadvantages of DNA vaccines include (a) incorporated CpG motifs can lead to overstimulation and toxicity (Zhao *et al.*, 2004; Heikenwalder *et al.*, 2004); (b) the required large amounts of highly characterized pDNA require processing with antibiotics and antibiotic resistance markers; (c) pDNA containing sequences meant to control gene expression (e.g. promoters, polyA signals, introns) that may deregulate gene expression after integration into the genome; and (d) genome-incorporated pDNA can result in uncontrolled duration and strength of antigen expression.

RNA vaccines offer the same advantages as DNA vaccines versus protein antigens but RNA vaccines have several added benefits over DNA vaccines (Gilboa and Vieweg, 2004). Since RNA does not require nuclear incorporation, expression of antigen in transfected cells occurs much faster than with pDNA. Potentially dangerous side effects are reduced since eukaryotic promoters needed for pDNA are not present in RNA constructs. mRNAs are relatively easy to manufacture in high amounts, to purify to homogeneity, and to characterize. Finally, RNA does not persist in the organism, and RNA is not incorporated into the genome.

One main drawback to both DNA and RNA vaccines is that the efficiency of naked oligonucleotide transfection is very low (compared with viral systems) owing to lack of protection from systemic nucleases, inability to migrate through cell membranes, or entrapment and degradation within endosomes. Consequently, non-viral gene constructs typically require some type of polymer delivery system. The reader is directed to a number of excellent reviews on the subject of polymer gene delivery published by Gao *et al.* (2007), Keegan and Saltzman (2006), Dang and Leong (2006), Pack *et al.* (2005), and Little and Langer (2005).

One distinct advantage of nucleotide transfection of dendritic cells (vs. direct antigen protein) is that one can modify the DNA or mRNA with targeting signals to better control MHC class I and class II presentation. For example, incorporation of ubiquitin mRNA with the target antigen will result in an enhanced formation of peptides for MHC class I presentation (Tobery and Siliciano, 1997; Tobery and Siliciano, 1999; Varshavsky *et al.*, 2000) whereas targeting sequences from the invariant chain (Ii) or lysosome-associated membrane protein (Lamp1) will lead to presentation of the antigen in the context of both MHC class I and class II, thus providing antigen-specific help (Wu *et al.*, 1995; Lin *et al.*, 1996; Thomson *et al.*, 1998; Bonini *et al.*, 2001). We are currently developing mRNA vaccines anti- to the *S. epidermidis* AtlE (autolysin adhesin) and AAP (accumulation-associated protein) used to colonize cardiovascular devices and anti- to the cell surface Arg-specific (RgpA) and Lys-specific (Kgp) proteinases (designated the RgpA–Kgp complex) used by the oral pathogen *Porphyromonas gingivalis*, the major cause of chronic periodontitis.

Concluding remarks

Parsek and Singh (2003) postulated diagnostic criteria that a disease must meet to be deemed a biofilm infection; further modified by Hall-Stoodley and Stoodley (2009) the common features of biofilm infections are: pathogenic bacteria are associated with a surface; direct examination of infected tissue or device demonstrates aggregated cells in cell clusters encased in a matrix, which may be of bacterial and host origin; infection is localized to a particular site in the host; recalcitrance to antibiotic treatment despite demonstrated susceptibility of planktonic bacteria (in the absence of culture, recalcitrance to antibiotic therapy may be inferred from the presence of live bacterial cells in the biofilm from *in situ* viability staining or RT-PCR); culture-negative results in spite of clinically documented high suspicion of infection (since localized bacteria in a biofilm may be missed in a conventional blood sample or aspirate); and ineffective host clearance evidenced by the location of bacterial cell clusters (macrocolonies) in discrete areas in the host tissue associated with host inflammatory cells.

As our population ages, there will be an increase in the number of people experiencing hospitalization and receiving short- or long-term biomedical implants. As engineered biomaterials and tissue regenerative medicine advance, an increasing portion of the population will receive one or multiple biomedical devices, ranging from disposable contact lenses, dental implants, orthopaedic implants, and vascular grafts to tissue engineered livers, small diameter vascular grafts that promote stem cells differentiation into endothelial cells, and polymer transfection systems that deliver micro-RNA knockout therapy to control chronic inflammation. The current health care approach to clean and sterilize has done little to prevent an epidemic in nosocomial infections. Biomaterials technologies employing disinfectant rinses, tethered or release antibiotics have also done little to reduce this epidemic and may even have contributed to the raise of antibiotic resistant bacteria.

Increasing scientific research over the past 10 years in biofilm formation has provided numerous possible targets with which to prevent or eradicate biofilm infections. Advances in the understanding of biofilm formation, coupled with emerging engineered biomaterials, provide many potential platforms and strategies to prevent or significantly reduce biofilm infections in susceptible populations.

References

Abrahams, H.M., and Stoller, M.L. (2003). Infection and urinary stones. Curr. Opin. Urol. *13*, 63–67.

Anderson, J.M. (2001). Biological responses to materials. Annu. Rev. Mater. Res. *31*, 81–110.

Ankenbauer, R., Sriyosachati, S., and Cox, C.D. (1985). Effects of siderophores on the growth of *Pseudomonas aeruginosa* in human serum and transferrin. Infect. Immun. *49*, 132–140.

Atkinson R.A., Salah El Din, A.L., Kieffer, B., Lefevre, J.F., and Abdallah, M.A. (1998). Bacterial iron transport: 1H NMR determination of the three-dimensional structure of the gallium complex of pyoverdin G4R, the peptidic siderophore of *Pseudomonas putida* G4R. Biochemistry 37, 15965–15973.

Bagge, N., Schuster, M., Hentzer, M., Ciofu, O., Givskov, M., Greenberg, E.P., and Høiby, N. (2004). *Pseudomonas aeruginosa* biofilms exposed to imipenem exhibit changes in global gene expression and β-lactamase and alginate production. Antimicrob. Agents Chemother. *48*, 1175–1187.

Bakaletz, L.O. (2007). Bacterial biofilms in otis media – Evidence and relevance. Ped. Infect. Dis. J. *26*, 517–519.

Bandara, B., Sankaridurg, P., and Willcox, M. (2004). Non-steroidal anti-inflammatory agents decrease bacterial colonisation of contact lenses and prevent adhesion to human corneal epithelial cells. Curr. Eye Res. 29, 245.

Barclay, R., and Ratledge, C. (1986). Participation of iron on the growth inhibition of pathogenic strains of *Mycobacterium avium* and *M. paratuberculosis* in serum. Zentralbl. Bakteriol. Mikrobiol. Hyg. *262*, 189–194.

Beaudoin, D., Bryers, J.D., Cunningham, A.B., and Peretti, S.W. (1998a). Mobilization of broad host range plasmid from *Pseudomonas putida* to established biofilm of *Bacillus azotoformans*. I. Experiments. Biotechnol. Bioeng. 57, 272–279.

Beaudoin, D., Bryers, J.D., Cunningham, A.B., and Peretti, S.W. (1998b). Mobilization of broad host range plasmid from *Pseudomonas putida* to established biofilm of *Bacillus azotoformans*. II. Modeling. Biotechnol. Bioeng. 57, 280–286.

Beikler, T., Peters, U., Prior, K., Ehmke, B., and Flemmig, T.F. (2005). Sequence variations in *rgpA* and *rgpB* of *Porphyromonas gingivalis* in periodontitis. J. Periodont. Res. *40*, 193–198.

Berendt, T., and Byren, I. (2004). Bone and joint infection. Clin. Med. 4, 510–518.

Bjarnsholt, T., Jensen, P.O., Fiandaca, M.J., Pedersen, J., Hansen, C.R., Andersen, C.B., Pressler, T., Givskov, M., and Hoiby, N. (2009a). *Pseudomonas aeruginosa* biofilms in the respiratory tract of cystic fibrosis patients. Pediatr. Pulm. *44*, 547–558.

Bjarnsholt, T., Tolker-Nielsen, T., Givskov, M., Janssen, M., and Chrsitensen, L.H. (2009b). Detection of bacteria by FISH in culture-negative soft tissue filler lesions. Dermatol. Surg. *35*, 1620–1624.

Boles, B.R, Thoendel, M., and Singh, P.K. (2005). Rhamnolipids mediate detachment of *Pseudomonas aeruginosa* from biofilms. Mol. Microbiol. *57*, 1210–1223.

Bonini, C., Lee, S.P., Riddell, S.R., and Greenberg, P.D. (2001). Targeting antigen in mature dendritic cells for simultaneous stimulation of CD4+ and CD8+ T cells. J. Immunol. *166*, 5250–5257.

Bothwell, M.R., Smith, A.L., and Phillips, T. (2003). Recalcitrant otorrhea due to *Pseudomonas* biofilm. Otolaryng. Head Neck *129*, 599–601.

Brown, L.J., Johns, B.A., and Wall, T.P. (2002). The economics of periodontal disease. Peridontology 2000 *29*, 233–234

Bryers, J.D., and Ratner, B.D. (2004). Bioinspired Implant materials befuddle bacteria. ASM News *70*, 232–237.

Bryers, J.D., Jarvis, R.A., Lebo, J., Prudencio, A., Kyriakides, T.R., and Uhrich, K. (2006). Biodegradation of poly(anhydride-esters) into non-steroidal anti-inflammatory drugs and their effect on *Pseudomonas aeruginosa* biofilms *in vitro* and on the foreign-body response *in vivo*. Biomaterials *27*, 5039–5048.

Burmølle, M, Thomsen, T.R., Fazli, M., Dige, I., Christensen, L., Homøe, P., Tvede, M., Nyvad, B., Tolker-Nielsen, T., Givskov, M., *et al.* (2010). Biofilms in chronic infections – a matter of opportunity – monospecies biofilms in multispecies infections. FEMS Immunol. Med. Microbiol. *59*, 324–336.

Cachia, P.J., and Hodges, R.S. (2003). Synthetic peptide vaccine and antibody therapeutic development: prevention and treatment of *Pseudomonas aeruginosa*. Biopolymers *71*, 141–168.

Calhoun, J.H., Cantrell, J., Cobos, J., Lacy, J., Valdez, R.R., Hokanson, J., and Mader, J.T. (1988). Treatment of diabetic foot infections: wagner classification, therapy, and outcome. Foot Ankle *9*, 101–106.

Casadevall, A., Dadachova, E., and Pirofski, L.A. (2004). Passive antibody therapy for infectious diseases. Nat. Rev. Microbiol. *2*, 695–703.

Castelli, P., Caronno, R., Ferrarese, S., Mantovani, V., Piffaretti, G., Tozzi, M., Lomazzi, C., Rivolta, N., and Sala, A. (2006). New trends in prosthesis infection in cardiovascular surgery. Surg. Infect. *7*, s45–s47.

Chandra, J., McCormick, T.S., Imamura, Y., Mukherjee, P.K., and Ghannoum, M.A. (2007). Interaction of *Candida albicans* with adherent human peripheral blood mononuclear cells increases *C. albicans* biofilm formation and results in differential expression of pro- and anti-inflammatory cytokines. Infect Immun. *75*, 2612–2620.

Chitambar, C.R., and Narasimhan, J. (1991). Targeting iron-dependent DNA synthesis with gallium and transferrin-gallium. Pathobiology *59*, 3–10.

Chole, R.A., and Faddis, B.T. (2002). Evidence for microbial biofilms in cholesteatomas. Arch. Otolaryngol. *128*, 1129–1133.

Chopra, T., and Kaatz, G.W. (2010). Treatment strategies for infective endocarditis. Expert Opin. Pharmacother. *11*, 345–360.

Cohen, N., Morisset, J., and Emilie, D. (2004). Induction of tolerance by *Porphyromonas gingivalis* on APCS: a mechanism implicated in periodontal infection. J. Dent. Res. *83*, 429–433.

Dang, J.M., and Leong, K.W. (2006). Natural polymers for gene delivery and tissue engineering. Adv. Drug Deliv. Rev. *58*, 487–499.

Davies, J.C., and Bilton, D. (2009). Bugs, biofilms, and resistance in cystic fibrosis. Respir. Care. *54*, 628–640.

De Smet, K., and Contreras, R. (2005). Human antimicrobial peptides: defensins, cathelicidins and histatins, Biotechnol. Lett. *27*, 1337–1347.

Domenico, P., Schwartz, S., and Cunha, B. (1989). Reduction of capsular polysaccharide production in *Klebsiella pneumoniae* by sodium salicylate. Infect. Immun. *57*, 3778–3782.

Domenico, P., Hopkins, T., and Cunha, B. (1990). The effect of sodium salicylate on antibiotic susceptibility and synergy in *Klebsiella pneumoniae*. J. Antimicrob. Chemother. *26*, 343–351.

Domenico, P., Straus, D., Woods, D., and Cunha, B. (1993). Salicylate potentiates amikacin therapy in rodent models of *Klebsiella pneumoniae* infection. J. Infect. Dis. *168*, 766–769.

Edelstein, B.L. (2002). Disparities in oral health and access to care: findings of national surveys. Ambul. Pediatr. *2*, 141–147.

Evans, R.T., Klausen, B., Sojar, H.T., Bedi, G.S., Sfintescu, C., Ramamurthy, N.S., Golub, L.M., and Genco, R.J. (1992). Immunization with *Porphyromonas (Bacteroides) gingivalis* fimbriae protects against periodontal destruction. Infect. Immun. *60*, 2926–2935.

Farber, B., and Wolff, A. (1992). The use of nonsteroidal anti-inflammatory drugs to prevent adherence of *Staphylococcus epidermidis* to medical polymers. J. Infect. Dis. *166*, 861–865.

Farber, B., Hsieh, H., Donnenfeld, E., Perry, H., Epstein, A., and Wolff, A. (1995). A novel antibiofilm technology for contact lens solutions. Ophthalmology *102*, 831–836.

Fichorova, R.N. (2009). Impact of *T. vaginalis* infection on innate immune responses and reproductive outcome. J. Reprod. Immunol. *83*, 185–189.

Finkelstein, R.A., Sciortino, C.V., and McIntosh, M.A. (1983). Role of iron in microbe–host interactions. Rev. Infect. Dis. *Suppl. 4*, S759–777.

Fischetti, V.A. (2005). Bacteriophage lytic enzymes: novel anti-infectives. Trends Microbiol. *13*, 491–496.

Fitzsimmons, K., Bamber, A.I., and Smalley, H.B. (2010). Infective endocarditis: changing aetiology of disease. Br. J. Biomed. Sci. *67*, 35–41.

Foster, T.J. (2005). Immune Evasion by *Staphylococci*. Nat. Rev. Microbiol. *3*, 948–958.

Gao, Y., Gu, W., Chen, L., Xu, Z., and Li, Y. (2007). A multifunctional nano device as non-viral vector for gene delivery: *In vitro* characteristics and transfection. J. Control Release *118*, 381–388.

Geddes, A. (2000). Infection in the twenty-first century: Predictions and postulates. J. Antimicrob. Chemother. *46*, 873–878.

George, A.M., Jones, P.M., and Middleton, P.G. (2009). Cystic fibrosis infections: treatment strategies and prospects. FEMS Microbiol. Lett. *300*, 153–164.

Gilbert, P., Allison, D.G., and McBain, A.J. (2002). Biofilms *in vitro* and *in vivo*: do singular mechanisms imply cross-resistance? J. Appl. Microbiol. *92*, 98S–110S.

Gilboa, E., and Vieweg, J. (2004). Cancer immunotherapy with mRNA transfected dendritic cells. Immunol. Rev. *199*, 251–263.

Govan J.R., and Deretic, V. (1996). Microbial pathogenesis in cystic fibrosis: mucoid *Pseudomonas aeruginosa* and *Burkholderia cepacia*. Microbiol. Rev. *60*, 539–574.

Hall-Stoodley, L., and Stoodley, P. (2009). Evolving concepts in biofilm infections. Cell. Microbiol. *11*, 1034–1043.

Hall-Stoodley, L., Hu, F.Z., Gieseke, A., Nistico, L., Nguyen, D., Hayes, J., Forbes, M., Greenberg, D.P., Dice, B., Burrows, A., *et al.* (2006). Direct detection of bacterial biofilms on the middle-ear mucosa of children with chronic otitis media. J. Am. Med. Assoc. *296*, 202–211.

Hassett, D.J., Sutton, M.D., Schurr, M.J., Herr, A.B., Caldwell, C.C., and Matu, J.O. (2009). *Pseudomonas aeruginosa* hypoxic or anaerobic biofilm infections within cystic fibrosis airways. Trends Microbiol. *17*, 130–138.

Hassett, D.J., Korfhagen, T.R., Irvin, R.T., Schurr, M.J., Sauer, K., Lau, G,W., Sutton. M.D., Yu, H., and Hoiby, N. (2010). *Pseudomonas aeruginosa* biofilm infections in cystic fibrosis: insights into pathogenic processes and treatment strategies. Expert Opin. Ther. Targets *14*, 117–130.

Hatt, J.K., and Rather, P.N. (2008). Role of bacterial biofilms in urinary tract infections. Curr. Top. Microbiol. Immunol. 322, 163–192.

Healy, B., and Freedman, A. (2006). Infections. BMJ 332, 838–841.

Healy, K.A., and Ogan, K. (2007). Pathophysiology and management of infectious staghorn calculi. Urol. Clin. North Am. *34*, 363–374.

Heikenwalder, M., Polymenidou, M., Junt, T., Sigurdson, C.,Wagner, H., Akira, S., Zinkernagel, R., and Aguzzi, A. (2004). Lymphoid follicle destruction and immunosuppression after repeated CpG oligodeoxynucleotide administration. Nat. Med. *10*, 187–192.

Hoffman, L.R., D'Argenio, D.A., MacCoss, M.J., Zhang, Z., Jones, R.A., and Miller, S.I. (2005). Aminoglycoside antibiotics induce bacterial biofilm Formation, Nature *436*, 1171–1175.

Høiby, N., Krogh Johansen, H., Moser, C., Song, Z., Ciofu, O., and Kharazmi, A. (2001). *Pseudomonas aeruginosa* and the *in vitro* and *in vivo* biofilm mode of growth. Microbes Infect. 3, 23–35.

Høiby, N., Bjarnsholt, T., Givskov, M., Molin, S., and Ciofu, O. (2010). Antibiotic resistance of bacterial biofilms. Int. J. Antimicrob. Agents 35, 322–332.

Hong, H.A., Ducle, H., and Cutting, S.M. (2005). The use of bacterial spore formers as probiotics. FEMS Microbiol. Rev. *29*, 813–835.

Hu, W., Eaton, J., and Tang, L. (2001). Molecular basis of biomaterial-mediated foreign body reactions. Blood *98*, 1231–1238.

Hubbard, J.A., Lewandowska, K.B., Hughes, M.N., and Poole, R.K. (1986). Effects of iron-limitation of *Escherichia coli* on growth, the respiratory chains and gallium uptake. Arch. Microbiol. *146*, 80–86.

Jain, A., Batista, E.L., Jr, Serhan, C., Stahl, G.L., and Van Dyke, T.E. (2003). Role for periodontitis in the progression of lipid deposition in an animal model. Infect. Immun. *71*, 6012–6018.

Jesaitis, A.J., Franklin, M.J., Berglund, D., Sasaki, M., Lord, C.I., Bleazard, J.B., Duffy, J.E., Beyenal, H., and Lewandowski, Z. (2003). Compromised host defense on *Pseudomonas aeruginosa* biofilms: characterization of neutrophil and biofilm interactions. J. Immunol. *171*, 4329–4339.

Jotwani, R., and Cutler, C.W. (2004). Fimbriated *Porphyromonas gingivalis* is more efficient than fimbria-deficient *P. gingivalis* in entering human dendritic cells *in vitro* and induces an inflammatory Th1 effector response. Infect. Immun. 72, 1725–1732.

Jotwani, R,, Pulendran, B., Agrawal, S., and Cutler, C.W. (2003). Human dendritic cells respond to *Porphyromonas gingivaux* LPS by promoting Th2 effector response *in vitro*. Eur. J. Immunol. *33*, 2980–2986.

Jotwani, R. and Cutler, C.W. (2004). Fimbriated *Porphyromonas gingivalis* is more efficient than fimbria-deficient *P. gingivalis* in entering human dendritic cells *in vitro* and induces an inflammatory Th1 effector response. Infect. Immun. 72, 1725–1732.

Jurado, R.L. (1997). Iron, infections, and anemia of inflammation. Clin. Infect. Dis. 25, 888–895.

Jurcisek, J., Greiner, L., Watanabe, H., Zaleski, A., Apicella, M.A., and Bakaletz, L.O. (2005). Role of sialic acid and complex carbohydrate biosynthesis in biofilm formation by nontypeable *Haemophilus influenzae* in the chinchilla middle ear. Infect. Immun. 73, 3210–3218.

Kadowaki, T., Nakayama, K., Okamoto, K., Abe, N., Baba, A., Shi, Y., Ratnayake, D., and Yamamoto, K. (2000). *Porphyromonas gingivalis* proteinases as virulence determinants in progression of periodontal diseases. J. Biochem. *128*, 153–159.

Kaneko, Y., Thoendel, M., Olakanmi, O., Britigan, B.E., and Singh. P.K. (2007). The transition metal gallium disrupts *Pseudomonas aeruginosa* iron metabolism and has antimicrobial and antibiofilm activity, J. Clin. Invest., *117*, 877–888.

Keegan, M.E., and Saltzman, W.M. (2006). Surface-modified biodegradable microspheres for DNA vaccine delivery. Methods Mol. Med. *127*, 107–113.

Kersulyte, D., Struelens, M.J., Deplano, A., and Berg, D.E. (1995). Comparison of arbitrarily primed PCR and macrorestriction (pulsed-field gel electrophoresis) typing of *Pseudomonas aeruginosa* strains from cystic fibrosis patients. J. Clin. Microbiol. 33, 2216–2219.

Kilty, S.J., and Desrosiers, M.Y. (2008). The role of bacterial biofilms and the pathophysiology of chronic rhinosinusitis. Curr. Allergy Asthma Rep. *8*, 227–233.

van Kooyk, Y., and Geijtenbeck, T.B. (2003). DC-SIGN: escape mechanism for pathogens. Nat. Rev. Immunol. *3*, 697–709.

van Kooyk, Y., Engering, A., Lekkerkerker, A.N., Ludwig, I.S., and Geijtenbeck, T.B. (2004). Pathogens use carbohydrates to escape immunity induced by dendritic cells. Curr. Opin. Immunol. *16*, 488–493.

Leid, J.G., Shirtliff, M.E., Costerton, J.W., and Stoodley, A.P. (2002). Human leukocytes adhere to, penetrate, and respond to *Staphylococcus aureus* biofilms. Infect. Immun. *70*, 6339–6345.

Leid, J.G., Willson, C.J., Shirtliff, M.E., Hassett, D.J., Parsek, M.R., and Jeffers, A.K. (2005). The exopolysaccharide alginate protects *Pseudomonas aeruginosa* biofilm bacteria from IFN-γ-mediated macrophage killing. J. Immunol. *175*, 7512–7518.

Lew, D.P., and Waldvogel, F.A. (2004). Osteomyelitis. Lancet *364*, 369–379.

Li, M., Lai, Y., Villaruz, A.E., Cha, D.J., Sturdevant, D.E., and Otto, M. (2007). Gram-positive three-component antimicrobial peptide-sensing system. Proc. Natl. Acad. Sci. U.S.A. *104*, 9469–9474.

Lin, D., Smith, M.A., Champagne, C., Elter, J., Beck, J., and Offenbacher, S. (2003). *Porphyromonas gingivalis* infection during pregnancy increases maternal tumor necrosis factor alpha, suppresses maternal interleukin-10, and enhances fetal growth restriction and resorption in mice. Infect. Immun. *71*, 5156–5162.

Lin, K.Y., Guarnieri, F.G., Staveley-O'Carroll, K.F., Levitsky, H.I., August, J.T., Pardoll, D.M., and Wu, T.C. (1996). Treatment of established tumors with a novel vaccine that enhances major histocompatibility class II presentation of tumor antigen. Cancer Res. *56*, 21–26.

Little, S.R., and Langer, R. (2005). Nonviral delivery of cancer genetic vaccines. Adv. Biochem. Eng. Biotechnol. *99*, 93–118.

Macassey, E., and Dawes, P. (2008). Biofilms and their role in otorhinolaryngological disease. J. Laryngol. Otol. *122*, 1273–1278.

Mader, J.T., Mohan. D., and Calhoun, J. (1997). A practical guide to the diagnosis and management of bone and joint infections. Drugs *54*, 253–264.

Marchisio, P., Bianchini, S., Capaccio, P., Esposito, S., Fusi, M., Nazzari, E., Torretta, S., and Principi, N. (2010). Insights into infectious otitis media. Int. J. Immunopathol. Pharmacol. *23*, 20–23.

Marcus, R.J., Post, J.C., Stoodley, P., Hall-Stoodley, L., McGill, R.L., Sureshkumar, K.K., and Gahlot, V. (2008). Biofilms in nephrology. Expert Opin. Biol. Ther. *8*, 1159–1166.

Martinez, L.R., and Casadevall, A. (2005). Specific antibody can prevent fungal biofilm formation and this effect correlates with protective efficacy. Infect. Immun. *73*, 6350–6362.

Mayrand, D., and Holt, S.C. (1988). Biology of asaccharolytic black-pigmented Bacteroides species. Microbiol. Rev. *52*, 134–152.

Mekori, Y., and Metcalfe, D. (2000). Mast cells in innate immunity. Immunol. Rev. *173*, 131–140.

Mercado, F.B., Marshall, R.I., Klestov, A.C., and Bartold, P.M. (2001). Relationship between rheumatoid arthritis and periodontitis. J. Periodontol. *72*, 779–787.

Mittal, R., Sharma, S., Chhibber, S., and Harjai, K. (2006). Effect of macrophage secretory products on elaboration of virulence factors by planktonic and biofilm cells of *Pseudomonas aeruginosa*, Comparative Immunology. Microbiol. Infect. Disease *29*, 12–26.

Miyakawa, H., Honma, K., Qi, M., and Kuramitsu, H.K. (2004). Interaction of *Porphyromonas gingivalis* with low density lipoproteins: implications for a role for periodontitis in atherosclerosis. J. Periodontal Res. *39*, 1–9.

Moreau-Marquis, S., Stanton, B.A., and O'Toole, G.A. (2008). *Pseudomonas aeruginosa* biofilm formation in the cystic fibrosis airway. Pulm. Pharmacol.Ther. *21*, 595–599.

Narasimhan, J., Antholine, W.E., and Chitambar, C.R. (1992). Effect of gallium on the tyrosyl radical of the iron-dependent M2 subunit of ribonucleotide reductase. Biochem. Pharmacol. *44*, 2403–2408.

O'Brien-Simpson, N.M., Veith, P.D., Dashper, S.G., and Reynolds, E.C. (2003). *Porphyromonas gingivalis* gingipains: the molecular teeth of a microbial vampire. Curr. Protein Pept. Sci. *4*, 409–426.

Offenbacher, S., Jared, H.L., O'Reilly, P.G., Wells, S.R., Salvi, G.E., Lawrence, H.P., Socransky, S.S., and Beck, J.D. (1998). Potential pathogenic mechanisms of periodontitis associated pregnancy complications. Ann. Periodontol. *3*, 233–250.

Oliver, R., L. Brown, and Loe, H. (1998). Periodontal diseases in the United States population. J. Periodontol. *69*, 269–278.

Otto, M. (2004). Quorum-sensing control in *Staphylococci* – a target for antimicrobial drug therapy? FEMS Microbiol. Lett. *241*, 135–141.

Pack, D.W., Hoffman, A.S., Pun, S., and Stayton, P.S. (2005). Design and development of polymers for gene delivery. Nat. Rev. Drug Discov. *4*, 581–593.

Pagella, F., Colombo, A., Gatti, O., Giourgos, G., and Matti, E. (2010). Rhinosinusitis and otitis media: the link with adenoids. Int. J. Immunopathol. Pharmacol. *23*, 38–40.

Parsek, M.R., and Singh, P.K. (2003). Bacterial biofilms: an emerging link to disease pathogenesis. Annu. Rev. Microbiol. *57*, 677–701.

Pawlowski, K.S., Wawro, D., and Roland, P.S. (2005). Bacterial biofilm formation on a human cochlear implant. Otol. Neurotol. *26*, 972–975.

Pedersen, S.S. (1992). Lung infection with alginate-producing, mucoid *Pseudomonas aeruginosa* in cystic fibrosis. APMIS Suppl. *28*, 1–79.

Persson, G.R. (2005). Immune responses and vaccination against periodontal infections. J. Clin. Periodontol. *32*, 39–53.

Post, J.C. (2001). Direct evidence of bacterial biofilms in otitis media. Laryngoscope *111*, 2083–2094.

Potempa, J., Sroka, A., Imamura, T., and Travis, J. (2003). Gingipains, the major cysteine proteinases and virulence factors of *Porphyromonas gingivalis*: structure, function and assembly of multidomain protein complexes. Curr. Protein Pept. Sci. *4*, 397–407.

Pucher, J., and Stewart, J. (2004). Periodontal disease and diabetes mellitus. Curr. Diab. Rep. *4*, 46–50.

Pulendran, B., Kumar, P., Cutler, C.W., Mohamadzadeh, M., Van Dyke, T., and Banchereau, J. (2001). Lipopolysaccharides from distinct pathogens induce different classes of immune responses *in vivo*. J. Immunol. *167*, 5067–5076.

Rajan, S., and Saiman, L. (2002). Pulmonary infections in patients with cystic fibrosis. Semin. Respir. Infect. *17*, 47–56.

Rayner, M.G., Zhang, Y., Gorry, M.C., Chen, Y., Post, J.C., and Ehrlich, G.D. (1998). Evidence of bacterial metabolic activity in culture negative otitis media with effusion. J. Am. Med. Assoc. *279*, 296–299.

Ratner, B.D., Hoffman, A.S., Frederick, J., Schoen, F.J., and Lemons, JE. (2004). Biomaterials Science, Second Edition: An Introduction to Materials in Medicine. Elsevier Academic Press, London, UK.

Reid, G., Dols, J., and Miller, W. (2009). Targeting the vaginal microbiota with probiotics as a means to counteract infections. Curr. Opin. Clin. Nutr. Metab. Care. *12*, 583–587.

Reid, S.D., Hong, W., Dew, K.E., Winn, D.R., Pang, B., Watt, J., Glover, D.T., Hollingshead, S.K., and Swords, W.E. (2009). *Streptococcus pneumoniae* forms surface-attached communities in the middle ear of experimentally infected chinchillas. J. Infect. Dis. *199*, 786–794.

Rooijakkers, S.H., and van Strijp, J.A. (2007). Bacterial complement evasion. Mol. Immunol. *44*, 23–32.

Saidi, I.S., Biedlingmaier, J.F., and Whelan, P. (1999). *In vivo* resistance to bacterial biofilm formation on tympanostomy tubes as a function of tube material. Otolaryng. Head Neck *120*, 621–627.

Savill, J. (1996). Phagocyte recognition of apoptotic cells. Biochem. Soc. Trans. *24*, 1065–1069.

Schwartz, B., Karsan, A., Bombeli, T., and Harlan, J. (1999). A novel β-1 integrin-dependent mechanism of leukocyte adherence to apoptotic cells. J. Immunol. *162*, 4842–4848.

Shankar, R., and Griesler, H. (1996). Inflammation, wound healing, and foreign body response. In Biomaterials Science: An Introduction to Materials in Medicine, Ratner, B.D., Hoffman, A.S., Schoen, F.J., and Lemons, J.E., eds. (Academic Press, New York), pp. 296–303.

Singh, P.K. (2004). Iron sequestration by human lactoferrin stimulates *P. aeruginosa* surface motility and blocks biofilm formation. Biometals *17*, 267–270.

Singh, P.K., Schaefer, A.L., Parsek, M.R., Moninger, T.O., Welsh, M.J., and Greenberg, E.P. (2000). Quorum-sensing signals indicate that cystic fibrosis lungs are infected with bacterial biofilms. Nature *407*, 762–64

Slots, J. (1982). Importance of black-pigmented Bacteroides in human periodontal disease. Host parasite interactions in periodontal diseases. In Host–parasite Interactions in Periodontal Diseases, Genco, R., and Merganhagan, S.E. (American Society for Microbiology, Washington, DC), pp. 27–45.

Smith, R.S., and Iglewski, B.H. (2003). *P. aeruginosa* quorum-sensing systems and virulence. Curr. Opin. Microbiol. *6*, 56–60.

Socransky, S.S., and Haffajee, A.D. (2000). Dental biofilms: difficult therapeutic targets. Periodontology *28*, 12–55.

Socransky, S.S., Haffajee, A.D., Cugini, M.A., Smith, C., and Kent, R.L., Jr. (1998). Microbial complexes in subgingival plaque. J. Clin. Periodontol. *25*, 134–144.

Stepanovic, S., Vukovic, D., Jesic, M., and Ranin, L. (2004). Influence of acetylsalicylic acid (aspirin) on biofilm production by *Candida* species. Chemotherapy *16*, 134–138.

Stewart, P.S. (2002). Mechanisms of antibiotic resistance in bacterial biofilms. Int. J. Med. Microbiol. *292*, 107–113.

Stewart, P.S., and Costerton. J.W. (2001). Antibiotic resistance of bacteria in biofilms. Lancet *358*, 135–138.

Stout, R., and Suttles, J. (2004). Functional plasticity of macrophages: reversible adaptation to changing microenvironments. J. Leukoc Biol. *76*, 509–513.

Stout, R.D., Jiang, C., Matta, B., Tietzel, I., Watkins, S.K., and Suttles, J. (2005). Macrophages sequentially change their functional phenotype in response to changes in microenvironmental influences. J. Immunol. *175*, 342–349.

Sulakvelidze, A., Alavidze, Z., and Morris, J.G., Jr. (2001). Bacteriophage therapy, Antimicrob Agents Chemother. *45*, 649–659.

Sun, D., Accavitti, M.A., and Bryers, J.D. (2005). Inhibition of biofilm formation by monoclonal antibodies against *Staphylococcus epidermidis* RP62A accumulation-associated protein, Clin. Diagn. Lab. Immunol. *12*, 93–100.

Swords, W.E., Moore, M.L., Godzicki, L., Bukofzer, G., Mitten, M.J., and VonCannon, J. (2004). Sialylation of lipooligosaccharides promotes biofilm formation by nontypeable *Haemophilus influenzae*. Infect. Immun. *72*, 106–113.

Tang, L., Jennings, T., and Eaton, J. (1998). Mast cells mediate acute inflammatory responses to implanted biomaterials. Proc. Natl. Acad. Sci. U.S.A. *95*, 8841–8846.

Tart A.H., and Wozniak, D.J. (2008). Shifting paradigms in *Pseudomonas aeruginosa* biofilm research. Curr. Top. Microbiol. Immunol. *322*, 193–206.

Tateda, K., Ishii, Y., Horikawa, M., Matsumoto, T., Miyairi, S., Pechere, J.C., Standiford, T.J., Ishiguro, M., and Yamaguchi, K. (2003). The *Pseudomonas aeruginosa* autoinducer N-3-oxododecanoyl homoserine lactone accelerates apoptosis in macrophages and neutrophils. Infect. Immun. *71*, 5785–93.

Thomson, S.A., Burrows, S.R., Misko, I.S., Moss, D.J., Coupar, B.E., and Khanna, R. (1998). Targeting a polyepitope protein incorporating multiple class II-restricted viral epitopes to the secretory/endocytic pathway facilitates immune recognition by CD4+ cytotoxic T lymphocytes: a novel approach to vaccine design. J Virol. 72, 2246–2252.

Timóteo, A.T., Branco, L.M., Pinto, M., Bico, P., and Ferreira, R.C. (2010). Nocardial endocarditis after mitral valve replacement: case report and review of the literature. Rev. Port. Cardiol. *29*, 291–297.

Tobery, T., and Siliciano, R.F. (1999). Cutting edge: induction of enhanced CTL-dependent protective immunity *in vivo* by N-end rule targeting of a model tumor antigen. J. Immunol. *162*, 639–642.

Tobery, T.W., and Siliciano, R.F. (1997). Targeting of HIV-1 antigens for rapid intracellular degradation enhances cytotoxic T lymphocyte (CTL) recognition and the induction of *de novo* CTL responses *in vivo* after immunization. J. Exp. Med. *185*, 909–920.

Varshavsky, A., Turner, G., Du, F., and Xie, Y. (2000). Felix Hoppe-Seyler Lecture (2000). The ubiquitin system and the N-end rule pathway. Biol. Chem. *381*, 779–789.

Verstraelen, H., and Verhelst, R. (2009). Bacterial vaginosis: an update on diagnosis and treatment. Expert Rev. Anti. Infect. Ther. *7*, 1109–1124.

Vuong, C., Kocianova, S., Voyich, J.M., Yao, Y., Fischer, E.R., DeLeo, F.R., and Otto, M. (2004). A crucial role for exopolysaccharide modification in bacterial biofilm formation, immune evasion, and virulence. J. Biol. Chem. *279*, 54881–54886.

Wagner, V.E., Frelinger, J.G., Barth, R.K., and Iglewski, B.H. (2006). Quorum sensing: dynamic response of *Pseudomonas aeruginosa* to external signals. Trends Microbiol. *14*, 55–58.

Wagner, V.E. and Iglewski, B.H. (2008). *P. aeruginosa* biofilms in CF infection. Clin. Rev. Allergy Immunol. *35*, 124–134.

Walker, T.S., Tomlin, K.L., Worthen, G.S., Poch, K.R., Lieber, J.G., Saavedra, M.T., Fessler, M.B., Malcolm, K.C., Vasil, M.L., and Nick, J.A. (2005). Enhanced *Pseudomonas aeruginosa* biofilm development mediated by human neutrophils. Infect. Immun. *73*, 3693–3701.

Wenzel, R.P. (2007). Health care-associated infections: major issues in the early years of the 21st century. Clin. Infect. Dis. *45*, S85–88.

Wilson, C., Clegg, R., Leavesley, D., and Pearcy, M. (2005). Mediation of biomaterial–cell interactions by adsorbed proteins: a review. Tissue Eng. *11*, 1–18.

van Winkelhoff, A.J., Loos, B.G., van der Reijden, W.A., and van der Velden, U. (2002). *Porphyromonas gingivalis*, *Bacteroides forsythus*, and other putative periodontal pathogens in subjects with and without periodontal destruction. J. Clin. Periodontol. *29*, 1023–1028.

Wizemann, T.M., Adamou, J.E., and Langermann, S. (1999). Adhesins as targets for vaccine development. Emerg. Infect. Disease. *5*, 395–403.

Wu, L., Estrada, O., Zaborina, O., Bains, M., Shen, L., Kohler, J.E., Patel, N., Musch, M.W., Chang, E.B., Fu, Y.X., *et al.* (2005). Recognition of host immune activation by *Pseudomonas aeruginosa*. Science *309*, 774–777.

Wu, T., Trevisan, M., Genco, R.J., Falkner, K.L., Dorn, J.P., and Sempos, C.T. (2000). Examination of the relation between periodontal health status and cardiovascular risk factors: serum total and high density lipoprotein cholesterol, C-reactive protein, and plasma fibrinogen. Am. J. Epidemiol. *151*, 273–282.

Wu, T.C., Guarnieri, F.G., Staveley-O'Carroll, K.F., Viscidi, R.P., Levitsky, H.I., Hedrick, L., Cho, K.R., August, J.T., and Pardoll, D.M. (1995). Engineering an intracellular pathway for major histocompatibility complex class II presentation of antigens. Proc. Natl. Acad. Sci. U.S.A. *92*, 11671–11675.

Zhao, H., Hemmi, H., Akira, S., Cheng, S.H., Scheule, R.K., and Yew, N.S. (2004). Contribution of Toll-like receptor 9 signaling to the acute inflammatory response to nonviral vectors. Mol. Ther. *9*, 241–248.

Zimmermann, S., Wagner, C., Müller, W., Brenner-Weiss, G., Hug, F., Prior, B., Obst, U., and Hänsch, G.M. (2006). Induction of neutrophil chemotaxis by the quorum-sensing molecule N-(3-oxododecanoyl)-L-homoserine lactone. Infect. Immun. *74*, 5687–5692.

The Ecological Significance of Plant-associated Biofilms

3

Venkatachalam Lakshmanan, Amutha Sampath Kumar and Harsh P. Bais

Abstract

Microorganisms have historically been studied as planktonic or free-swimming cells, but most exist as sessile communities attached to surfaces, in multicellular assemblies known as biofilms. In the process of coping with both the pathogenic and beneficial interactions, the rhizosphere of plant roots encourages formation of sessile communities that begins with the attachment of free-floating microorganisms to a surface. Certain bacteria such as plant growth promoting rhizobacteria not only induce plant growth but also protect plants from soil-borne pathogens in a process known as biocontrol. By contrast, other rhizobacteria in a biofilm matrix may cause pathogenesis in plants. Although research suggests that biofilm formation on plants is associated with biological control and pathogenic response, little is known about how plants regulate this association. The scope of this chapter is restricted to biofilm-forming bacteria and their interactions with terrestrial plants, specifically emphasizing recent work. After an overview of documented interactions between bacteria and plant tissues, we examine some of the more prominent mechanisms of biofilm formation on and around plant surfaces.

Introduction

A bacterial biofilm, is generally described as a highly structured, dense, sessile population settled on an inert surface. However, sessile populations are also present in association with various biotic surfaces. In the case of plants, surface attached bacteria can be found on seeds, leaves, and roots (Bais *et al.*, 2004; Rudrappa *et al.*, 2008a). Plant-associated communities are usually not as dense and structured as biofilms formed in other environmental conditions. However, they have an added element of complexity, since in this case the surface is not only an inert support for bacterial growth, but also the main source of nutrients for the microorganisms. Plants influence their environment by synthesizing complex organic compounds from inorganic matter and sunlight, releasing a substantial portion of this organic material as exudates in the rhizosphere (Bais *et al.*, 2006). As much as 40% of the carbon fixed by the land plants is released into the soil as root exudates, including compounds rich in nitrogen and phosphorus (Lynch and Whipps, 1990). These nutrients promote bacterial growth in the rhizosphere, which in turn allows other organisms, most prominently protozoans, to feed on the bacteria. The nitrogen released in the process leads to increased branching of the roots and thus more exudates for bacteria to feed upon. The effect of plant root secreted chemicals on microbial populations in the zone of soil influenced by the root, known as 'the rhizosphere,' has been well established (Grayston *et al.*, 1998; Bais *et al.*, 2006; Rudrappa *et al.*, 2008a). As a result, a number of important beneficial (Dong *et al.*, 2001; Biswas *et al.*, 2000; Ryu *et al.*, 2003, 2004; Rudrappa *et al.*, 2008a, 2010) and harmful interactions (Hawes and Smith, 1989; Timmusk *et al.*, 2005) between plants and microorganisms, localized in the rhizosphere have been studied (Table 3.1).

Several other types of beneficial soil-borne microbes which are non-pathogenic, such as plant

Table 3.1 Bacteria known to colonize various plant part surfaces of different plant species

Interaction	Organism	Plant	Plant part	Reference
Biocontrol	*Bacillus subtilis*	Crop plants	Root	Bais *et al.* (2004)
	Pseudomonas fluorescens	Crop plants	Root	Silby and Levy (2004)
	Pseudomonas putida	Crop plants	Root	Espinosa-Urgel *et al.* (2002)
	Pseudomonas chlororaphis	Crop plants	Root	Chin-A-Woeng *et al.* (2000)
	Pseudomonas aureofaciens	Crop plants	Root	Sigler *et al.* (2001)
	Enterobacter agglomerans	Cotton	Root	Chernin *et al.* (1995)
	Enterobacter cloacae	Crop plants	Seeds	Kageyama and Nelson (2003)
	Bacillus polymyxa	Cucumber	Root	Bertin *et al.* (2003)
Symbiosis	*Rhizobium*	Legumes	Root	Fujishige *et al.* (2006)
	Sinorhizobium	Legumes	Root	Fujishige *et al.* (2006)
Beneficial	*Azoarcus strain BH72*	Grasses	Vasculature	Hurek *et al.* (1994)
	Herbaspirillum seropedicae	Wheat, rice	Endophytic	Baldani *et al.* (1986)
	Azorhizobium caulinodans	Rice	Root	Van Nieuwenhove *et al.* (2004)
	R. leguminosarum	Rice	Root	Yanni *et al.* (1997
	Azospirillum brasilense	Wheat	Root	Kim *et al.* (2005)
	Azospirillum lipoferum	Maize	Seed	Jacoud *et al.* (1999)
	Gluconacetobacter diazotrophicus	Sugar cane	Endophytic	James *et al.* (2001)
	Serratia entomophila	Grass species	Shoot	Johnson *et al.* (2001)
	Klebsiella pneumoniae	Wheat	Root	Dong *et al.* (2004)
	Salmonella enterica	Alfalfa	Seed sprouts	Puohiniemi *et al.* (1997)
Pathogenic	*Agrobacterium tumefaciens*	Pea	Root	Hawes and Smith (1989)
	Escherichia coli	Leafy vegetables	Leaves	Delaquis *et al.* (2007)
	Xylella fastidiosa	Grapes	Vasculature	Marques *et al.* (2002
	Xylophylus ampelinus	Grapes	Vasculature	Grall and Manceau (2003)
	Pseudomonas syringae	Cabbage	Leaf	Park *et al.* (2005)
	Erwinia amylovora	Apple/pear	Vasculature	Zhao *et al.* (2005)
	Clavibacter michigenesis	Potato	Leaf	Wolf *et al.* (2005)
	Pantoea stewartii	Sweet maize	Vasculature	Ahmad *et al.* (2001)
	Ralstonia solanacearum	Potato	Vasculature	Williamson *et al.* (2002)
	Xanthomonas axonopodis	Bean	Leaf	Jacques *et al.* (2005)
	Burkholderia cepacia	Wheat	Root	Balandreau *et al.* (2001)
	Enterococcus faecalis	Arabidopsis	Leaves/root	Jha *et al.* (2005)
	Erwinia chrysanthemi	Crop plants	Leaves	Hugouvieux-Cotte-Pattat *et al.* (1996)

Adapted and modified from Rudrappa *et al.* (2008b).

growth promoting rhizobacteria and fungi, can be associated with plant root surfaces and are known to contribute towards increases in plant yields by mechanisms such as improved mineral uptake, phytohormone production (Biswas *et al.*, 2000; Ryu *et al.*, 2003, 2004; Fujishige *et al.*, 2006), competitive suppression of pathogens by production of antibiotics (Mazzola *et al.*, 1992) and induction of secondary metabolite-mediated systemic resistance (Ryu *et al.*, 2004). This biological control activity is exerted either directly through antagonism of soil-borne pathogens *or* indirectly by eliciting a plant-mediated resistance response (Van Loon *et al.*,1998; Pozo and Azcon-Aguilar, 2007).

Microbial diversity in soil

Soil represents a highly heterogeneous environment for the microbiota and the number of prokaryotic cells on earth is estimated to exceed ~10^{10}, of which the biggest fraction is found abundant in the soil (Whitman *et al.*, 1998). The different components of the solid fractions in soil (sand, silt, clay, and organic matter) provide a myriad of different microhabitats (van Elsas and Trevors, 1997). But the community structure of the indigenous bacterial microflora in soil is determined by many variables, including geographic location and soil structure (Gelsomino *et al.*, 1999), soil particle size (Postma and Van veen, 1990; Ranjard *et al.*, 2001; Sessitsch *et al.*, 2001), mineral composition (Carson *et al.*, 2009), and agricultural practices (Benizri *et al.*, 2007; Rooney and Clipson, 2009). The term biodiversity has been defined in various ways. In microbial terms, it describes the number of different types of microbial species and their relative abundance in a given community in a given habitat. In molecular-ecological terms, it can be defined as the number and distribution of different sequence types present in the DNA extracted from the community in the habitat. However, the term 'community structure' implies the information on the number of individuals of different recognizable taxa (Liesack *et al.*, 1997). These divergent terms are often used interchangeably in publications on soil microbial diversity. With respect to microbial diversity, the number of types present and the evenness of their distribution are both important. A habitat with a larger number of species is more diverse, whereas an evenly distributed community is more diverse than an unevenly distributed community with the same number of species (Hedrick *et al.*, 2000).

Root secretions and rhizosphere microbial diversity

Soil organisms have to compete for nutrients and other resources that are sparsely available in soil. Because of these limiting circumstances, bacterial proliferation in soil is typically slow. However, microbial activity in soil is greatly influenced by plant root exudates (Bais *et al.*, 2006). Around, 5–60% (Marschner, 1995) of the entirety of photosynthetic carbon fixed by plants can be transferred into the soil as a rhizo-deposition. Rhizo-deposition includes shedding of root cells, water soluble compounds such as amino acids, sugars, hormones and vitamins that leak from the root surfaces, or actively secreted polymeric carbohydrates and enzymes. Products of the roots may also include gases such as CO_2 and ethylene, lysates released when cells autolyse, solid materials including cell walls, sloughed cells and root border cells, and eventually large parts such as root hairs or roots themselves (Bertin *et al.*, 2003; Bais *et al.*, 2004, 2006). Plants collectively produce a diverse array of ≥ 100,000 compositionally different secondary metabolites, each with different functions in the rhizosphere (Bais *et al.*, 2004, 2006). Microorganisms can use these compounds as substrates, resulting in an increased microbial biomass and activity around the roots, the so called rhizosphere effect (Table 3.2).

Bacterial biofilms in the rhizosphere

Root biofilm initiation and development is complex and not well understood owing to the dynamic nature of plant root surfaces. Soil is the main reservoir of bacterial immigrants for the potential rhizosphere community (Normander and Prosser, 2000; De Ridder-Dunie *et al.*, 2005; Berg and Smalla, 2009). In addition to physicochemical variations throughout the root surface, it is probable that other abiotic factors such as nutrient availability, temperature and relative humidity influence root biofilm associations (Stanley and Lazazzera, 2004). Yet, despite these challenges, diverse bacterial species have adapted to these ever-changing conditions and are capable of starting colonization by forming micro colonies on different parts of the roots from tip to elongation

Table 3.2 Components of root exudates, their functions in the rhizosphere, and compounds identified in root exudates of different plant species, including knapweeds, flax, tomatoes, turnip, peanut, lupine, pine, rape, and other agronomically significant species

Exudates	Specific compounds	Rhizospheric effect
Organic acids	Citric, oxalic, malic, fumeric, succinic, acetic, butyric, valeric, glycolic, piscidic, formic, aconitic, lactic, pyruvic, glutaric, malonic, aldonic, erythronic, ferulic, botanic, syringic, rosmarinic, trans-cinnamic, vanillic, tetronic	Major carbon source, chemoattractant signal for microbes, chelators of poorly soluble mineral nutrients, soil acidifiers, Al detoxifiers and *nod* gene inducers
Amino acids	α- and β-alanine, proline, asparagine, valine, aspartate, tryptophan, cysteine, ornithine, cysteine, histidine, glutamate, arginine, glycine, homoserine, isoleucine, phenylalanine, lysine, α-aminoadipic acid, methionine, serine, threonine	Nutrient source, chemoattractant signal for microbes, chelators of poorly soluble mineral nutrients
Sugars and vitamins	Glucose, fructose, galactose, maltose, ribose, xylose, rhamnose, arabinose, raffinose, deoxyribose, oligosaccharides, biotin, thiamin, niacin, pantothenate and riboflavin	Plant and microbial growth promoters; nutrient sources for microbes
Enzymes	Acid/alkaline phosphatase, invertase, amylase; protease	Catalysts for P release from organic molecules; biocatalysts for organic matter transformations
Inorganic ions and gases	HCO_3^-, OHH^+, CO_2, H_2	Chemoattractant signals to microbes
Phenolics	Liquiritigenin, luteolin, Genistein, Coumetrol, eriodictyol, naringenin, isoliquiritigenin, umbelliferone, formononetin, quercetin 4′,7-dihydroxyflavone 4,4′-dihydroxy-2′-methoxychalcone 4′,7-dihydroxyflavone 3,5,7,3′-tetrahydroxy-4′-methoxyflavone 7,3′-dihydroxy-4′-methoxyflavone (+)– and (–)– catechin	Nutrient source, chemoattractant signals to microbial growth promoters, *nod* gene inducers in rhizobia resistance, inducers against phytoalexins, chelators of poorly soluble mineral nutrients, phytoalexins against soil pathogens, detoxifiers of Al
Root border cells	Produce signals controlling gene expression; stimulate microbial growth; release chemo attractant; synthesize defence molecules for the rhizosphere; act as decoys that keep root cap infection-free; release mucilage and proteins	

Adapted and modified from Dakora and Phillips (2002) and Bais *et al*. (2004).

zone. Evidence is increasing that plants are able to actively select for their bacterial rhizosphere microflora, thereby creating a habitat which is most favourable for the plant. Such micro colonies eventually grow into large population sizes on roots to form mature biofilms (Bais *et al.*, 2004; Garbeva *et al.*, 2004; Rudrappa *et al.*, 2008, 2010).

The colonization of the rhizosphere by plant beneficial bacteria has received increasing attention in the past two decades. Plant roots and the surrounding soil areas (the rhizosphere) are preferential sites for colonization by soil bacteria. Bacterial establishment in the rhizosphere is perhaps the best studied aspect. Following the trail of the numerous studies on the association between symbiotic microorganisms (*Rhizobium*, *Sinorhizobium*, etc.) and legume roots, colonization of the rhizosphere by other plant beneficial bacteria has received increasing attention owing to the fact that their effectiveness is directly dependent on their efficiency to colonize the roots (Bull *et al.*, 1991; Thomashow, 1996; Chin-A-Woeng *et al.*, 2000).

Chemotaxis during bacterial biofilm formation

Bacteria can move by different mechanisms, including flagellar swimming, swarming, twitching, and gliding movements. Bacterial motility towards the root is not a random phenomenon, but results from chemotactic responses to root exudates. Flagellar motility provides access to attachment sites on the plant, and possibly drives initial attachment, but is also involved in biofilm maturation. Plant exudates are released from specific sites and many bacteria preferentially move towards and colonize positions on the plant surface with optimal nutrient availability. Chemotaxis towards sugars present in root exudates has been demonstrated in *Azospirillum brasilense* (Van Bastelaere *et al.*, 1999). In this organism, as well as in *Pseudomonas fluorescens*, chemotaxis has been shown to constitute an important factor in root colonization (Vande Broek and Vanderleyden, 1995; Vande Broek *et al.*, 1998). Thus, as is the case for the initiation of biofilm formation on abiotic surfaces, environmental and nutritional cues are important for biofilm formation on root surfaces, with the difference that in the later case nutrients come from the surface itself. This explains why chemotaxis plays a significant role in bacterial attachment to plant roots, and not in biofilm formation on abiotic surfaces, which at least in *Escherichia coli* has been shown to be chemotaxis independent (Pratt and Kolter 1998). *A. brasilense* is thought to use aero taxis to identify optimal oxygen concentrations that allow respiration but do not inhibit nitrogenase (Greer-Phillips *et al.*, 2004). Taxis towards oxygen and other attractants promote *A. brasilense* root colonization. Motility is also involved in biofilm formation of many different microbes, and although its importance is often conditional, motility mutants tend to be much less competitive when compared with motile strains over time (Turnbull *et al.*, 2001; Yao and Allen, 2006). For example, *S. meliloti* flagellar mutants showed reduced biofilm formation and delayed in nodule formation (Fujishige *et al.*, 2006). Twitching motility conferred by retractable type IV pili is used by *Pseudomonas* spp. to spread across the surface and form mushroom like structures in biofilms (Klausen *et al.*, 2003). The rate of chemotaxis by different bacteria towards different nutrients is highly dependent upon the species and strain, as well as on the attractant. Generally, bacterial strains show a stronger reaction to their native host plants and endophytes often respond more strongly than rhizobacteria (Bacilio-Jiménez *et al.*, 2003).

Motility is involved not only in the initiation and development of biofilms, but also in their dispersal, an essential mechanism for spreading and colonizing new habitats (Sauer *et al.*, 2002). For vascular pathogens such as *P. stewartii* subsp. *stewartii* and *Xylella fastidiosa*, migration from an initial colonization site can strongly influence the nature of disease, and motility drives this emigration. Dispersion from plant biofilms is also certain to play a role in the ecology and disease progression of other plant-associated microbes.

Nutrient source for biofilm formation

Root exudates serve as a major plant-derived factor responsible for triggering root colonization (Lugtenberg *et al.*, 1999) and biofilm associations (Walker *et al.*, 2004). These organic compounds may provide a nutrient source for the encroaching microorganisms (Lugtenberg *et al.*, 1999). The carboxylic acids malate and citrate are suitable carbon (C) sources for many microorganisms (López-Bucio *et al.*, 2000; Neumann and Martinoia, 2002). Carboxylate excretion and acidification at the root surface might thus have a strong general impact on microbial communities (Marschner *et al.*, 2002). As previously mentioned, plant roots exude photosynthates which include many small molecular weight compounds (Bais *et al.*, 2006). Bacteroids living in symbiosis with plants receive their energy supply through the peribacteroidal membrane largely in the form of malate. Similarly, free-living beneficial bacteria preferentially use malate as C-sources (Kleijn *et al.*, 2010). The root surfaces of plants are continually subjected to the two-way traffic of solutes from plants to the soil and vice versa (Lugtenberg *et al.*, 1999; Dakora and Phillips, 2002). A broad range of environmental factors cause fluctuations in root surface properties and this dynamic environment may therefore make it challenging for two-way communication between plants and microbial communities in the rhizosphere (Bais *et al.*, 2006). This interaction becomes more complicated when more than one bacterium is involved, as observed in the case of multispecies microbial associations (An *et al.*, 2006). The concept of heterogeneous bacterial colonization has been reported for the rhizosphere (Beattie and Lindow, 1999; Joseph *et al.*, 2007), controlled in part by the pH gradient that exists along the root surface from the root tip to basal regions (Fischer *et al.*, 1989). Additionally, it is believed that metabolite requirements and organic compound secretions fluctuate between the root tip and mature root regions; therefore the nutrients available for bacterial use will not be distributed homogeneously throughout the rhizoplane (Hofland *et al.*, 1989; Eshel and Waisel, 1996). It is known that the root tip is engaged in maximum carbon turnover, in terms of root secretions and cell debris formation, because of the frequent 'decapping' of the root cap (Hawes *et al.*, 2000). Such localized differences in nutrient availability may influence the structural formation of biofilms. For example, lower nutrient levels have been observed to influence the swarming of *P. aeruginosa*, resulting in 'flatter' biofilm structures (Shrout *et al.*, 2006). Similarly, our studies have shown that *Bacillus subtilis* strain FB17, a beneficial bacterium, exhibits differential biofilm formation on *A. thaliana* roots (Rudrappa *et al.*, 2008, 2010; Venkatachalam *et al.*, 2008a) (Fig. 3.1).

The discussion of biofilms thus far has been supported by studies regarding bacterial aggregation on homogeneous abiotic surfaces, in sewage treatment plants and a few plant and animal tissues, but complexity increases for bacterial colonization in the rhizosphere. This leads to the speculation that plants could be using root-exuded compounds in the rhizosphere to take advantage of this bacterial communication system and influence colonizing communities. Discovery and characterization of these plant secreted compounds could have important biological implications in agriculture. Among such critical processes, regulation of microbial recruitment and dynamics are most vital. It has been reported that plants regulate microbial processes by deterring pathogenic microorganisms and selectively attracting beneficial microorganisms (Ramey *et al.*, 2004). How these microorganisms establish themselves as communities on the root surface is a critical question because, like their plant counterpart, microorganisms are equally dynamic and employ various mechanisms to cope with changed conditions (Langer *et al.*, 2004).

Recent studies have detected specific root-secreted compounds in the rhizosphere that act as plant quorum sensing (QS) mimics (Teplitski *et al.*, 2000; Mathesius *et al.*, 2003; Keshavan *et al.*, 2005). For example, evidence for the ability of roots to secrete acyl hormoserine lactones (AHLs) has been shown by engineering tobacco and potato to secrete AHLs in the rhizosphere (Toth *et al.*, 2004; Scott *et al.*, 2006). Apart from secretion of AHLs and AHL mimics, plant-derived compounds also influence biofilm formation by interfering with the bacterial QS mechanism (Rasmussen *et al.*, 2005). Indeed, a fair body of evidence suggests that cross-talk between plants and bacteria may occur through QS signal mimics.

Figure 3.1 Differential biofilm of *Bacillus subtilis* (strain FB17) on *Arabidopsis thaliana* (Col-0) root surface with respect to the intra-plant signalling. (A) Schematic of intra-plant signalling (Malate transporter – ALMT1) to recruit beneficial rhizobacteria through secretion of malic acid (MA), which acts as a chemoattractant for *Bacillus subtilis* FB17 after leaf infection *by Pseudomonas syringae pv. tomato (Pst)*. ISR in the schematic refers to systemic induced resistance. (B) The figure shows less or no colonization in the uninfected plant, followed by minimal attachment in the roots of an *atalmt1* mutant (plants deficient in malic acid secretion) (C) Higher colonization and biofilm formation 5 days post treatment in response to aerial infection with *P. syringae* pv. Tomato (*Pst*DC3000). (D) In plants ove-expressing ALMT1 (plant lines secreting more malic acid) as observed through representative root colonization experiments by confocal microscopy. (E) in plants overexpressing AtALMT1 (plant lines secreting more malic acid) and infected aerially with with *P. syringae* pv. Tomato (*Pst*DC3000) as observed through representative root colonization experiments in plants by confocal microscopy. Scale bar = 50 μm.

Thus, the rhizosphere may be considered as a region of overlapping, communicating bacterial populations, each defined by mutual recognition of specific QS signals, and each affected in different ways by the secretion of host plant QS mimics. In addition to their associations with QS mechanisms, plants may also regulate bacterial associations by influencing the structure of biofilms attached to their root surface by varying rhizosphere nutrient status, as suggested by abiotic surface studies (Shrout *et al.*, 2006).

Biofilms in response to phosphorus availability

Phosphorus, is an important fitness factor for plant-associated bacteria, controlling aggregation and surface attachment. In many bacteria the *Pho*R-*Pho*B two component regulatory system, similar to that identified in *E. coli*, activates the response to phosphorus limitation. The impact of limiting phosphorus on biofilms, however, depends on the organism and its role in the ecosystem. *Pseudomonas aureofaciens* PA147-2 shows a *phoB*-dependent inhibition of biofilm formation under phosphate starvation (Monds *et al.*, 2001; Danhorn *et al.*, 2004). In contrast, phosphorus limitation and *phoB* overexpression increase the density of *A. tumefaciens* biofilms (Danhorn *et al.*, 2004).

Bacterial signals in biofilm formation

Bacterial biofilm formation is a highly regulated process. Each species responds to its own set of environmental cues via a distinct set of molecular mechanisms. Despite this mechanistic diversity, there are several stimuli that are generally important for plant-associated bacteria and act through conserved regulatory systems. QS mediates several plant–microbe interactions, both pathogenic and beneficial. The first described examples of plant-secreted QS mimics were halogenated furanones produced by the marine red algae *Delisea pulchra* (Yoon *et al.*, 2006), which bind competitively to AHL receptors, instigating proteolytic degradation and inhibition of associated QS signals (Teplitski *et al.*, 2004).

The initial stages of root colonization by *P. putida* appear to be very dynamic, and the bacterial population reaches its maximum size with respect to plant biomass 1–2 days after seed inoculation (Espinosa-Urgel *et al.*, 2002). At that time, growth of the bacterial population on the root appears to couple with plant growth, so that the number of bacteria that are recovered per gram of root remains essentially constant afterwards. Thus, the initial steps may be key for the success of root colonization. Microbial colonization in general is highly regulated in a population density dependent manner by quorum sensing (QS) (Whitehead *et al.*, 2001). It has been reported that the cells, use auto-inducers (AI) to measure limitations to mass transfer or diffusion, termed diffusion sensing (DS) (Redfield, 2002). However, the more recent efficiency sensing (ES) theory (Hense *et al.*, 2007), which unifies QS and DS, is most applicable in the context of plant-root-associated bacteria, which will experience spatial distribution and mass transfer limitations under natural settings. In accordance with this hypothesis, root-associated bacteria might form small aggregates of clonal cells (microcolonies). Once established, there will be ES between microcolonies especially in the case of sessile bacteria. This leads to expression of specific genes that allows microcolonies/bacteria to coordinate with each other in a manner reminiscent of a multicellular organism (Whitchurch *et al.*, 2002; Kearns *et al.*, 2005; Kolter, 2005) referred to as QS. QS activation is mediated by small autoinducer (AI) molecules, which are responsible for cell–cell communication, owing to the coordinated action of many bacteria, including plant-associated bacteria. The most commonly reported type of autoinducer signals are N-acyl homoserine lactones (AHLs) (von Bodman *et al.*, 2003), although half a dozen or so other molecules, including diketopiperazines in several Gram-negative bacteria (Holden *et al.*, 1999), furanosyl borate diester in *Vibrio harveyi* (Chen *et al.* 2002), and γ-butyrolactone in *Streptomyces* (Yamada and Nihira, 1998), have also been implicated in density-dependent signalling. Typically, a basal level of AHLs are constitutively synthesized until a threshold population of bacteria has been achieved, which point out that these molecules serve as ligands to a global transcription regulator (LuxR or LuxR-like proteins) that activates many QS controlled genes, including virulence factors. The rhizosphere contains a higher proportion of AHL-producing bacteria as compared with bulk soil, suggesting that they play a role in colonization (Elasri *et al.*, 2001). Diverse polysaccharide and autoinducers produced in different strains of organisms allows for variations between biofilm structures (Mack *et al.*, 1996; Miller and Bassler, 2001; Waters and Bassler, 2005).

While considering natural encounters between plants and bacteria, the disruption of QS regulation by other rhizosphere bacteria may be as important as disruption by QS mimics from the host plant. The best example is the production of AHL lactonases (which degrade AHLs) by the soil bacterium *Bacillus thuringiensis* (Dong *et al.*, 2001).

Rhizospheric biofilms and plant defence

Plant defences have the potential to affect bacterial populations in the rhizosphere by either recruiting beneficial bacteria or actively repressing pathogen proliferation. One of the best studied examples is the biological control of the fungus *Gaeumannomyces graminis* var. *tritici* (*Ggt*) the causal agent of 'take all' in wheat. When wheat is continuously grown in the same field, a build-up of the pathogen may occur. However, after several years of wheat monoculture and a severe outbreak of the disease, a decrease in take-all is observed, a phenomenon known as take-all decline (TAD). TAD is associated with the proliferation of specific strains of fluorescent *Pseudomonas* spp. in the wheat rhizosphere that produce the antibiotic 2,4-diacetylphloroglucinol (DAPG) and successfully suppress *Ggt* (Weller *et al.*, 2002).

A recent study from our group showed that inoculation of *Arabidopsis thaliana* leaves with the foliar pathogen *Pseudomonas syringae pv. tomato* (*Pst*DC3000) induced malate excretion in roots (Rudrappa *et al.*, 2008, 2010). The study revealed that *Pst*DC3000 infected shoots relay chemical signal(s) underground through root malate secretion, resulting in specific chemotaxis to recruit rhizobacteria *Bacillus subtilis* strain FB17. The authors further demonstrated that infection of *A. thaliana* leaves up-regulated AtALMT1, a plasma membrane-located malate channel, in roots. Although various groups have shown mechanisms of *B. subtilis* biocontrol of root infections, the studies on the effect of *B. subtilis*-induced systemic resistance (ISR) on aerial bacterial pathogenic infections are limited (Bais *et al.*, 2004; Rudrappa *et al.*, 2008, 2010) (Figs. 3.1 and 3.2). Therefore, it was postulated that root exudates have direct defensive qualities and diseased plants 'signal for help' in the rhizosphere.

Microbial control of plant diseases is a complex process involving not only the biocontrol microbe, the pathogen, and the plant, but also the indigenous microflora, macrobiota such as nematodes and protozoa, and the plant growth substrate such as soil, stone wool, or vermiculite. Many reviews on biocontrol have been written (Chin-A-Woeng *et al.*, 2003; Compant *et al.*, 2005; Haas and Defago, 2005). To act efficiently, the controlling microbe should remain active under a large range of conditions, such as varying pH, temperature, and concentrations of different ions. These requirements are not easy to fulfil.

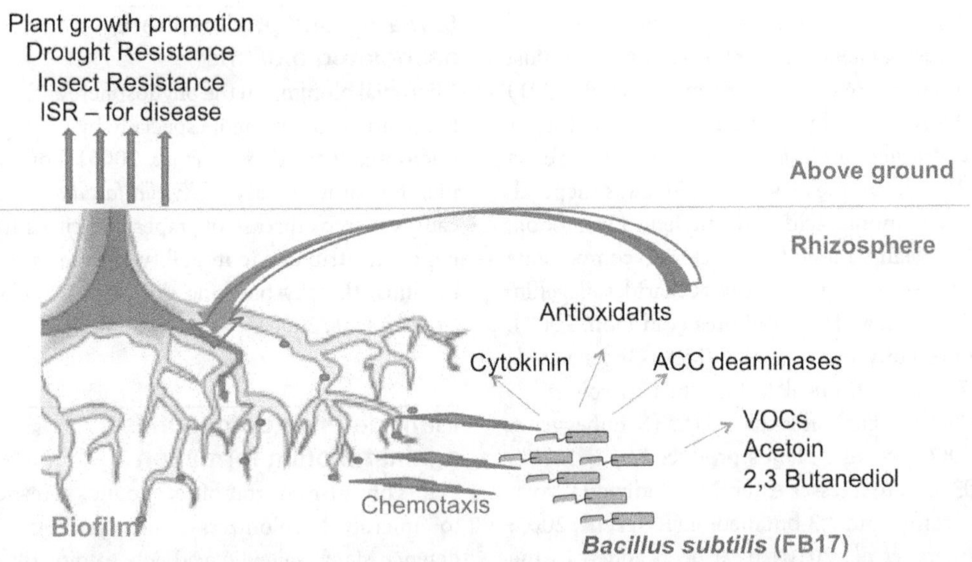

Figure 3.2 Model for the induced systemic resistance (ISR) signalling pathway. Biofilm formation by beneficial rhizobacteria *Bacillus subtilis* FB17 leads to activation of innate defence responses in plants. Systemically, induction of ISR is associated with priming for enhanced expression of a set of jasmonic acid (JA)-responsive and/or ethylene (ET)-responsive genes leading to increased plant resistance against invading pathogens.

Therefore, it is not surprising that the efficacy of several first-generation commercial biocontrol products (Copping, 2004) is not always sufficient. However, as our understanding of mechanisms of biocontrol and selection procedures for active strains increases, biocontrol products will improve and we predict that biocontrol has a good future.

Plant immune responses triggered by beneficial microbes

Only few plant–beneficial microbe interactions leading to enhanced systemic resistance have been studied for locally induced changes in plant gene expression or metabolism. In most cases only weak, transient, or strictly localized defence-associated responses were elicited, which differs greatly from the massive induction of defence responses triggered during plant–pathogen interactions (Gianinazzi-Pearson *et al.*, 1996; Verhagen *et al.*, 2004), but, interaction of some bacteria with the plant roots can result in plants resistant to some pathogenic bacteria, fungi, and viruses. This phenomenon is called induced systemic resistance (ISR). ISR shares many properties with innate immunity in humans (Bittel and Robatzek, 2007). ISR was discovered by the findings that resistance can be induced by the rhizobacterium *Pseudomonas* sp. strain WCS417r against *Fusarium* wilt of carnation (Van Peer *et al.*, 1991) and by selected rhizobacteria against the fungus *Colletotrichum orbiculare* in cucumber (Wei *et al.*, 1991). In these studies ISR was dependent on jasmonic acid and ethylene (Van Loon, 2007). Many individual bacterial components induce ISR, such as lipopolysaccharides, flagellin, salicylic acid and siderophores (van loon, 2007). More recently, cyclic lipopeptides (Ongena *et al.*, 2007), the antifungal factor Phl (Iavicoli *et al.*, 2003), the signal molecule AHLs (Schuhegger *et al.*, 2006), volatile blends produced by *B. subtilis* GB03 and, to a lesser extent, the individual volatiles acetoin and 2,3-butanediol (Ryu *et al.*, 2003; Rudrappa *et al.*, 2010) have been added to the list. In contrast to many biocontrol mechanisms, extensive colonization of the root system is not required for ISR, as shown by Kamilova *et al.* (2005) and also Dekkers *et al.* (2000), using root colonization mutants of *P. fluorescens* WCS365. It is unlikely that a poor colonizer acts through antibiosis, since colonization is the delivery system for antifungal components along the root system (Chin-A-Woeng *et al.*, 2000). Therefore, the finding that certain strains of *B. cereus*, which are poor colonizers, are good biocontrol agents was surprising (Gilbert *et al.*, 1994). However, the observation that certain antifungal metabolites (AFMs) can induce ISR explains this phenomenon. We therefore speculate that many of the *Bacillus* strains that can act as biocontrol agents, act through ISR rather than antibiosis. Rudrappa *et al.* (2008) reported that infection of leaves of *A. thaliana* seedlings with the foliar pathogen *P. syringae* pv. *tomato Pst*DC3000 results in enhanced secretion of l-malic acid by the roots, and that the enhanced level of l-malic acid selectively signals and recruits the beneficial rhizobacterium *B. subtilis* FB17, which is a biocontrol bacterium that protects the plant through ISR. Previously, De Weert *et al.* (2002) reported that another biocontrol bacterium, *P. fluorescens* WCS365, which also acts through ISR (Kamilova *et al.*, 2005), shows strong chemotaxis towards the major tomato root exudate component, citric acid.

In planta and phyllosphere-associated biofilms

Microbial biofilms on the phyllosphere and vasculature are also reported, especially by pathogenic microorganisms (Torres *et al.*, 2006). For example, the pathogenicity of *Xylella fastidiosa*, which causes Pierce's disease of grapes and citrus fruits, has been attributed to its ability to form extensive biofilms, thereby blocking the plant vasculature (Torres *et al.*, 2006).

Plant defence compounds against biofilm formation

Plants produce an array of compounds in response to microbial colonizers. The systemic plant defence signal salicylic acid acts to directly suppress virulence genes in *P. aeruginosa*, inhibiting attachment and biofilm formation on *A. thaliana* roots (Prithiviraj *et al.*, 2005). Conversely, rosmarinic acid is effective against free-swimming *P. aeruginosa* but far less effective on biofilms

(Walker *et al.*, 2004). These studies demonstrate the complex interplay between plant defence, biofilms, and infection.

Phenazine

Although antibiotics, including antifungal compounds, play important roles in biocontrol, they are usually not implicated in colonization of the plant per se. An exception is phenazine produced by *Pseudomonas chlororaphis* strain 30–84, which is regulated by the PhzR-PhzI quorum-sensing system. Both *phzR* and *phzI* regulatory mutants and *phzB* phenazine biosynthetic mutants formed significantly reduced and immature biofilms *in vitro* but were rescued by heterologous *phzB* expression or exogenous phenazine. The *phzR/I* mutants were also deficient in colonization of wheat seeds and roots (Maddula *et al.*, 2006). Two quorum sensing (QS) systems of regulation were identified: Phz/R and CsaI/R. Genes *phzI* and *csaI* were cloned and sequenced (Veselova *et al.*, 2008). Phenazine apparently contributes significantly to biofilm maturation *in vitro* and on plants, in addition to its antibiotic activity.

Surfactants

Rhamnolipids shape *P. aeruginosa* biofilm architecture by maintaining open fluid channels between cell clusters (Davey *et al.*, 2003), and the role of surfactants in biofilm formation in other species has also been examined. Cyclic lipopeptides (CLPs) are another class of surfactants produced by several pseudomonads (Raaijmakers *et al.*, 2006). CLPs can increase virulence in plant pathogens and are of great importance for biocontrol species because of their antimicrobial properties and their role in motility and biofilm formation. *P. putida* strain PCL1445 forms two CLP surfactants, both of which inhibit biofilm formation of several *Pseudomonas* species and can even dissolve existing biofilms. The CLPs appear to have a similar function in PCL1445, because a CLP-less mutant shows increased aggregation and formation of thicker biofilms (Kuiper *et al.*, 2004). *B. subtilis* 6051 produces extensive biofilms on inert surfaces and roots and secretes an antimicrobial CLP surfactant called surfactin. Loss of surfactin biosynthesis leads to a deficiency in flagella-independent surface motility and reduced colonization of roots and inert surfaces. The surfactin mutant is outcompeted by *P. syringae* pv. *tomato* in the rhizosphere and fails to protect *A. thaliana* from infection (Bais *et al.*, 2004). Swarming motility in *Rhizobium etli* is dependent on the *cin* quorum-sensing system. Long-chain 3-hydroxy AHLs synthesized by *cinI* are signals recognized by the transcriptional regulator CinR, which in turn induces *cinI*. These long-chain AHLs reduce the viscosity of the extracellular slime produced by *R. etli* during swarming, indicating that they serve not only as quorum-sensing signals but also as biosurfactants (Daniels *et al.*, 2006). The coordinated development of differentiated subpopulations suggests a potential need for cell to cell signalling systems, which facilitate communication among and across differentiating subpopulations (Ilka *et al.*, 2009).

Biofilm inhibitors

A major concern in developing an antimicrobial therapy against opportunistic pathogens like *P. aeruginosa* is their inherent potential to develop antibiotic resistance through biofilm formation. Disturbingly, recent evidence also suggests that some aminoglycoside antibiotics induce biofilm formation in *P. aeruginosa* at sub inhibitory concentrations (O'Toole and Stewart, 2005). Conversely, plants combat these opportunistic pathogens by secreting antimicrobials through roots (Walker *et al.*, 2004). To counter these antimicrobials, microorganisms in the biofilms synthesize compounds which can bind and sequester antibiotics in the periplasm. The sequestering of antibiotics prevents their entry through the periplasmic space to their site of action in the cytoplasm as reported for *P. aeruginosa* (Mah *et al.*, 2003). Alternatively, if these compounds diffuse in a controlled manner or can degrade the established biofilm matrix, they could potentially be more effective. Rosmarinic acid (RA) secreted by *Ocimum basilicum* (sweet basil), was potent against the planktonic form of *P. aeruginosa*, therefore preventing biofilm initiation, but lacked the ability to penetrate mature biofilms on the root surfaces (Walker *et al.*, 2004). This could be because the pathogen might acquire resistance to the secretions either by altered lipopolysaccharide production, enhanced expression of antibiotic efflux pumps or by detoxification of the antimicrobial compound (Chuanchuen *et al.*, 2001). The above-mentioned lines of research encourage the concept that specific root secretions may function as natural plant pathogenic biofilm inhibitors. Screening of root exudates for potential biofilm inhibitors against several plant pathogens may lead to the discovery of the regulatory pathways controlling biofilm formation in plant pathogens by identifying compounds that inactivate bacterial binding and/or QS.

Human pathogens and associated biofilms

Human food-borne pathogens can also colonize plant surface and tissues and cause enteric illness upon consumption of the contaminated plant product. According to the World Health Organization (Fact sheet No. 237, http://www.who.int/mediacentre/factsheets/fs237/en/), 76 million cases of food borne diseases occur annually in the United States. Several of the pathogens responsible for these epidemics form biofilms on aerial plant surfaces, and these studies may provide insight on how they would infect root tissues. Enteric bacteria such as *Salmonella enterica* and *Escherichia coli* are most often associated with food crops including sprouts, fruits, and leafy vegetables (Hogan and Kolter, 2002). Their ability to form biofilms on the plant surfaces prevents removal of the bacteria during sanitizing procedures and allows colonization of interior plant surfaces (Cooley *et al.*, 2003; Brandl, 2006; Kutter *et al.*, 2006). The recent issue of spinach contaminated with *E. coli* O157:H7 is one of the best examples (Boxrud *et al.*, 2010). Several of the genes in *S. enterica* responsible for biofilm formation and virulence in humans are those that cause attachment to plant surfaces (Cooley, 2007). With all of the aforementioned human pathogens, extensive research will be needed to fully understand how these bacteria operate in diverse hosts and inflict virulence through biofilm formation. Continued research into plant-associated bacteria, identification of genes responsible for biofilm formation and possible strategies for preventing or curing human illness should spearhead future research.

In recent years increased attention has been paid to human pathogens that are spread *via* association with plants (Aruscavage *et al.*, 2006). A subset of animal pathogens only come into contact with plants coincidentally (e.g. through contaminated irrigation water or in processing facilities), for others the plant is an alternative host. Removal of pathogenic *Escherichia coli* and *Salmonella* strains formed biofilms on cantaloupe melons, and *Salmonella* from alfalfa and mung bean seeds had limited success, even with harsh treatments (Fett and Cooke, 2003). Furthermore, biofilm formation has been shown to play a role in the persistence of *Salmonella* after chlorination treatment of parsley (Fett and Cooke, 2003).

Some of the virulence factors are important for animal infection by *Salmonella enterica* function during attachment to plant tissues as well (Bais *et al.*, 2004). In addition to animal pathogens that are merely transmitted through plants, there are those that infect both plants and animals, including several *P. aeruginosa* PA14 (He *et al.*, 2004) and *Enterococcus faecalis* strains (Jha *et al.*, 2005). Horizontal gene transfer probably plays a significant role in the acquisition of the pathogenicity islands of *P. aeruginosa* PA14 and is facilitated by the extended and intimate contact of cells within biofilms (Espinosa-Urgel, 2004).

Perspective and prospects

Biofilms grow on and around plant tissues and an increasing number of studies have begun to reveal their importance in plant–microbe interactions. While much work remains to be done, insights have been gained at the molecular level into the processes that drive the interaction between biofilm forming microbes and plants. Defined model systems focusing on a few select plant and bacterial species have been and will continue to be important to understand the interactions mediated by biofilm forming microbes on plants. Field studies have been conducted to acquire an improved understanding of the microbes that inhabit specific plant environments. These studies focus on the need of the microbial participants in plant–bacteria interactions rather than the mechanisms through which they are involved. It will not be a trivial task to integrate these ecological findings with the molecular mechanisms and the genetic requirements involved. H.P.B. acknowledges the support from University of Delaware Research Foundation (UDRF) and NSF Award 105-0814477.

References

Ahmad, M., Majerczak, D.R., Pike, S., Hoyos, M.E., Novacky, A., and Coplin, D.L. (2001). Biological activity of harpin produced by *Pantoea stewartii* subsp. stewartii. Mol. Plant Microbe Interact. *14*, 1223–1234.

An, D., Danhorn, T., Fuqua, C., and Parsek, M.R. (2006). Quorum sensing and motility mediate interactions between *Pseudomonas aeruginosa* and *Agrobacterium tumefaciens* in biofilm cocultures. Proc. Natl. Acad. Sci. U.S.A. *103*, 3828–3833.

Aruscavage, D., Lee, K., Miller, S., and LeJeune, J.T. (2006). Interactions affecting the proliferation and control of human pathogens on edible plants. J. Food Sci. *71*, R89–99.

Bacilio-Jiménez, M., Aguilar-Flores, S., Ventura-Zapata, E., Pérez-Campos, E., and Bouquelet, S. (2003). Chemical characterization of root exudates from rice (*Oryza sativa*) and their effects on the chemotactic response of endophytic bacteria. Plant Soil *249*, 271–277

Bais, H.P., Fall, R., and Vivanco, J.M. (2004). Biocontrol of *Bacillus subtilis* against infection of *Arabidopsis* roots by *Pseudomonas syringae* is facilitated by biofilm formation and surfactin production. Plant Physiol. *134*, 307–319.

Bais, H.P., Weir, T.L., Perry, L.G., Gilroy, S., and Vivanco, J.M. (2006). The role of root exudates in rhizosphere interactions with plants and other organisms. Annu. Rev. Plant Biol. *57*, 233–266.

Balandreau, J., Viallard, V., Cournoyer, B., Coenye, T., Laevens, S., and Vandamme, P. (2001). *Burkholderia cepacia* Genomovar III is a common plant-associated bacterium. Appl. Environ. Microbiol. *67*, 982–985.

Baldani, J.I., Baldani, V.L.D., Seldin, L., and Döbereiner, J. (1986). Characterization of *Herbaspirillum seropedicae* a root associated nitrogen-fixing bacterium. Int. J. Syst. Bacteriol. *36*, 86–93.

Beattie, G.A., and Lindow, S.E. (1999). Bacterial colonization of leaves: a spectrum of strategies. Phytopathol. *89*, 353–359.

Benizri, E., Nguyen, C., Piutti, S., Slezack-Deschaumes, S., and Philippot, L. (2007). Additions of maize root mucilage to soil changed the structure of the bacterial community. Soil Biol. Biochem. *39*, 1230–1233.

Berg, G., and Smalla, K. (2009). Plant species and soil type cooperatively shape the structure and function of microbial communities in the rhizosphere. FEMS Microbiol. Ecol. *68*, 1–13.

Bertin, C., Yang, X.H., and Weston, L.A. (2003). The role of root exudates and allelochemicals in the rhizosphere. Plant Soil *256*, 67–83.

Biswas, J.C., Ladha, J.K., and Dazzo, F.B. (2000). Rhizobial inoculation improves nutrient uptake and growth of lowland rice. Soil Sci. Soc. Am. J. *64*, 1644–1650.

Bittel, P., and Robatzek, S. (2007). Microbe-associated molecular patterns (MAMPs) probe plant immunity. Curr. Opin. Plant Biol. *10*, 335–341.

Brandl, M.T. (2006). Fitness of human enteric pathogens on plant and implications for food safety. Annu. Rev. Phytopathol. *44*, 367–392.

von Bodman, S.B., Bauer, W.D., and Coplin, D.L. (2003). Quorum-sensing in plant pathogenic bacteria. Annu. Rev. Phytopathol. *41*, 455–482.

Boxrud, D., Monson, T., Stiles, T., and Besser, J. (2010). The role, challenges, and support of pulsenet laboratories in detecting foodborne disease outbreaks. Public Health Rep. *2*, 57–62.

Bull, C.T., Weller, D.M., and Thomashow, L.S. (1991). Relationship between root colonisation and suppression of *Gaeumannomyces graminis* var. *tritici* by *Pseudomonas fluorescens* strain 2–79. Phytopathology *81*, 954–959.

Carson, J.K., Campbell L., Rooney, D., Clipson, N., and Gleeson, D.B. (2009). Minerals in soil select distinct bacterial communities in their microhabitats. FEMS Microbiol. Ecol. *67*, 381–388.

Chen, X., Schauder, S., Potier, N., van Drosselaer, A., and Pelczer, I. (2002). Structural identification of a bacterial quorum-sensing signal containing boron. Nature *415*, 545–549.

Chin-A-Woeng, T.F.C., Bloemberg, G.V., Mulders, I.H.M., Dekkers, L.C., and Lugtenberg, B.J.J. (2000). Root colonization is essential for biocontrol of tomato foot and root rot by the phenazine-1-carboxamide-producing bacterium *Pseudomonas chlororaphis* PCL1391. Mol. Plant–Microbe Interact. *13*, 1340–1345.

Chin-A-Woeng, T.F.C., Bloemberg, G.V., and Lugtenberg, B.J.J. (2003). Mechanisms of biological control of phytopathogenic fungi by *Pseudomonas* spp. In Plant–Microbe Interactions, Stacey, G., and Keen, N.T., eds. (The American Phytopathological Society, St. Paul, MN), pp. 173–224.

Chernin, L., Ismailov, Z., Haran, S., and Chet, I. (1995). Chitinolytic *Enterobacter agglomerans* antagonistic to fungal plant pathogens. Appl. Environ. Microbiol. *61*, 1720–1726.

Chuanchuen, R., Beinlich, K., Hoang, T.T., Becher, A., Karkoff-Schweizer, R.R., and Schweizer, H.P. (2001). Cross-resistance between triclosan and antibiotics in *Pseudomonas aeruginosa* is mediated by multidrug efflux pumps: exposure of a susceptible mutant strain to triclosan selects nfxB mutants over expressing MexCD-Opr. Antimicrob. Chemother. *45*, 428–432.

Compant, S., Duffy, B., Nowak, J., Cl′ement, C., and Barka, E.A. (2005). Use of plant growth-promoting bacteria for biocontrol of plant diseases: principles, mechanisms of action, and future prospects. Appl. Environ. Microbiol. *71*, 4951–4959.

Cooley, M. (2007). Incidence and tracking of *Escherichia coli* O157:H7 in a major produce production region in California. PLoS ONE 2, e1159.

Cooley, M.B., Miller, W.G., and Mandrell, R.E. (2003). Colonization of *Arabidopsis thaliana* with *Salmonella enterica* and enterohemorrhagic *Escherichia coli* O157:

H7 and competition by *Enterobacter asburiae*. Appl. Environ. Microbiol. *69*, 4915–4926.

Copping, L.G. (2004). The Manual of Biocontrol Agents. (Br. Crop Prot. Counc. Publ., Alton, Hampshire, UK).

Dakora, F.D., and Phillips, D.A. (2002). Root exudates as mediators of mineral acquisition in low-nutrient environments. Plant Soil *245*, 35–47.

Danhorn, T., Hentzer, M., Givskov, M., Parsek, M.R., and Fuqua, C. (2004). Phosphorus limitation enhances biofilm formation of the plant pathogen *Agrobacterium tumefaciens* through the PhoR-PhoB Regulatory System. J. Bacteriol. *186*, 4492–4501.

Daniels, R., Reynaert, S., Hoekstra, H., Verreth, C., and Janssens, J. (2006). Quorum signal molecules as biosurfactants affecting swarming in *Rhizobium etli*. Proc. Natl. Acad. Sci. U.S.A. *103*, 14965–14970.

Davey, M.E., Caiazza, N.C., and O'Toole, G.A. (2003). Rhamnolipid surfactant production affects biofilm architecture in *Pseudomonas aeruginosa* PAO1. J. Bacteriol. *185*, 1027–1036.

De Ridder-Duine, A.S., Kowalchuk, G.A., Klein Gunnewiek, P.J.A., Smant, W., Van Veen, J.A., and De Boer, W. (2005). Rhizosphere bacterial community composition in natural stands of Carex arenaria (sand sedge) is determined by bulk soil community composition. Soil Biol. Biochem. *37*, 349–357.

Dekkers, L.C., Mulders, C.H.M., Phoelich, C.C., Chin-A-Woeng, T.F.C., Wijfjes, A.H.M., and Lugtenberg, B.J.J. (2000). The *sss* colonization gene of the tomato-*Fusarium* f.sp. *radicis-lycopersici* biocontrol strain *Pseudomonas fluorescens* WCS365 can improve root colonization of other wild type *Pseudomonas* spp. bacteria. Mol. Plant–Microbe Interact. *13*, 1177–1183.

Delaquis, P., Bach, S., and Dinu, L.D. (2007). Behavior of *Escherichia coli* O157:H7 in leafy vegetables. J. Food Protect. *70*, 1966–1974.

DeWeert, S., Vermeiren, H., Mulders, I.H.M., Kuiper, I., and Hendrickx, N. (2002). Flagella-driven chemotaxis towards exudate components is an important trait for tomato root colonization by *Pseudomonas fluorescens*. Mol. Plant Microbe. Interact. *15*, 1173–1180.

Dong, Y.H., Wang, L.H., Xu, J.L., Zhang, H.B., Zhang, X.F., and Zhang, L.H. (2001). Quenching quorum-sensing-dependent bacterial infection by an N-acyl homoserine lactonase. Nature *411*, 813–817.

Dong, Y.M., Iniguez, A.L., and Triplett, E.W. (2004). Nitrogen fixation in wheat provided by *Klebsiella pneumoniae* 342. Mol. Plant–Microbe Interact. *17*, 1078–1085.

Elasri, M., Delorme, S., Lemanceau, P., Stewart, G., and Laue, B. (2001). Acyl-homoserine lactone production is more common among plant-associated *Pseudomonas* spp. than among soil borne *Pseudomonas* spp. Appl. Environ. Microbiol. *67*, 1198–1209.

van Elsas, J.D., and Trevors, J.T. (1997). Modern Soil Microbiology. (Marcel Dekker, New York).

Eshel, A., and Waisel, Y. (1996). Multiform and multifunction of various constituents of one root system. In Plant Roots: The Hidden Half, Waisel, Y., Eshel, A., and Kafkafi, U., eds. (Marcel Dekker, New York), pp. 175–192.

Espinosa-Urgel, M. (2004). Plant-associated *Pseudomonas* populations: molecular biology, DNA dynamics, and gene transfer. Plasmid *52*, 139–150.

Espinosa-Urgel, M., Kolter, R., and Ramos, J.L. (2002). Root colonization by *Pseudomonas putida*: love at first sight. Microbiology *148*, 341–343.

Fett, W.F., and Cooke, P.H. (2003). Reduction of *Escherichia coli* O157:H7 and *Salmonella* on laboratory-inoculated alfalfa seed with commercial citrus-related products. J. Food Prot. *66*, 1158–1165.

Fischer, K.S., Edmeades, G.O., and Johnson, E.C. (1989). Selection for the improvement of maize yield under moisture-deficits. Field Crop Res. *22*, 227–243.

Fujishige, N.A., Kapadia, N.N., De Hoff, P.L., and Hirsch, A.M. (2006). Investigations of *Rhizobium* biofilm formation. FEMS Microbiol. Ecol. *56*, 195–206.

Garbeva, P., van Veen J.A., and van Elsas, J.D. (2004). Microbial diversity in soil: selection of microbial populations by plant and soil type and implications for disease suppressiveness. Annu. Rev. Phytopathol. *42*, 243–270.

Gelsomino, A., Keijzer-Wolters, A.C., Cacco, G., and van Elsas, J.D. (1999). Assessment of bacterial community structure in soil by polymerase chain reaction and denaturing gradient gel electrophoresis. J. Microbiol. Meth. *38*, 1–15.

Gianinazzi-Pearson, V. (1996). Plant cell responses to arbuscular mycorrhizal fungi: getting to the roots of the symbiosis. Plant Cell *8*, 1871–1883.

Gilbert, G.S., Handelsman, J., and Parke, J.L. (1994). Root camouflage by disease control. Phytopathology *84*, 222–225.

Grall, S., and Manceau, C. (2003). Colonisation of *Vitis vinifera* by a green fluorescence protein-labeled GFP-marked strain of *Xylophus ampelinus*, the causal agent of bacterial necrosis of grapevine. Appl. Environ. Microbiol. *69*, 1904–1912.

Grayston, S.J., Wang, S.Q., Campbell, C.D., and Edwards, A.C. (1998). Selective influence of plant species on microbial diversity in the rhizosphere. Soil Biol. Biochem. *30*, 369–378.

Greer-Phillips SE, Stephens BB, Alexandre G. (2004). An energy taxis transducer promotes root colonization by *Azospirillum brasilense*. J. Bacteriol. *186*, 6595–6604.

Haas, D., and Defago, G. (2005). Biological control of soil-borne pathogens by fluorescent pseudomonads. Nat. Rev. Microbiol. 3, 307–319.

Hawes, M.C., and Smith, L.Y. (1989). Requirement for chemotaxis in pathogenicity of *Agrobacterium tumefaciens* on roots of soil grown pea plants. J. Bacteriol. *171*, 5668–5671.

Hawes, M.C., Gunawardena, U., Miyasaka, S., and Zhao, X. (2000). The role of root border cells in plant defense. Trends Plant Sci. *5*, 128–133.

He, J.X., Baldini, R.L., Deziel, E., Saucier, M., and Zhang, Q.H. (2004). The broad host range pathogen *Pseudomonas aeruginosa* strain PA14 carries two pathogenicity islands harboring plant and animal virulence genes. Proc. Natl. Acad. Sci. U.S.A. *101*, 2530–2535.

Hedrick, D.B., Peacock, A., Stephen, J.R., Macnaughton, S.J., Bruggemann, J., and White, D.C. (2000). Measuring soil microbial community diversity using polar lipid fatty acid and denaturing gradient gel electrophoresis data. J. Microbiol. Methods *41*, 235–248.

Hense, B.A., Kuttler, C., Johannes, M., Hartmann, R.A., and Kreft, J.U. (2007). Does Efficiency sensing unify diffusion and quorum sensing? Nat. Rev. Microbiol. *5*, 230–239.

Hoffland, E., Findenegg, G.R., and Nelemans, J.A. (1989). Solubilization of rock phosphate by rape. II. Local root exudation of organic acids as a response to P-starvation. Plant Soil *113*, 161–165.

Hogan, D.A., and Kolter, R. (2002). *Pseudomonas–Candida* interactions: an ecological role for virulence factors. Science *296*, 2229–2232.

Holden, M.T.G., Chhabra, S.R., de Nys, R., Stead, P., and Bainton, N.J. (1999). Quorum-sensing cross-talk: isolation and chemical characterization of cyclic dipeptides from *Pseudomonas aeruginosa* and other Gram-negative bacteria. Mol. Microbiol. 33, 1254–1266.

Hugouvieux-Cotte-Pattat, N., Condemine, G., Nasser, G., and Reverchon, S. (1996). Regulation of pectinolysis in *Erwinia chrysanthemi*. Annu. Rev. Microbiol. *50*, 213–257.

Hurek, T., Reinhold-Hurek, B., van Montagu, M., and Kellenberger, E. (1994). Root colonization and systemic spreading of *Azoarcus* strain BH72 in grasses. J. Bacteriol. *176*, 1913–1923.

Iavicoli, A., Boutet, E., Buchala, A., and M´etraux, J.P. (2003). Induced systemic resistance in *Arabidopsis thaliana* in response to root inoculation with *Pseudomonas fluorescens* CHA0. Mol. Plant–Microbe Interact. *16*, 851–858.

Ilka, B.B., Joshua, A.H., Aiwen, W.L., Denise, M.W., and Adam, P.A. (2009). Complexity in bacterial cell–cell communication: Quorum signal integration and subpopulation signaling in the *Bacillus subtilis* phosphorelay. Proc. Natl. Acad. Sci. U.S.A. *106*, 6459–6464.

Jacoud, C., Job, D., Wadoux, P. and Bally, R. (1999). Initiation of root growth stimulation by *Azospirillum lipoferum* CRT1 during maize seed germination. Can. J. Microbiol. *45*, 339–342.

Jacques, M.A., Josi, K., Darrasse, A., and Samson, R. (2005). *Xanthomonas axonopodis* pv. *phaseoli* var. *fuscans* is aggregated in stable biofilm population sizes in the phyllosphere of fieldgrown beans. Appl. Environ. Microbiol. *71*, 2008–2015.

James, E.K., Olivares, F.L., de Oliveira, A.L.M., dos Reis, F.D., Jr., da Silva, L.G., and Reis, V.M. (2001). Further observations on the interaction between sugar cane and *Gluconacetobacter diazotrophicus* under laboratory and greenhouse conditions. J. Exp. Bot. *52*, 747–760.

Jha, A.K., Bais, H.P., and Vivanco, J.M. (2005). *Enterococcus faecalis* mammalian virulence-related factors exhibit potent pathogenicity in the *Arabidopsis thaliana* plant model. Infect. Immun. 73, 464–475.

Johnson, V.W., Pearson, J.F., and Jackson, T.A. (2001). Formulation of *Serratia entomophila* for biological control of grass grub. N.Z. Plant Protec. *54*, 125–127.

Joseph, B., Patra, R.R., and Lawrence, R. (2007). Characterization of plant growth promoting rhizobacteria associated with chickpea (*Cicer arietinum* L.). Int. J. Plant. Prod. *2*, 141–152.

Kageyama, K., and Nelson, E.B. (2003). Differential inactivation of seed exudate stimulation of *Pythium ultimum* sporangium germination by *Enterobacter cloacae* influences biological control efficacy on different plant species. Appl. Environ. Microbiol. *69*, 1114–1120.

Kamilova, F., Validov, S., Azarova, T., Mulders, I., and Lugtenberg, B. (2005). Enrichment for enhanced competitive plant root tip colonizers selects for a new class of biocontrol bacteria. Environ. Microbiol. *7*, 1809–1817.

Kearns, D.B., Chu, F., Branda, S.S., Kolter, R. and Losick, R. (2005). A master regulator for biofilm formation by *Bacillus subtilis*. Mol. Microbiol. *55*, 739–749.

Keshavan, N.D., Chowdhary, P.K., Haines, D.C., and Gonzalez, J.E. (2005). L-canavanine made by Medicago sativa interferes with quorum sensing in *Sinorhizobium meliloti*. J .Bacteriol. *187*, 8427–8436.

Kim, C., Kecskes, M.L., Deaker, R.J., Gilchrist, K., New, P.B., Kennedy, I.R., Kim, S., and Sa, T. (2005). Wheat root colonization and nitrogenase activity by *Azospirillum* isolates from crop plants in Korea. Can. J. Microbiol. *51*, 948–956.

Klausen, M., Heydorn, A., Ragas, P., Lambertsen, L., Aaes-Jørgensen, A., Morin, S., and Tolker-Nielson, T. (2003). Biofilm formation by *Pseudomonas aeruginosa* wild type, flagella and type IV pili mutants. Mol. Microbiol. *48*, 1511–1124.

Kleijn, R.J., Buescher, J.M., Le Chat, L., Jules, M., Aymerich, S., and Sauer, U. (2010). Metabolic fluxes during strong carbon catabolite repression by malate in *Bacillus subtilis*. J. Biol. Chem. *285*, 1587–1596.

Kolter, R. (2005). Surfacing views of biofilm biology. Trends Microbiol. *13*, 1–2.

Kuiper, I., Lagendijk, E.L., Pickford, R., Derrick, J.P., Lamers, G.E., Thomas-Oates, J.E., Lugtenberg, B.J.J., and Bloemberg, G.V. (2004). Characterization of two *Pseudomonas putida* lipopeptide biosurfactants, putisolvin I and II, which inhibit biofilm formation and break down existing biofilms. Mol. Microbiol. *51*, 97–113.

Kutter, S., Hartmann, A., and Schmid, M. (2006). Colonization of barley (*Hordeum vulgare*) with *Salmonella enterica* and *Listeria* spp. FEMS Microbiol. Ecol. *56*, 262–271.

Langer, U., Bohme, L., and Bohme, F. (2004). Classification of soil microorganisms based on growth properties: a critical view of some commonly used terms. J. Plant Nutr. Soil Sci. *167*, 267–269.

Liesack, W., Janssen, P.H., Rainey, F.A., Ward-Rainey, N.L., and Stackebrandt, E. (1997). Microbial diversity in soil: the need for a combined approach using molecular and cultivation techniques. In Modern Soil

Microbiology, van Elsas, J.D., and Trevors, J.T., eds. (MarcelDekker, New York)

Lopez-Bucio, J., de la Vega, O.M., Guevara-Garcia, A., and Herrera-Estrella, L. (2000). Enhanced phosphorus uptake in transgenic tobacco plants that overproduce citrate. Nat. Biotechnol. *18*, 450–453.

Lugtenberg, B.J.J., Kravchenko, L.V., and Simons, M. (1999). Tomato seed and root exudate sugars: composition, utilization by *Pseudomonas* biocontrol strains and role in rhizosphere colonization. Environ. Microbiol. *1*, 439–446.

Lynch, J.M., and Whipps, J.M. (1990). Substrate flow in the rhizosphere. Plant Soil *129*, 1–10.

Mack, D., Fischer, W., Krokotsch, A., Leopold, K., Hartmann, R., Egge, H., and Laufs, R. (1996). The intercellular adhesin involved in biofilm accumulation of *Staphylococcus* epidermis is a linear beta-1,6-linked glucosaminoglycan: purification and structural analysis. J. Bacteriol. *178*, 175–183.

Maddula, V., Zhang, Z., Pierson, E.A., and Pierson, L.S. (2006). Quorum sensing and phenazines are involved in biofilm formation by *Pseudomonas chlororaphis* (*aureofaciens*) strain 30–84. Microb. Ecol. *52*, 289–301.

Mah, T.F., Pitts, B., Pellock, B., Walker, G.C., and Stewart, P.S. (2003). A genetic basis for *Pseudomonas aeruginosa* biofilm antibiotic resistance. Nature *426*, 306–310.

Marques, L.L.R., Ceri, H., Manfio, G.P., Reid, D.M., and Olson, M.E. (2002). Characterization of biofilm formation by *Xylella fastidiosa in vitro*. Plant Dis. *86*, 633–638.

Marschner, P., Neumann, G., Kania, A., Weiskopf, L., and Lieberei, R. (2002). Spatial and temporal dynamics of the microbial community structure in the rhizosphere of cluster roots of white lupin (*Lupinus albus* L.). Plant Soil *246*, 167–174.

Marschner, H. (1995). Mineral Nutrition of Higher Plants, 2nd edition. (Academic Press, London).

Mathesius, U., Mulders, S., Gao, M., Teplitski, M., Caetano-Anollés, G., Rolfe, B.G., and Bauer, W.D. (2003). Extensive and specific responses of a eukaryote to bacterial quorum-sensing signals. Proc. Natl. Acad. Sci. U.S.A. *100*, 1444–1449.

Mazzola, M., Cook, R.J., Thomashow, L.S., Weller, D.M., and Pierson, L.S. (1992). Contribution of phenazine antibiotic biosynthesis to the ecological competence of fluorescent pseudomonads in soil habitats. Appl. Environ. Microbiol. *58*, 2616–2624.

Miller, M.B., and Bassler, B.L. (2001). Quorum sensing in bacteria. Annu. Rev. Microbiol. *55*, 165–199.

Monds, R.D., Silby, M.W., and Mahanty, H.K. (2001). Expression of the Pho regulon negatively regulates biofilm formation by *Pseudomonas aureofaciens* PA147-2. Mol. Microbiol. *42*, 415–426.

Neumann, G., and Martinoia, E. (2002). Cluster roots – an underground adaptation for survival in extreme environments. Trends Plant Sci. *7*, 162–167.

Normander, B., and Prosser, J.I. (2000). Bacterial origin and community composition in the barley phytosphere as a function of habitat and presowing conditions. Appl. Environ. Microbiol. *66*, 4372–4377.

O'Toole, G.A., and Stewart, P.S. (2005). Biofilms strike back. Nat. Biotech. *23*, 1378–1379.

Ongena, M., Jourdan, E., Adam, A., Paquot, M., Brans, A., Joris, B., Arpingy, J.-L., and Thonart, P. (2007). Surfactin and fengycin lipopeptides of *Bacillus subtilis* as elicitors of induced systemic resistance in plants. Environ. Microbiol. *9*, 1084–1090.

Park, Y.-S., Jeon, M.H., Lee, S.-H., Moon, J.S., Cha, J.-S., Kim, H.Y., and Cho, T.-J. (2005). Activation of defense responses in Chinese cabbage by a nonhost pathogen, *Pseudomonas syringae* pv. Tomato. J. Biochem. Mol. Biol. *38*, 748–754.

Postma, J., and van Veen, J.A. (1990). Habitable pore space and survival of *Rhizobium leguminosarum* biovar *trifolii* introduced into soil. Microb. Ecol. *19*, 149–161.

Pozo, M.J., and Azcon-Aguilar, C. (2007). Unraveling mycorrhiza-induced resistance. Curr. Opin. Plant Biol. *10*, 393–398.

Pratt, L.A., and Kolter, R. (1998). Genetic analysis of *Escherichia coli* biofilm formation: Roles of flagella, motility, chemotaxis and type I pili. Mol. Microbiol. *30*, 285–293.

Prithiviraj, B., Bais, H.P., Weir, T., Suresh, B., Najarro, E.H., Dayakar, B.V., Schweizer, H.P., and Vivanco, J.M. (2005). Down regulation of virulence factors of *Pseudomonas aeruginosa* by salicylic acid attenuates its virulence on *Arabidopsis thaliana* and *Caenorhabditis elegans*. Infect. Immun. *73*, 5319–5328.

Puohiniemi, R., Heiskanen, T., and Siitonen, A. (1997). Molecular epidemiology of two international sprout-borne *Salmonella* outbreaks. J. Clin. Microbiol. *35*, 2487–2491.

Raaijmakers, J.M., de Bruijn, I., and de Kock, M.J.D. (2006). Cyclic lipopeptide production by plant-associated *Pseudomonas* spp.: diversity, activity, biosynthesis, and regulation. Mol. Plant–Microbe Interact. *19*, 699–710.

Ramey, B.E., Koutsoudis, M., von Bodman, S.B., and Fuqua, C. (2004). Biofilm formation in plant–microbe associations. Curr. Opin. Microbiol. *7*, 602–609.

Ranjard, L., Poly, F., Lata, J.C., Mougel, C., Thioulouse, J., and Nazaret, S. (2001). Characterization of bacterial and fungal soil communities by automated ribosomal intergenic spacer analysis fingerprints: biological and methodological variability. Appl. Environ. Microbiol. *67*, 4479–4487.

Rasmussen, T.B., Bjarnsholt, T., Skindersoe, M.E., Hentzer, M., Kristoffersen, P., K¨ote, M., Nielsen, J., Eberl, L., and Givskov, M. (2005), Screening for quorum-sensing inhibitors (QSI) by use of a novel genetic system, the QSI selector. J. Bacteriol. *187*, 1799–1814.

Redfield, R.J. (2002). Is quorum sensing a side effect of diffusion sensing? Trends Microbiol. *10*, 365–370.

Rooney, D.C., and Clipson, N.J.W. (2009). Phosphate addition and plant species alters microbial community structure in acidic upland grassland soil. Microb. Ecol. *57*, 4–13.

Rudrappa, T., Czymmek, K.J., Pare, P.W., and Bais, H.P. (2008a). Root-secreted malic acid recruits beneficial soil bacteria. Plant Physiol. *148*, 1547–1556.

Rudrappa, T., Biedrzycki, M.L., and Bais, H.P. (2008b). Causes and consequences of plant associated biofilm. FEMS Microbiol. Ecol. *64*, 153–166.

Rudrappa, T., Biedrzycki, M.L., Kunjeti, S.G., Donofrio, N.M., Czymmek, K.J., Paré, P.W., and Bais, H.P. (2010). The rhizobacterial elicitor acetoin induces systemic resistance in *Arabidopsis thaliana*. Comm. Integrative Biol. 3, 1–9.

Ryu, C.-M., Farag, M.A., Hu,C.-H., Reddy, M.S., Wei, H.X., Paré, P.W., and Kloepper, J.W. (2003). Bacterial volatiles promote growth in *Arabidopsis*. Proc. Natl. Acad. Sci. U.S.A. *100*, 4927–4932.

Ryu, C.M., Farag, M.A., Hu, C.H., Reddy, M.S., Kloepper, J.W., and Paré, P.W. (2004). Bacterial volatiles induce systemic resistance in *Arabidopsis*. Plant Physiol. *134*, 1017–1026.

Sauer, K., Camper, A.K., Ehrlich, G.D., Costerton, J.W., and Davies, D.G. (2002). *Pseudomonas aeruginosa* displays multiple phenotypes during development as a biofilm. J. Bacteriol. *184*, 1140–1154.

Scott, J., Weil, P., Le, T., Williams, P., Fray, R.G., von Bodman, S.B., and Savka, M.A. (2006). Long- and short-chain plant-produced bacterial N-Acyl-homoserine lactones become components of phyllosphere, rhizosphere, and soil. Mol. Plant Microbe Interact. *19*, 227–239.

Sessitsch, A., Weilharter, A., Gerzabek, M.H., Kirchmann, H., and Kandeler, E. (2001). Microbial population structures in soil particle size fractions of a longterm fertilizer field experiment. Appl. Environ. Microbiol. *67*, 4215–4224.

Shrout, J.D., Chopp, D.L., Just, C.L., Hentzer, M., Givskov, M., and Parsek, M.R. (2006). The impact of quorum sensing and swarming motility on *Pseudomonas aeruginosa* biofilm formation is nutritionally conditional. Mol. Microbiol. *62*, 1264–1277.

Schuhegger, R., Ihring, A., Gantner, S., Bahnweg, G., Knappe, C., Vogg, G., Hutzler, P., Schmid, M., Van Breusegem, F., Eberl, L., *et al.* (2006). Induction of systemic resistance in tomato by *N*-acyl-L-homoserine lactone-producing rhizosphere bacteria. Plant Cell Environ. *29*, 909–918.

Sigler, W.V., Nakatsu, C.H., Reicher, Z.J., and Turco, R.F. (2001). Fate of the biological control agent *Pseudomonas aureofaciens* TX-1 after application to Turfgrass. Appl. Environ. Microbiol. *67*, 3542–3548.

Silby, M.W., and Levy, S.B. (2004). Use of *in vivo* expression technology to identify genes important in growth and survival of *Pseudomonas fluorescens* Pf0–1 in soil: discovery of expressed sequences with novel genetic organization. J. Bacteriol. *186*, 7411–7419.

Stanley, N.R., and Lazazzera, B.A. (2004). Environmental signals and regulatory pathways that influence biofilm formation. Mol. Microbiol. *52*, 917–924.

Teplitski, M., Robinson, J.B., and Bauer, W.D. (2000). Plants secrete substances that mimic bacterial N-acyl homoserine lactone signal activities and affect population density dependent behaviors in associated bacteria. Mol. Plant Microbe Int. *13*, 637–648.

Teplitski, M., Chen, H., Rajamani, S., Gao, M., Merighi, M., Sayre, R.T., Robinson, J.B., Rolfe, B.G., and Bauer, W.D. (2004). *Chlamydomonas reinhardtii* secretes compounds that mimic bacterial signals and interfere with quorum sensing regulation in bacteria. Plant Physiol. *134*, 137–146.

Thomashow, L.S., and Weller, D.M. (1996). Current concepts in the use of introduced bacteria for biologicaldisease control: mechanisms and antifungal metabolites. In Plant–Microbe Interact, Stacey, G., and Keen, N.T., eds. (Chapman and Hall, New York), pp.187–235.

Timmusk, S., Grantcharova, N., and Wagner, E.G.H. (2005). *Paenibacillus polymyxa* invades plant roots and forms biofilms. Appl. Environ. Microbiol. *71*, 7292–7300.

Torres, M.A., Jones, J.D.G., and Dangl, J.L. (2006). Reactive oxygen species signaling in response to pathogens. Plant Physiol. *141*, 373–378.

Toth, I.K., Newton, J.A., Hyman, L.J., Lees, A.K., Daykin, M., Ortori, C., Williams III, P., and Frayl, R.G. (2004). Potato plants genetically modified to produce N-acylhomoserine lactones increase susceptibility to soft rot erwiniae. Mol. Plant Microbe Int. *17*, 880–887.

Turnbull, G.A., Morgan, J.A.W., Whipps, J.M., and Saunders, J.R. (2001). The role of bacterial motility in the survival and spread of *Pseudomonas fluorescens* in soil and in the attachment in bacterial–plant interactions. FEMS Microbiol. Ecol. *35*, 57–65.

Van Bastelaere, E., Lambrecht, M., Vermeiren, H., Van Dommelen, A., Keijers, V., Proost, P., and Vanderleyden, J. (1999). Characterization of a sugar-binding protein from *Azospirillum brasilense* mediating chemotaxis to and uptake of sugars. Mol. Microbiol. *32*, 703–714.

Van Loon, L.C. (2007). Plant responses to plant growth-promoting bacteria. Eur. J. Plant Pathol. *119*, 243–54.

Van Loon, L.C., Bakker, P.A,H.M., and Pieterse, C.M.J. (1998). Systemic resistance induced by rhizosphere bacteria. Annu. Rev. Phytopathol. *36*, 453–483.

Van Nieuwenhove, C., Van Holm, L., Kulasooriya, S.A., Vlassak, K., and Van Holm, L. (2004). Establishment of *Azorhizobium caulinodans* in the rhizosphere of wetland rice (*Oryza sativa* L.). Biol. Fert. Soils *31*, 143–149.

Van Peer, R., Niemann, G.J., and Schippers, B. (1991). Induced resistance and phytoalexin accumulation in biological control of *Fusarium* wilt of carnation by *Pseudomonas* sp. strain WCS417r. Phytopathology *81*, 728–734.

Vande Broek, A., and Vanderleyden, J. (1995). The role of bacterial motility, chemotaxis, and attachment in bacteria–plant interactions. Mol. Plant–Microbe Interact. *8*, 800–810.

Vande Broek, A., Lambrecht, M., and Vanderleyden, J. (1998). Bacterial chemotactic motility is important for the initiation of wheat root colonisation by *Azospirillum brasilense*. Microbiology *144*, 2599–2606.

Verhagen, B.W.M., Glazebrook, J., Zhu, T., Chang, H.-S., Van Loon, L.C., and Pieterse, C.M.J. (2004). The transcriptome of rhizobacteria-induced systemic resistance in *Arabidopsis*. Mol. Plant–Microbe Interact. *17*, 895–908.

Veselova, M.A., Klein, S.H., Bass, I.A., Lipasova, V.A., Metlitskaya, A.Z., Ovadis, M.I., Chernin, L.S., and Khmel, I.A. (2008). Quorum sensing systems of regulation, synthesis of phenazine antibiotics, and antifungal activity in rhizospheric bacterium *Pseudomonas chlororaphis* 449. Russ. J. Genet. *44*, 1400–1408.

Walker, T.S., Bais, H.P., Déziel, E., Schweizer, H.P., Rahme, L.G., Fall, R., and Vivanco, J.M. (2004). *Pseudomonas aeruginosa*-plant root interactions. Pathogenicity, biofilm formation, and root exudation. Plant Physiol. *134*, 320–331.

Waters, C.M., and Bassler, B.L. (2005). Quorum sensing: cell-to-cell communication in bacteria. Annu. Rev. Cell Dev. Biol. *21*, 319–346.

Weller, D.M., Raaijmakers, J.M., McSpadden Gardener, B.B., and Thomashow, L.S. (2002). Microbial populations responsible for specific soil suppressiveness to plant pathogens. Annu. Rev. Phytopathol. 40, 309–348.

Wei, G., Kloepper, J.W., and Tuzun, S. (1991). Induction of systemic resistance of cucumber to *Colletotrichum orbiculare* by select strains of plant growth-promoting rhizobacteria. Phytopathology *81*, 1508–1512.

Whitchurch, C.B., Tolker-Nielsen, T., Ragas, P.C., and Mattick, J.S. (2002), Extracellular DNA required for bacterial biofilm formation. Science *295*, 487.

Whitehead, N.A., Barnard, A.M., Slater, H., Simpson, N.J., and Salmond, G.P. (2001). Quorum sensing in Gram-negative bacteria. FEMS Microbiol Rev. *25*, 365–404.

Whitman, W.B., Coleman, D.C., and Wiebe, W.J. (1998). Prokaryotes: The unseen majority. Proc. Natl. Acad. Sci. U.S.A. *95*, 6578–6583.

Williamson, L., Nakoho, K., Hudelson, B., and Allen, C. (2002). *Ralstonia solanacearum* race 3, biovar 2 strains isolated from geranium are pathogenic on potato. Plant Dis. *86*, 987–991.

Wolf, J.M., Beckhoven, J.R.C.M., Hukkanen, A., Karjalainen, R., and Muller, P. (2005). Fate of *Clavibacter michiganensis* ssp. *sepedonicus*, the causal organism of bacterial ring rot of potato, in weeds and field crops. J. Phytopathol. *153*, 358–365.

Yamada, Y., and Nihira, T. (1998). Microbial hormones and microbial chemical ecology. In Comprehensive Natural Products Chemistry, Vol. 8, Barton, D.H.R., and Nakanishi, K., eds. (Elsevier, Oxford), pp. 377–413.

Yanni, Y.G., Rizk, R.Y., Corich, V., Squartini, A., Ninke, K., Philip-Hollingsworth, S., Orgambide, G., de Bruijn, F., Stoltzfus, R., Buckley, D., *et al.* (1997). Natural endophytic association between *Rhizobium leguminosarum* bv. trifolii and rice roots and assessment of its potential to promote rice growth. Plant Soil. *194*, 99–114.

Yao, J., and Allen, C. (2006). Chemotaxis is required for virulence and competitive fitness of the bacterial wilt pathogen *Ralstonia solanacearum*. J. Bacteriol. *188*, 3697–3708.

Yoon, H.S., Millers, K.M., Sheath, S.R., Ott, F.D., and Bhattacharya, D. (2006). Defining the major lineages of red algae (Rhodophyta). J. Phycol. *42*, 482–492.

Zhao, Y., Blumer, S.E., and Sundin, G.W. (2005). Identification of *Erwinia amylovora* genes induced during infection of immature pear tissue. J. Bacteriol. *187*, 8088–8103.

An Invisible Workforce: Biofilms in the Soil

4

Mette Burmølle, Annelise Kjøller and Søren J. Sørensen

Abstract

Biofilms in soil are composed of multiple species microbial consortia attached to soil particles and biotic surfaces including roots, fungal hyphae and decomposing organic material. The bacteria present in these biofilms gain several advantages including protection from predation, desiccation and exposure to antibacterial substances, and optimized acquisition of nutrients released in the mycosphere. Studies of soil biofilms are complicated by the composite structure of the soil environment; therefore, various simplified model systems have been applied to study succession and bacterial interactions in soil biofilms. Model system observations indicate an increased efficiency to degrade and decompose organic material and xenobiotic compounds by these multispecies bacterial communities. Consequently, soil biofilms may be valuable tools for bioremediation and biocontrol. However, soil biofilms may also provide survival sites for opportunistic pathogenic bacteria, providing enhanced protection and increasing their potential to survive and evolve in the soil environment. In this review, we provide evidence that biofilms are of major importance for the fitness of individual bacteria and the wider soil ecology, owing to the accumulated selective advantage provided to bacteria by the biofilm mode of life.

Introduction

It has been estimated that '99% of all microbial activity in open ecosystems is stuck (in biofilms) to surfaces' (Potera, 1996). Bacteria are believed to obtain several advantages by living in biofilms (Jefferson, 2004) and our understanding of biofilm structure, dynamics and function has become a crucial part not only of microbiology but many related research areas. Despite this, little remains known about biofilms in soil, which may be due to difficulties associated with studying this structurally complex environment at the microscopic scale without disturbing the soil and its associated biofilm structures. Several studies, however, indicate a dominant role for biofilms in soil ecology, where they are involved in the degradation of organic matter and other compounds, including pollutants. This highlights the bioremediation potential of soil biofilms, as well as the promising role of some biofilm-associated organisms to be candidates as biological sensors and biocontrol agents.

It is clear that biofilms are crucial components in soil ecology and represent a vital, but largely invisible workforce in the complex and hostile soil environment. In this chapter, we discuss soil biofilms from a bacterial perspective, and highlight their function and roles in the protection and evolution of pathogenic bacteria.

Bacterial advantages of the biofilm mode of life in soil

Adaptation and self-preservation are crucial for soil bacterial survival as the physical and chemical conditions characteristic of soil may change dramatically over very short time periods. Embedment of biofilm-associated cells in the biofilm may help to 'buffer' them from variations in surrounding environmental conditions (e.g. in the soil water content or pH). For example, soil water content can vary greatly, causing desiccation of bacterial cells. However, a soil *Pseudomonas* sp. has been shown to produce more exopolysaccharides when subjected to soil drying (Roberson and Firestone, 1992) and biofilm forming (mucoid) strains of other species have been observed to survive desiccation better than non-mucoid mutants (Ophir and Gutnick, 1994). Increased desiccation survival indicates a protective effect of the biofilm, probably caused by the moisture retaining nature of the biofilm matrix (Sutherland, 2001; Jass *et al.*, 2002).

Soil bacteria are often challenged with exposure to antibiotics, which are naturally produced by microorganisms in soil (Hansen *et al.*, 2001), or to heavy metals and other toxic compounds (Verbruggen *et al.*, 2009). Generally, bacteria are more resistant to antibacterial compounds when present in biofilms rather than the free-living, planktonic state. Resistance to antibacterial compounds has been attributed to reduced permeability and/or a destabilizing effect of the extracellular polymeric substance (EPS) matrix to some compounds, or altered gene and protein expression patterns up-regulating stress responses (efflux pumps, EPS production and other stress factors), leading to the formation of more resistant 'persister cells' (Mah and O'Toole, 2001; Gilbert *et al.*, 2002; Jefferson, 2004). Biofilms in soil have shown similar protective effects. For example in a study by Almås *et al.* (2005), it was reported that tightly bound bacterial cells (i.e. those present in very small soil pores or biofilms) had increased susceptibility to trace metals, which may indicate that they were previously spatially protected from metals present in the soil. In contrast, the loosely bound cells (in the pore water or large pores) were less susceptible to the metals, which was attributed to their previous exposure to these metals, altering the bacterial community to better resist such exposure. Little is known about the importance of protozoan–bacterial interactions for biofilm formation in soil but there is evidence from aquatic systems that grazing by protozoa favours the establishment of biofilms. Given the high importance of protozoan grazing as a mortality factor for bacteria in soil (Ekelund and Rønn, 1994) it is probable that similar mechanisms operate here.

Soil bacterial biofilms consist of multiple species mixtures interacting metabolically and socially, conferring several selective advantages to the constituent bacteria (Jefferson, 2004). The presence of different species in a high cell density environment increases the possibility of horizontal gene transfer (Sørensen *et al.*, 2005), a process that may be further stabilized and thereby enhanced by the structural nature of the biofilm. Thus, biofilm organization increases the possibility of cells receiving genes from other bacteria. Such genes often encode traits conferring selective advantages (Burmølle *et al.*, 2008). Subsequently, the cells adaptation to the versatile soil environment increases.

The biofilm environment may also offer optimized conditions for quorum sensing (further discussed in Chapter 1), whereby bacteria regulate gene expression according to the cell density. Intraspecies communication based on *N*-acyl homoserine lactones has been demonstrated in soil (Burmølle *et al.*, 2005) and interspecies communication based on universal signal compounds, such as AI-2 (Xavier and Bassler, 2003), is likely to play a role in the development and physiology of multispecies biofilms. Additionally, multispecies biofilms are more resistant to antibacterial compounds than single species biofilms (Burmølle *et al.*, 2006) and the biofilm environment may also serve to protect species that would otherwise be outcompeted (Stewart *et al.*, 1997; Klayman *et al.*, 2009).

Thus, the organization of biofilm communities confers many selective advantages to soil dwelling bacteria. From an evolutionary perspective, these may result in an improved fitness for the biofilm-associated bacteria. It seems probable that the selective pressure in soil has favoured bacteria capable of biofilm formation.

Succession and diversity of soil biofilm microbial communities

Natural biofilms comprise bacteria attached to surfaces in soil or water, but fungi, algae and protozoa are also present as common inhabitants. Investigations to date have primarily concentrated on bacterial biofilms, but other taxonomic groups also contribute to biofilm functions. A general model for the development of bacterial biofilms was suggested by Jackson (2003) based on experimental work. In this model, biofilm species diversity is hypothesized to initially increase, but decrease at later stages when less competitive species are no longer present. With time, resource diversity increases and the significance of competition is reduced. Interaction between bacteria and predation by protozoa becomes increasingly important at later stages and has been found to subsequently affect biofilm diversity (Jackson *et al.*, 2001; Jackson, 2003). By use of a flow biofilm system, we followed community succession in an early stage multispecies biofilm. The observed succession pattern could not be explained by the 'Jackson model'; however, it was possible to determine early, intermediate and late colonizers in the represented species (Burmølle *et al.*, 2007). We also showed that the ability of single species to form biofilm was reflected in the time taken for that species to appear in a multispecies biofilm. For example, the species forming a considerable amount of biofilm after 1–2 days when grown under monoculture conditions also appeared early in the multispecies consortium. In contrast, the late consortium-colonizer demanded more time for monospecies biofilm formation. However, the planktonic growth rate of the single species did not seem to influence the time of appearance in the multispecies biofilm. Thus, it seems that the ability and rate at which cells formed single-species biofilms some way reflect their ability to colonize multispecies biofilms (Burmølle *et al.*, 2007).

A dominance of Gram-negative species was observed in biofilms in soil (Burmølle *et al.*, 2007) and in aquatic systems (Hu *et al.*, 2003; Martiny *et al.*, 2003). However, some of the results obtained may be artefacts since the methods used are typically more effective for growth and DNA extraction of Gram-negative bacteria. In a recent mini-review, the ecology of hitherto-uncultured bacteria in the rhizosphere is presented (da Rocha *et al.*, 2009). In these studies, *Proteobacteria* were equally represented in both culture-dependent and culture-independent approaches, whilst the *Actinobacteria* were more frequent in the culture-dependent group. The highest density representatives among the groups included *Acidobacteria*, *Verrucomicrobia* and *Planctomycetes*. The functions of these groups in the rhizosphere are not yet known, as no isolates have been obtained from this environment.

Bacterial distribution in soil

Bacterial distribution in the soil environment has been found to vary and is dependent on several factors. The highest bacterial densities are found near nutrient sources such as on plant roots or decaying organic material (Foster, 1988; Pearce *et al.*, 1995; Nunan *et al.*, 2003). In the bulk soil, bacteria are found in patches known as microcolonies that contain low cell numbers. These microcolonies are often composed of several different bacterial species (Nunan *et al.*, 2003; Grundmann, 2004). On exposure to nutrient sources, such microcommunities have the potential to develop into high bacterial density multispecies biofilms (Nunan *et al.*, 2003). The biofilm-associated fraction of soil bacteria may represent the stage of the bacterial life cycle characterized by metabolic activity, whereas the microcolony stage represents a dormant or survival stage of low metabolic activity. If so, cell division and interactions with other bacteria in the active biofilm stage must be of major importance for bacterial adaptation and proliferation in soil.

Important biotic surfaces for biofilm formation in soil are free-living fungal hyphae and the mycelium around plant roots. The attachment takes place via flagellar movement, chemotaxi and possibly by shutting down the defence mechanism of the fungi (e.g. production of antibiotics). The biofilm formation on fungal hyphae includes diverse bacterial populations. The benefits for the attached bacteria are obvious; acquisition of nutrients released by the fungus (e.g. mannitol, trehalose) and the potential to colonize new habitats in the soil. These biofilms were recently named *'the hyphal highway'* (Nazir *et al.*, 2010b). Two kinds of microhabitats for establishing bacterial biofilms in soil were also described; (i) the mycorrhizosphere; mycorrhizal hyphae on roots, characterized as a stable habitat and (ii) the mycosphere; a dense hyphal network underneath mushrooms in the soil, characterized as an unstable habitat that changes with the age of the hyphae. Finally, bacterial biofilms also form on free-living hyphae of saprophytic microfungi present in soil.

Bacteria able to thrive in special fungal environments are known as mycosphere competent. These include specific fungiphilic bacteria (Warmink *et al.*, 2009). When soil from the dense mycosphere networks under 'the feet' of a range of different basidiomycetes was examined for various bacterial community characteristics, some isolates were found to be universal fungiphiles while others were adapted to only one or few fungi. The bacteria were shown to utilize the specific carbon compounds excreted by the fungi, a phenomenon that may be responsible for selection of bacterial species present in the mycosphere. This was further demonstrated for *Variovorax paradoxus*-like bacteria in the mycosphere of the Basidomycota *Lyophyllum* sp. (Boersma *et al.*, 2010; Nazir *et al.*, 2010a). In a microcosm experiment, it was found that these bacteria were strongly selected for and able to grow on compounds released by the fungus (e.g. glycerol). It was also shown that *Lyophyllum* sp. was able to modulate pH in the microenvironment and raise the pH to approximately 5, in order to permit *V. paradoxus* to grow, also providing a suitable habitat for other pH-sensitive bacteria.

Bacterial migration via the fungal hyphae was recently investigated. In a microcosm experiment with a *Lyophyllum* spp., it was shown how an inoculated bacteria, *Burkholderia* spp., developed a biofilm surrounding the hyphae utilizing nutrients released by the fungus. The bacteria only migrate when introduced at the apex of the hyphae and this migration takes place in the direction of the growth of the hyphae. The bacterial mobility occurs in the thin water film around the hyphae (Warmink and van Elsas, 2009) and has importance for translocation and colonization to new microhabitats. Plant roots represent another very important biotic surface commonly subject to bacterial biofilm formation, as further detailed in Chapter 3.

Function of soil biofilms

Biofilms play a dominant role in the degradation of decaying organic material in soil. Such degradation is usually dependent on the presence of extracellular enzymes, and it is therefore advantageous for the degrading organisms to cover the material as well as possible. Nutrient induced establishment of biofilm in soil on or near a degradable source (litter, roots, etc.) imparts a great advantage to the bacteria involved when compared with unattached bacteria, as evidenced by their increased growth rates, physiological capabilities and metabolic activity (Jass *et al.*, 2002). In the biofilm, bacteria and microfungi act simultaneously in the degradation process. Both bacteria and fungi are able to produce a variety of extracellular enzymes that degrade organic matter in soil (Nannipieri *et al.*, 2002), although only fungi are known to attack and degrade recalcitrant polymeric lignin compounds (Baldrian, 2006). The resulting soluble nutrients are later available both for bacteria and other microorganisms. A recent review lists all enzymes identified in aquatic biofilms (Flemming and Wingender, 2010). This encompasses all the enzymes known to act in degrading organic matter (e.g. cellulases, chitinases, phenol oxidases, proteases, lipases), which are also produced by soil microbial biofilms. One important question arising is how this enzymatic activity contributes to carbon cycling on a global scale, as enormous quantities of dissolved polymers are produced for further decomposition.

The effect of bacterial biofilms in the mycorrhizosphere on chemical weathering, denudation and plant nutrition has been investigated in pine/mycorrhiza growth experiments (Balogh-Brunstad *et al.*, 2008; Koele *et al.*, 2009). Bacteria in the biofilm accelerated mineralization, released nutrients for the plants and reduced denudation. These observations were based on isolation of the root–hyphal-mineral interface in the rhizosphere from the bulk soil solutions. Establishment of a biofilm in the mycorrhizosphere can be beneficial to both parties (Deveau *et al.*, 2010). The basidiomycete *Laccaria bicolor* in association with biofilms of *Pseudomonas fluorescens* accumulated trehalose in the fungal hyphae, which was found to be released both from live and dead hyphae and may serve to attract *P. fluorescens* to the hyphae. The interaction between the two microorganisms was shown to be beneficial as *P. fluorescens* produces thiamine and other metabolites that in turn improve the growth of the fungus.

Given the current state of investigation, research is needed to better understand the complexity of biofilm–mycorrhiza interactions. Pilus-mediated interactions are important for attachment and establishment of biofilm in the mycorrhizosphere (Hogan *et al.*, 2007) and the full significance of their impact needs further elucidation. Furthermore, the response of the fungus to biofilm formation is important. The fungal factors that may block or eradicate biofilm formation must be investigated before initiation of potential applications can begin. The molecular analysis of both bacterial and fungal factors involved in biofilm formation and development is important and requires further investigation.

Soil bioremediation applications

The various activities of biofilms may be employed to aid the breakdown of a variety of pollutants and other unwanted compounds. As previously described, soil biofilms are involved in the degradation of organic matter, but xenobiotic compounds and pollutants can also be efficiently degraded by these diverse communities. An example of this is polycyclic aromatic hydrocarbons (PAHs), which are common pollutants in the environment, generated from incomplete combustion of organic matter produced by home heating, traffic, etc. (Johnsen *et al.*, 2005). Degradation of PAHs by biofilm bacteria has been evaluated in several studies (Stach and Burns, 2002; Johnsen *et al.*, 2005). The efficiency with which biofilm bacteria degrade PAH's has been associated with various factors including enhancement of PAH flux into cells by biofilm formation on PAH-containing sorbents. The arrangement results in a reduced diffusion path length and thus raised degradation efficiency (Johnsen *et al.*, 2005). Furthermore, higher species diversity in soil biofilms may contribute to elevated levels of PAH degradation compared with monospecies biofilms. It has been demonstrated that a PAH-degrading biofilm culture, composed of soil bacteria, was more diverse, both at the species and molecular levels, when compared with PAH exposed enrichment cultures. In the biofilm cultures, there were more different PAH-degrading species present and a higher diversity of genes involved in PAH degradation could be observed (Stach and Burns, 2002). In line with this, bacterial degradation of fenemiphos was only found to take place by a mixed consortium of six Gram-negative bacteria and with the presence of soil particles in the growth medium (Ou and Thomas, 1994).

Biofilms associated with fungi have also been demonstrated to enhance degradation of pollutants and these biofilms may spread widely as a consequence of the hyphal association. The importance of biofilm association with the ecto-mycorrhizal hyphae of two basidiomycetes on Scots pine grown in petroleum hydrocarbon (PHC) contaminated soil has been elucidated (Sarand *et al.*, 1998). A *Pseudomonas fluorescens* isolate able to degrade *m*-xylene and *m*-toluate harbouring transmissible, plasmid borne catabolic marker genes was isolated from the PHC contaminated soil. Microbial biofilm on the ecto-mycorrhizosphere hyphae was estimated to consist of 10^6 aerobic bacteria per gram fresh weight. By growing on mycorrhizal hyphae, degrading bacteria could reach a wide area of a polluted site and therefore be capable of enhancing the degradation process.

These observations point in the direction of the efficient pollutant degradation and bioremediation potential of soil biofilms, based on their tight community associations and the high bacterial and genetic diversity. In fact, it has been suggested that biotreatment of soil is improved by immobilization of soil and bacteria in a soil bioreactor, further encouraging the bacterial cells to form biofilms (Karamanev *et al.*, 1997). In this study, it was demonstrated how the pentachlorophenol mineralization was increased by one to three orders of magnitude in the immobilized soil bioreactor when compared with previously reported values. The increase in mineralization was attributed to the particular hydrodynamics and mass transfer properties of the system. In addition to the physical properties resulting in bioremediation potential, it should also be noted that, as mentioned previously, biofilms are hot-spots for horizontal gene transfer and genes involved in degradation of pollutants or other xenobiotic compounds may therefore easily spread in the biofilm, increasing the degradation potential of the bacterial community.

Biocontrol applications

The interactions between bacteria and fungi are not always beneficial for the fungi, but such 'non-beneficial' interactions may also have practical applications for use in biocontrol. Biofilms can protect fungal hyphae against changing environments, but living hyphae may also be subject to attacks by the bacteria. During the establishment of biofilms, bacteria may attach to living hyphae via polar interactions (de Boer *et al.*, 2005). In some cases the established biofilms may include bacteria that are able to lyse fungal hyphae, resulting in a condition known as mycophagy. The mycolytic bacteria produce hydrolytic enzymes (e.g. chitinase, glucanase, protease) that may subsequently degrade the hyphae. This knowledge has applicable potentials for the biocontrol of plant pathogenic fungi. In a recent mini-review, it was described how the mycophagy of the chitinolytic bacteria *Collimonas* (*Oxalobacteriaceae*) was able to suppress root fungal pathogens (Leveau *et al.*, 2010). Additionally, it has been shown that a *Pseudomonas fluorescens* strain produced antimicrobial metabolites inhibiting the growth of phytopathogenic fungi (Sharma and Das, 2010).

Another application has been suggested after review of the information on the microbial interactions in the mycorrhizosphere (Johansson *et al.*, 2004). Biocontrol suitable bacteria able to attach to roots and hyphae can form protective biofilms against pathogens on these biotic surfaces making them of possible importance for sustainable agriculture. Recently, microbial communities colonizing the hyphae of extraradical arbuscular mycorrhizal fungi (AMF) have been characterized using terminal restriction fragment length polymorphism (TRFLP) and 16S rRNA analysis (Scheublin *et al.*, 2010). Results show that bacteria in the biofilm on the hyphae were different from both the non-attached bacterial community and from bacteria living on a non-biological (glass wool) surface placed in the same soil. One particular bacterial species, identified as a member of the *Oxalobacteriaceae* family, was abundant on the AMF hyphae after 20 hours incubation. This family probably has a special interaction with the fungus, which is in agreement with several recent reports.

Seneviratne and co-workers investigated colonization of saprophytic fungal hyphae with different *Rhizobia* strains forming dense biofilms (Seneviratne and Jayasinghearachchi, 2003; Seneviratne *et al.*, 2006). The importance of the observation was that the hyphae in soil may serve as a survival site for the otherwise free-living nitrogen fixing bacteria, establishing a nitrogen fixing system. *Rhizobia* are not spore forming and it has been unclear how they survive under adverse soil conditions in the absence of their host. Survival in a biofilm on fungal hyphae may be a reasonable way to deal with extreme environments. The approach has been further developed by addition of nitrogen-fixing bacteria to fungal-bacterial biofilms in soil and thus forming fungal–rhizobial biofilms. This application could be used in agricultural and environmental settings but has not been introduced in practice so far. The biofilm formation on fungal hyphae may also affect the legume nodulation process. *Rhizobia* survive unfavourable environments in the biofilm around the roots, which may ensure success of the nodulation process and effective symbiosis with the host plant (Rinaudi and Giordano, 2010). More research is needed to elucidate how this may happen.

How to study biofilms in soil

As previously mentioned, *in situ* studies of soil biofilms are problematic because of the complex structure of the soil environment. The complex structure, combined with an astonishingly high bacterial cell number (~10^9 cells/g soil) and diversity (~10^6 species/g soil) (Torsvik *et al.*, 1990; Gans *et al.*, 2005), imposes challenges making such studies difficult. Microscopic analysis of ultra-thin soil slices provides information on the distribution of bacteria in the soil (Foster, 1981, 1988; Nunan *et al.*, 2003), but not on the bacterial nature or function. One way to circumvent this is to study aspects of biofilm formation and physiology in the laboratory with model systems composed of one or more soil isolates in a similar way to the bacterial response to desiccation studies mentioned previously (Roberson and Firestone, 1992; Ophir and Gutnick, 1994). Flow cells are a very commonly used model system in which biofilms are established on glass surfaces and continuously studied by confocal scanning laser microscopy (Møller *et al.*, 1998). Flow cells have been adapted for studies of soil conditions by addition of sand grains to the medium (Rodriguez and Bishop, 2007). Biofilm establishment in flow cells and in culture tubes allows high-resolution microscopy of structural aspects of biofilm growth (Baum *et al.*, 2009).

Attempts have been made to better mimic the conditions in the soil through establishing biofilms in different types of bioreactors (Holden *et al.*, 1997; Karamanev *et al.*, 1997; von Canstein *et al.*, 2002; Dunsmore *et al.*, 2004; Jean *et al.*, 2004). Analysis of the composition and function of the community is then performed by microscopic examination, molecular methods or by detection of the production or removal of various compounds. von Canstein *et al.* (2002) used a bioreactor to demonstrate that multispecies biofilms were more efficient at reducing mercury than those consisting of only one species. In our own group, we have developed a flow biofilm system, based on the isolation and growth of soil bacteria on agar covered glass slides supplied with autoclaved soil water nutrients. This avoids plate cultivation dependent isolation and the use of artificial growth media (Fig. 4.1). This system allowed identification of the dominant species present and facilitated studies of biofilm succession (Burmølle *et al.*, 2007). Lünsdorf and co-workers isolated biofilm forming bacteria and showed by microscopy that the cells were organized into small 'clay hutches' that may serve different functions including protection against protozoan grazing (Lünsdorf *et al.*, 2000, 2001).

These model systems provide important knowledge on biofilm structure, function, diversity and physiology that are essential in understanding soil biofilms. However, many aspects of how soil complexity and species diversity affect biofilms are still missing. Regardless of how well the model system is refined, it is impossible to recreate conditions that perfectly reflect those in soil. Nevertheless,

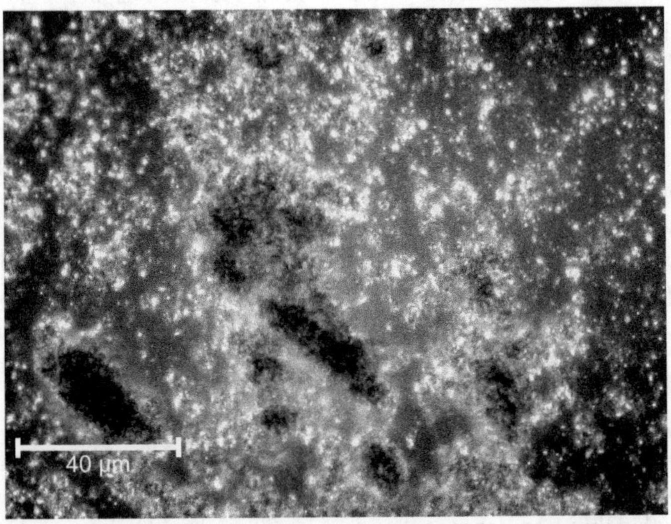

Figure 4.1 Morphology of a multispecies biofilm composed of soil bacteria. The biofilm was established by immersion of agar coated glass slides into soil slurry for 7 days. The attached biofilm was rinsed in a saline solution, stained with crystal violet and rinsed again, followed by examination by light microscopy with the ×40 objective (see Burmølle *et al.*, 2007, for further details). The micrograph illustrates how the bacteria organize into microcolonies of varying sizes. These microcolonies may be initiated by attachment of bacteria onto small soil particles; however, such soil particles are not visualized by the applied staining method. The scale bar represents 40 µm.

we still depend on these systems to improve our knowledge of soil biofilms.

Pathogenicity and biofilms in soil

Many opportunistic pathogenic bacteria including *Burkholderia, Pseudomonas, Salmonella, Listeria, Campylobacter* and *Legionella* occur naturally in the soil environment (Winfield and Groisman, 2003; Berg *et al.*, 2005; Gray *et al.*, 2006). However, characterization of these bacteria has focused on their biology and population dynamics in human hosts or in hospital environments and has ignored the impact of the external (non-host) environment. Several studies suggest that bacterial pathogenicity is greatly affected by the time spent and conditions faced by the bacteria in these external environments, of which soil is of major importance (Wilson and Salyers, 2003; Walther and Ewald, 2004). As stated, soil conditions may select for biofilm formation and attachment, unfortunately also enhancing the potential pathogenicity of the bacteria involved. Additionally, species incapable of surviving in the soil environment on their own may become part of soil biofilms in which they persist and evolve. This is of particular interest – and concern – with respect to pathogenic bacteria.

The colonization and survival of human pathogens in soil and rhizosphere has been studied recently (Kutter *et al.*, 2006; Opelt *et al.*, 2007; Warriner and Namvar, 2010). Warriner and Namvar (2010) found that the human pathogens *Escherichia coli* and *Salmonella* were adapted to survive in the rhizosphere as part of their life cycle and that they compete effectively with the rhizosphere flora. The survival time for *E. coli* in the rye or alfalfa rhizosphere was approximately 95 days, but only 24–41 days in plant-less fallow soil. Kutter *et al.* (2006) investigated the colonization of barley roots with the food borne pathogens *Salmonella* and *Listeria*. The survival was evaluated after 1–4 weeks, based on both cultivation-dependent and cultivation-independent (i.e. molecular) methods. The results showed that not only was the rhizosphere colonized but so were the root interiors. The problem arose when manure or sewage were used as fertilizers. A very special reservoir for potentially human pathogenic bacteria was also found by Opelt *et al.* (2007). Bacteria were identified from bryophyte genus *Sphagnum* peat bog vegetation with low pH and very low nutrient concentration. Among the bacteria a high proportion of opportunistic human pathogens such as *Staphylococcus, Hafnia, Yersinia* and *Pantoea* were identified that probably survive in biofilms on the 'roots' of the bryophyte.

Information on the effect of climate change on the survival, persistence, movement and virulence of pathogenic bacteria in soil has recently been reviewed (French *et al.*, 2009). When manure is amended to agricultural soil, potential pathogenic bacteria can be added to areas where they have previously not been present. Pathogenic bacteria will, as most indigenous soil bacteria, become concentrated in biofilms in the rhizosphere and may, with higher temperature and elevated CO_2 levels, survive for extended time periods. The extended time periods will cause enhanced spread of virulence factors and genes encoding antibiotic resistance determinants, owing to bacterial dispersal by water flow through soil. The impact of higher temperature and elevated CO_2 levels on the survival of pathogenic bacteria has been shown in the laboratory, but has not yet been studied in the field. Predicted higher winter temperatures are especially expected to improve pathogen survival despite competition with the indigenous soil community. Manure application also increases risks imparted by horizontal gene transfer from the pathogens to soil bacteria. Urban effluents spread on agricultural land can similarly carry bacteria expressing antibiotic resistance determinants that are transferred to wildlife (Blanco *et al.*, 2009). Therefore, biofilms may ultimately serve as reservoirs of antibiotic resistance.

Conclusions and future trends

There are many reasons that may cause soil bacteria to organize into biofilm communities. Biofilm-dwelling bacteria obtain increased protection from the harsh conditions faced by bacteria living in the bulk soil and the stability of their surrounding environment is improved. Additionally, their potential for fast adaptation and maximum nutrient gain is increased. We therefore hypothesize that the biofilm mode of growth is the preferred life style for active bacteria in soil as a result of selective evolutionary factors. However, research in soil biofilm ecology is still in its infancy and in many areas, we are just beginning to reveal and understand the complex interactions involved. This is especially true of bacterial–fungal interactions in biofilms. Progress is necessary to further understand the complicated relationship that may rely on factors such as co-migration, communication and horizontal gene transfer between bacteria in the biofilm and also between bacteria and their fungal hosts.

Studies of soil biofilms are complicated and our understanding depends on ecologically relevant model systems. Model systems are fundamental tools for expanding our knowledge of the bacterial life in soil – knowledge that is useful for exploiting the potential of bacterial bioremediation and biocontrol. Studies performed so far point in the direction of biofilms as being efficient tools for removing pollutants and biologically controlling plant pathogens. Model systems of soil biofilms may provide insights into long-term aspects of bacterial evolution in soil, which could be valuable for understanding and surveying the development of new aggressive pathogenic bacteria that may be less harmful if their arrival is prepared for.

Acknowledgement

The authors would like to thank Timothy Evison for careful and critical correction.

References

Almås, A.R., Mulder, J., and Bakken, L.R. (2005). Trace metal exposure of soil bacteria depends on their position in the soil matrix. Environ. Sci. Technol. *39*, 5927–5932.

Baldrian, P. (2006). Fungal laccases – occurrence and properties. FEMS Microbiol. Rev. *30*, 215–242.

Balogh-Brunstad, Z., Keller, C.K., Gill, R.A., Bormann, B.T., and Li, C.Y. (2008). The effect of bacteria and fungi on chemical weathering and chemical denudation fluxes in pine growth experiments. Biogeochemistry *88*, 153–167.

Baum, M.M., Kainovic, A., O'Keeffe, T., Pandita, R., McDonald, K., Wu, S., and Webster, P. (2009). Characterization of structures in biofilms formed by a *Pseudomonas fluorescens* isolated from soil. BMC Microbiol. 9, 103.

Berg, G., Eberl, L., and Hartmann, A. (2005). The rhizosphere as a reservoir for opportunistic human pathogenic bacteria. Environ. Microbiol. 7, 1673–1685.

Blanco, G., Lemus, J.A., and Grande, J. (2009). Microbial pollution in wildlife: Linking agricultural manuring and bacterial antibiotic resistance in red-billed choughs. Environ. Res. *109*, 405–412.

de Boer, W., Folman, L.B., Summerbell, R.C., and Boddy, L. (2005). Living in a fungal world: impact of fungi on soil bacterial niche development. FEMS Microbiol. Rev. *29*, 795–811.

Boersma, H., Otten, R., Nazir, R., and van Elsas, J.D. (2010). Selection of *Variovorax paradoxus*-like bacteria in the mycosphere and the role of fungal-released compounds. Soil Biol. Biochem. *42*, 2737–2745.

Burmølle, M., Hansen, L.H., and Sørensen, S.J. (2005). Use of a whole-cell biosensor and flow cytometry to detect AHL production by an indigenous soil community during decomposition of litter. Microb. Ecol. *50*, 221–229.

Burmølle, M., Webb, J.S., Rao, D., Hansen, L.H., Sørensen, S.J., and Kjelleberg, S. (2006). Enhanced biofilm formation and increased resistance towards antimicrobial agents and bacterial invasion are caused by synergistic interactions in multispecies biofilms. Appl. Environ. Microbiol. 72, 3916–3923.

Burmølle, M., Hansen, L.H., and Sørensen, S.J. (2007). Establishment and early succession of a multispecies biofilm composed of soil bacteria. Microb. Ecol. *54*, 352–362.

Burmølle, M., Bahl, M.I., Jensen, L.B., Sørensen, S.J., and Hansen, L.H. (2008). Type 3 fimbriae, encoded by the conjugative plasmid pOLA52, enhance biofilm formation and transfer frequencies in *Enterobacteriaceae* strains. Microbiology *154*, 187–195.

von Canstein, H., Kelly, S., Li, Y., and Wagner-Dobler, I. (2002). Species diversity improves the efficiency of mercury-reducing biofilms under changing environmental conditions. Appl. Environ. Microbiol. *68*, 2829–2837.

Deveau, A., Brule, C., Palin, B., Champmartin, D., Rubini, P., Garbaye, J., Sarniguet, A., and Frey-Klett, P. (2010). Role of fungal trehalose and bacterial thiamine in the improved survival and growth of the ectomycorrhizal fungus *Laccaria bicolor* S238N and the helper bacterium *Pseudomonas fluorescens* BBc6R8. Environ. Microbiol. Rep. 2, 560–568.

Dunsmore, B.C., Bass, C.J., and Lappin-Scott, H.M. (2004). A novel approach to investigate biofilm accumulation and bacterial transport in porous matrices. Environ. Microbiol. *6*, 183–187.

Ekelund, F., and Rønn, R. (1994). Notes on protozoa in agricultural soil with emphasis on heterotrophic flagellates and naked amoebae and their ecology. FEMS Microbiol. Rev. *15*, 321–353.

Flemming, H.C., and Wingender, J. (2010). The biofilm matrix. Nat. Rev. Microbiol. *8*, 623–633.

Foster, R.C. (1981). Polysaccharides in soil fabrics. Science *214*, 665–667.

Foster, R.C. (1988). Microenvironments of soil-microorganisms. Biol. Fert. Soils 6, 189–203.

French, S., Levy-Booth, D., Samarajeewa, A., Shannon, K.E., Smith, J., and Trevors, J.T. (2009). Elevated temperatures and carbon dioxide concentrations:

effects on selected microbial activities in temperate agricultural soils. World J. Microb. Biot. *25*, 1887–1900.

Gans, J., Wolinsky, M., and Dunbar, J. (2005). Computational improvements reveal great bacterial diversity and high metal toxicity in soil. Science *309*, 1387–1390.

Gilbert, P., Allison, D.G., and McBain, A.J. (2002). Biofilms *in vitro* and *in vivo*: do singular mechanisms imply cross-resistance? J. Appl. Microbiol. *92*, 98S–110S.

Gray, M.J., Freitag, N.E., and Boor, K.J. (2006). How the bacterial pathogen *Listeria monocytogenes* mediates the switch from environmental Dr. Jekyll to pathogenic Mr. Hyde. Infect. Immun. *74*, 2505–2512.

Grundmann, G.L. (2004). Spatial scales of soil bacterial diversity – the size of a clone. FEMS Microbiol. Ecol. *48*, 119–127.

Hansen, L.H., Ferrari, B., Sørensen, A.H., Veal, D., and Sørensen, S.J. (2001). Detection of oxytetracycline production by *Streptomyces rimosus* in soil microcosms by combining whole-cell biosensors and flow cytometry. Appl. Environ. Microbiol. *67*, 239–244.

Hogan, D.A., Wargo, M.J., and Beck, N. (2007). Bacterial biofilms on fungal surfaces. In The Biofilm Mode of Life: Mechanisms and Adaptations, Kjelleberg, S., and Givskov, M., eds. (Wymondham, Horizon Bioscience), pp. 235–245.

Holden, P.A., Hunt, J.R., and Firestone, M.K. (1997). Toluene diffusion and reaction in unsaturated *Pseudomonas putida* biofilms. Biotechnol. Bioeng. *56*, 656–670.

Hu, J.Y., Fan, Y., Lin, Y.H., Zhang, H.B., Ong, S.L., Dong, N., Xu, J.L., Ng, W.J., and Zhang, L.H. (2003). Microbial diversity and prevalence of virulent pathogens in biofilms developed in a water reclamation system. Res. Microbiol. *154*, 623–629.

Jackson, C.R. (2003). Changes in community properties during microbial succession. Okios *101*, 444–448.

Jackson, C.R., Churchill, P.F., and Roden, E.E. (2001). Successional changes in bacterial assemblage structure during epilithic biofilm development. Ecology *82*, 555–566.

Jass, J., Roberts, S.K., and Lappin-Scott, H.M. (2002). Microbes and enzymes in biofilms. In Enzymes in the Environment Activity, Ecology and Applications, Burns, R.G., and Dick, R.D., eds. (Marcel Dekker Inc., New York, USA), pp. 307–326.

Jean, J., Tsao, C., and Chung, M. (2004). Comparative endoscopic and SEM analyses and imaging for biofilm growth on porous quartz sand. Biogeochemistry *70*, 427–445.

Jefferson, K.K. (2004). What drives bacteria to produce a biofilm? FEMS Microbiol. Lett. *236*, 163–173.

Johansson, J.F., Paul, L.R., and Finlay, R.D. (2004). Microbial interactions in the mycorrhizosphere and their significance for sustainable agriculture. FEMS Microbiol. Ecol. *48*, 1–13.

Johnsen, A.R., Lukas, W.Y., and Harms, H. (2005). Principles of microbial PAH-degradation in soil. Environ. Pollut. *133*, 71–87.

Karamanev, D.G., Chavarie, C., and Samson, R. (1997). Soil immobilization: New concept for biotreatment of soil contaminants. Biotechnol. Bioeng. *57*, 471–476.

Klayman, B.J., Volden, P.A., Stewart, P.S., and Camper, A.K. (2009). *Escherichia coli* O157:H7 requires colonizing partner to adhere and persist in a capillary flow cell. Environ. Sci. Technol. *43*, 2105–2111.

Koele, N., Turpault, M.P., Hildebrand, E.E., Uroz, S., and Frey-Klett, P. (2009). Interactions between mycorrhizal fungi and mycorrhizosphere bacteria during mineral weathering: Budget analysis and bacterial quantification. Soil. Biol. Biochem. *41*, 1935–1942.

Kutter, S., Hartmann, A., and Schmid, M. (2006). Colonization of barley (*Hordeum vulgare*) with *Salmonella enterica* and *Listeria* spp. FEMS Microbiol. Ecol. *56*, 262–271.

Leveau, J.H.J., Uroz, S., and de Boer, W. (2010). The bacterial genus *Collimonas*: mycophagy, weathering and other adaptive solutions to life in oligotrophic soil environments. Environ. Microbiol. *12*, 281–292.

Lünsdorf, H., Strömpl, C., Osbourn, A.M., Bennasar, A., Moore, E.R.B., Abrahams, W.R., and Timmis, K.N. (2001). Approach to analyse interactions of microorgniams, hydrophobic substrates, and soil colloids leading to formation of composite biofilms, and to study initial events in microbiogeological processes. Methods Enzymol. *336*, 317–331.

Lünsdorf, H., Erb, R.W., and Timmis, K.W. (2000). 'Clay hutches': a novel interaction between bacteria and clay minerals. Environ. Microbiol. *2*, 161–168.

Mah, T.F., and O'Toole, G.A. (2001). Mechanisms of biofilm resistance to antimicrobial agents. Trends Microbiol. *9*, 34–39.

Martiny, A.C., Jørgensen, T.M., Albrechtsen, H.J., Arvin, E., and Molin, S. (2003). Long-term succession of structure and diversity of a biofilm formed in a model drinking water distribution system. Appl. Environ. Microbiol. *69*, 6899–6907.

Møller, S., Sternberg, C., Andersen, J.B., Christensen, B.B., Ramos, J.L., Givskov, M., and Molin, S. (1998). *In situ* gene expression in mixed-culture biofilms: evidence of metabolic interactions between community members. Appl. Environ. Microbiol. *64*, 721–732.

Nannipieri, P., Kandeler, E., and Ruggiero, P. (2002). Enzyme activities and microbiological and biochemical processes in soil. In Enzymes in the Environment: Activity, Ecology, and Applications, Burns, R.G., and Dick, R.P., eds. (Marcel Dekker, Inc., New York, USA), pp. 1–33.

Nazir, R., Boersma, H., Warmink, J.A., and van Elsas, J.D. (2010a). *Lyophyllum* sp. strain Karsten alleviates pH pressure in acid soil and enhances the survival of *Variovorax paradoxus* HB44 and other bacteria in the mycosphere. Soil. Biol. Biochem. *42*, 2146–2152.

Nazir, R., Warmink, J.A., Boersma, H., and van Elsas, J.D. (2010b). Mechanisms that promote bacterial fitness in fungal-affected soil microhabitats. FEMS Microbiol. Ecol. *71*, 169–185.

Nunan, N., Wu, K.J., Young, I.M., Crawford, J.W., and Ritz, K. (2003). Spatial distribution of bacterial communities and their relationships with the microarchitecture of soil. FEMS Microbiol. Ecol. *44*, 203–215.

Opelt, K., Berg, C., and Berg, G. (2007). The bryophyte genus *Sphagnum* is a reservoir for powerful and extraordinary antagonists and potentially facultative human pathogens. FEMS Microbiol. Ecol. *61*, 38–53.

Ophir, T., and Gutnick, D.L. (1994). A role for exopolysaccharides in the protection of microorganisms from desiccation. Appl. Environ. Microbiol. *60*, 740–745.

Ou, L.T., and Thomas, J.E. (1994). Influence of soil organic-matter and soil surfaces on a bacterial consortium that mineralizes fenamiphos. Soil. Sci. Soc. Am. J. *58*, 1148–1153.

Pearce, D., Bazin, M.J., and Lynch, J.M. (1995). The rhizosphere as a biofilm. In Microbial Biofilms, Lappin-Scott, H.M., and Costerton, J.W., eds. (Cambridge University Press, New York, USA), pp. 207–220.

Potera, C. (1996). Biofilms invade microbiology. Science *273*, 1795–1797.

Rinaudi, L.V., and Giordano, W. (2010). An integrated view of biofilm formation in *Rhizobia*. FEMS Microbiol. Lett. *304*, 1–11.

Roberson, E.B., and Firestone, M.K. (1992). Relationship between desiccation and exopolysaccharide production in a soil *Pseudomonas sp*. Appl. Environ. Microbiol. *58*, 1284–1291.

da Rocha, U.N., van Overbeek, L.S., and van Elsas, J.D. (2009). Exploration of hitherto-uncultured bacteria from the rhizosphere. FEMS Microbiol. Ecol. *69*, 313–328.

Rodriguez, S.J., and Bishop, P.L. (2007). Three-dimensional quantification of soil biofilms using image analysis. Environ. Eng. Sci. *24*, 96–103.

Sarand, I., Timonen, S., Nurmiaho-Lassila, E.L., Koivula, T., Haahtela, K., Romantschuk, M., and Sen, R. (1998). Microbial biofilms and catabolic plasmid harbouring degradative fluorescent pseudomonads in Scots pine mycorrhizospheres developed on petroleum contaminated soil. FEMS Microbiol. Ecol. *27*, 115–126.

Scheublin, T.R., Sanders, I.R., Keel, C., and van der Meer, J.R. (2010). Characterisation of microbial communities colonising the hyphal surfaces of arbuscular mycorrhizal fungi. ISME J. *4*, 752–763.

Seneviratne, G., and Jayasinghearachchi, H.S. (2003). Mycelial colonization by bradyrhizobia and azorhizobia. J. Biosci. *28*, 243–247.

Seneviratne, G., Zavahir, J.S., Bandara, W.M.M.S., and Weerasekara, M.L.M.A.W. (2006). Fungal-bacterial biofilms: their development for novel biotechnological applications. World J. Microbiol. Biotechnol. *24*, 739–743.

Sharma, D., and Das, M. (2010). Efficacy of antimicrobial metabolites of *Pseudomonas fluorescens* PFG against phytopathogenic fungi. P. Natl. A. Sci. India. B. *80*, 68–71.

Sørensen, S.J., Bailey, M., Hansen, L.H., Kroer, N., and Wuertz, S. (2005). Studying plasmid horizontal transfer *in situ*: a critical review. Nat. Rev. Microbiol. *3*, 700–710.

Stach, J.E.M., and Burns, R.G. (2002). Enrichment versus biofilm culture: a functional and phylogenetic comparison of polycyclic aromatic hydrocarbon-degrading microbial communities. Environ. Microbiol. *4*, 169–182.

Stewart, P.S., Camper, A.K., Handran, S.D., Huang, C., and Warnecke, M. (1997). Spatial distribution and coexistence of *Klebsiella pneumoniae* and *Pseudomonas aeruginosa* in biofilms. Microb. Ecol. *33*, 2–10.

Sutherland, I. (2001). Biofilm exopolysaccharides: a strong and sticky framework. Microbiology *147*, 3–9.

Torsvik, V., Goksöyr, J., and Daae, F.L. (1990). High diversity in DNA of soil bacteria. Appl. Environ. Microbiol. *56*, 782–787.

Verbruggen, N., Hermans, C., and Schat, H. (2009). Mechanisms to cope with arsenic or cadmium excess in plants. Curr. Opin. Plant Biol. *12*, 364–372.

Walther, B.A., and Ewald, P.W. (2004). Pathogen survival in the external environment and the evolution of virulence. Biol. Rev. Camb. Philos. Soc.*79*, 849–869.

Warmink, J.A., and van Elsas, J.D. (2009). Migratory response of soil bacteria to *Lyophyllum* sp strain Karsten in soil microcosms. Appl. Environ. Microbiol. *75*, 2820–2830.

Warmink, J.A., Nazir, R., and van Elsas, J.D. (2009). Universal and species-specific bacterial 'fungiphiles' in the mycospheres of different basidiomycetous fungi. Environ. Microbiol. *11*, 300–312.

Warriner, K., and Namvar, A. (2010). The tricks learnt by human enteric pathogens from phytopathogens to persist within the plant environment. Curr. Opin. Biotech. *21*, 131–136.

Wilson, B.A., and Salyers, A.A. (2003). Is the evolution of bacterial pathogens an out-of-body experience? Trends Microbiol. *11*, 347–350.

Winfield, M.D., and Groisman, E.A. (2003). Role of nonhost environments in the lifestyles of *Salmonella* and *Escherichia coli*. Appl. Environ. Microbiol. *69*, 3687–3694.

Xavier, K.B., and Bassler, B.L. (2003). LuxS quorum sensing: more than just a numbers game. Curr. Opin. Microbiol. *6*, 191–197.

Biofilms: Applications in Bioremediation

5

Gabriele Pastorella, Giulio Gazzola, Seratna Guadarrama and Enrico Marsili

Abstract

Bioremediation uses microorganisms to remove, detoxify, or immobilize pollutants, and does not require addition of harmful chemicals. Bioremediation is particularly suitable for large areas where contaminant concentrations are relatively low and the hydrology of the soil does not support an aggressive chemical remediation strategy. In the last few years, researchers have described the mechanisms of bioremediation for numerous priority pollutants, including chlorinated hydrocarbons, polyaromatic hydrocarbons, and heavy metals. However, most studies published to date have dealt with planktonic cultures grown under controlled laboratory conditions.

Microorganisms in the environment occur mostly as biofilms, whose development is encouraged by the presence of solid surfaces and the limited amounts of organic carbon. Therefore, optimization of bioremediation processes in the field requires a thorough knowledge of biofilm structure, dynamic, and interaction with pollutants and other environmental factors.

In this chapter, we describe the recent advances in bioremediation, with particular regard to the role of microbial biofilms. We discuss emerging technologies, such as bioelectroremediation and microbially produced surfactants. We also show how genetic engineering technologies may be employed to improve bioremediation effectiveness, both in laboratory and in field applications.

Introduction

'Bio-remediate' is a term used to explain how biological organisms transform, remove, or immobilize pollutants. Bioremediation methods can be adopted for clean-up of all environmental compartments and interfaces, such as soil, groundwater and air. This chapter deals mostly with bioremediation of soil and groundwater. The bioremediation procedure and the pollutant(s) removal mechanism change according to the characteristics of the contaminated area, the nature of the pollutant(s) and the microorganisms adopted in the treatment. In principle, bioremediation is a method with low-environmental impact. It does not strictly require the addition of harmful chemicals to mobilize, transform, or remove contaminants. Bioremediation is particularly suitable for large areas if time is not a constraint and the characteristics of the soil do not support an aggressive chemical remediation strategy. For example, bioremediation is the optimal solution for clean-up of diluted pollutants in large areas, where conventional methods would be exceedingly expensive and low pollutants concentrations do not pose an immediate threat to surrounding water bodies or urban communities. In the last 20 years, researchers have described the fundamental mechanisms of bioremediation for most of the priority pollutants, including chlorinated hydrocarbons, polyaromatic hydrocarbons, and heavy metals. Most mechanisms have been elucidated for planktonic cultures under controlled laboratory conditions. However, bacteria in the environment occur mostly as biofilms. The development of biofilms is encouraged by the presence of solid surfaces, such as rocks, soil, metal particles, and limited concentrations of organic carbon. *In situ* bioremediation processes are carried out by native microbial biofilms, or through bioaugmentation, where bacteria that have better tolerance to the pollutant are instead added to the system, thereby increasing rates of removal or immobilization. *In situ* processes may be enhanced through engineering measures, such as air venting, biobarriers, addition of chemicals. *Ex situ* processes, on the contrary, require removal of the contaminated soil or water, and treatment inside a bioreactor.

In all *in situ* and *ex situ* bioremediation processes, biofilm structures and activity are dictated by the surrounding environment. For example, highly toxic chemicals may trigger the production of polymeric capsulae that reduce direct contact between cells and pollutants. Also, the concentration and availability of nutrients and electron donor/acceptor determine biofilm microstructure. In multistep degradation processes, typical of hydrocarbon bioremediation in complex microbial communities, the distribution of microorganisms within the biofilm changes constantly. Furthermore, the gradient of the pollutant(s) across the thickness of the biofilms affects the oxygen profile causing biofilms to stratify into several niches. A thorough understanding of these complex correlations between biofilm microstructure and activity is needed for the development of more efficient bioremediation processes, both *in situ* and *ex situ*. When the pollutant is elemental and therefore cannot be degraded, biofilms can change its redox state and accumulate it in a less toxic form within their extracellular capsulae. This process is particularly important to metal bioremediation. The localization of metals within the biofilm, the crystal structure and the size of the metal particles produced during bioreduction determine the long-term efficiency of the bioremediation process.

Microorganisms living in polluted sites can acquire bio-degradative capabilities and increased resistance to pollutants through evolution and horizontal gene transfer. The rate of these adaptation processes is low, especially for highly toxic contaminants that require complex degradative machinery. Numerous priority pollutants are artificial and their release in the environment is relatively recent. Therefore, microorganisms in biofilms exposed to novel contaminants have not yet optimised their metabolism to use the novel pollutants as carbon or energy source, or to degrade them through their co-metabolism. The toxicity resistance and the pollutant degradation rate of the microorganisms required for bioremediation can be improved by molecular biology techniques, as well as by those from genetics, ecology, metabolic and protein engineering, and '-omics' technologies. For instance, recombinant DNA technology can be used to construct a microorganism with the desirable catabolic or resistance pathways, while site-directed mutagenesis can be used to modify enzymes with more efficient bio-degradative characteristics. From an environmental bioremediation perspective, accelerated-evolution techniques such as genome shuffling can potentially contribute to a better understanding of the bio-remediation processes, as well as generate fast-evolving and well-adapted microorganisms.

In this chapter, we describe the recent advances in bioremediation for chosen priority contaminants, with particular regard to the role of microbial biofilms. We also show how emerging genetic engineering technologies and the use of microbially produced surfactants may be employed to improve bioremediation effectiveness, both in laboratory and in field applications.

Remediation versus bioremediation

Bioremediation of a contaminated environment is a process in which viable microorganisms or their by-products transform the pollutant(s) into less harmful forms (Bragg *et al.*, 1994; N'Guessan *et al.*, 2010). It is an alternative to conventional physicochemical remediation treatments, such as advanced oxidation, chelation and precipitation (Bragg *et al.*, 1994; Kulik *et al.*, 2006; Jacques *et al.*, 2008; N'Guessan *et al.*, 2010). Such treatments require expensive chemical agents and raise multiple environmental concerns. Current policy requires the near-complete recovery and/or recycling of these agents at the end of treatment. Physicochemical decontamination methods are recommended when rapid treatment is crucial, and when the high concentration of pollutants can potentially reduce the effectiveness of any biological decontamination methods (e.g. large oil spills, concentrated toxic plumes close to the pollution source). Also, these physicochemical methods are chosen when the contaminated water or soil can be easily removed and treated in a chemical reactor, as in the so-called pump and treat approach (Jacques *et al.*, 2008). In such cases, the chemical reactor is set to maximize the rate of contaminant removal. However. bioremediation is the optimal strategy for long-term detoxification of large areas, for decontamination of inaccessible subsurface environments such as groundwater, and for the removal of priority contaminants at a low concentration (Rittman and McCarty, 2000).

Laboratory studies demonstrated that most toxic chemicals or mixtures thereof may be degraded into harmless forms by single species or mixed bacterial communities under specific reaction conditions. The reader is referred to published reviews for an extensive explanation of biodegradation pathways for organic and inorganic pollutants (Bopp *et al.*, 2010; Erable *et al.*, 2010; Williams *et al.*, 2010).

The application of a detoxification strategy in the field is a complex task and no general rules can be drawn. Since soils are very complex media, their composition and morphology affect detoxification rates and reaction pathways. Soil characterization is a time-consuming process in most cases, with the exception of sandy subsurface soils which possess a relatively simple structure. Despite these shortcomings, bioremediation proves to be a sustainable treatment of contaminated environments.

Biofilm in subsurface environments

Biofilm is the preferred growth mode among microorganisms in the environment where surfaces are amenable for cell attachment (Calvo *et al.*, 2009; Conn *et al.*, 2010; Conrad *et al.*, 2010; Williams *et al.*, 2010). Diverse biofilm communities can be found in a wide-range of environments, including those facing challenging conditions (e.g. high temperature, salinity, pressure). Subsurface environments are frequently the target zones for bioremediation, as they have large porosity and a low organic carbon concentration (Prpich *et al.*, 2006; Conn *et al.*, 2010). As mentioned in earlier chapters, these conditions also promote the development of thin and strong biofilm. Bioremediation in subsurface environments is, inherently therefore, a biofilm-mediated process.

Subsurface biofilms comprise mixed microbial consortia. Such communities are much more complex than the single species biofilms commonly encountered in laboratory research. The structure of subsurface biofilms depends on several factors such as the nutrient availability, solid phase morphology, and physicochemical parameters of the microenvironment. The heterogeneity of the subsurface environment results in a large variety of biofilm growth forms, including cluster- and protrusion-type structures, cell aggregate bridging, and thick bioweb-type growths (Rodriguez and Bishop, 2007). When planning a bioremediation process, the process designer should assess the native microbial consortia and the existing metabolic network, in order to create the optimal bioremediation strategy. For example, in case of *in situ* bioremediation, the composition of native microbial consortia will help determine if biostimulation is preferred over the addition of novel strains into the soil system. While the use of a specific strain may be very effective in the laboratory set-up, in the field, such bacteria may be competitively excluded by native microorganisms, thereby reducing the effectiveness of the treatment.

Biofilm formation in contaminated sites

Initial biofilm cell attachment occurs via physicochemical interactions and protein adhesion-secretion to form a cell monolayer (Prpich *et al.*, 2006). Surface recognition proceeds through chemotaxis, and then swirling and swimming microorganisms form the initial biofilm as they attach by secreting extracellular proteins to create a matrix from extracellular polymeric substances (EPS). EPS promotes the

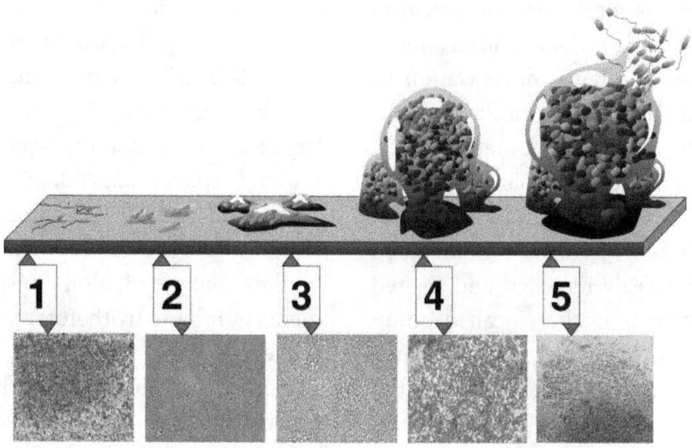

Figure 5.1 Schematic of the five-stages of biofilm development. Stage 1, initial attachment; stage 2, irreversible attachment; stage 3, maturation I; stage 4, maturation II; stage 5, dispersion. Each stage of development in the diagram is paired with a photomicrograph of a developing *P. aeruginosa* biofilm. All photomicrographs are shown to same scale (image credit: D. Davis). Reproduced with permission from Monroe (2007).

attachment of other microorganisms (Fig. 5.1). Biofilm thickness is limited by shear stress, nutrient availability and occurrence of toxic chemicals (Joannis-Cassan *et al.*, 2007; Mukherjee *et al.*, 2009). Biofilm microstructure is also affected by the morphology of the solid phase or attaching surface. For instance, in sandy or fine particle soils, where shear stress is very low, thick biofilms grow around the mineral particles. Biofilms on rock surfaces grow thinner because of higher shear stress and lower specific area of the support (Cunningham *et al.*, 1991; Shafahi and Vafai, 2009). As cells die, active bioconversion and/or biodegradation lead to solute transfer to or from the bulk liquid. This eventually results in partial cell detachment from the surface. Detachment may be also caused by changes in nutrient availability and/or hydrodynamic conditions, and it enables biofilm to spread along the groundwater flow. Detachment is particularly relevant to bioremediation and bioimmobilization processes that involve engineered surfaces, such as biobarriers. Large biofilm sloughing may impair the ability of the biobarrier to remove the contaminant(s) at the needed rate. As a result, the contaminant may migrate downstream and reach other groundwaters and soils. Furthermore, in biosorption and bioimmobilization, sloughing remobilises environmental pollutants. In order to avoid these problems, engineered bioremediation devices need to be regularly monitored and biofilm removal scheduled (Kulik *et al.*, 2006; Sabater *et al.*, 2007).

In certain environments, such as undisturbed, organic rich sediments, biofilm can grow further and form thick layers, also known as microbial mats. However, these complex microbial communities are not commonly encountered in bioremediation processes. The presence of contaminants affects biofilm in several ways. The existent microbial physiology may change owing to acute contamination (Martínková *et al.*, 2009) altering biofilm microstructure (i.e. homogeneous vs. irregular) and the pollutant(s) solubility and concentration (Salinero *et al.*, 2009; Yoshida *et al.*, 2009). Hydrophobicity/hydrophilicity of contaminants also affects the biofilm structure. Poorly soluble pollutants typically lead to the formation of a more dispersed and thin biofilm, while hydrophilic pollutants cause formation of a thicker biofilm (Antizar-Ladislao and Galil, 2010). Similar effects were observed in a biofilm grown in presence of high Cr(VI) concentration: the biofilm cells were elongated and grouped in clusters (Morel *et al.*, 2009).

Roles of biofilm in bioremediation

Early research

In 1981, Akinson described for the first time the use of biofilm in water bioremediation (Akinson, 1981). Since then, many researchers have outlined the correlation between biofilm microstructure and bioremediation activity. The reader may find an excellent account of this research in two reviews, one concerning the microbial metabolism of xenobiotics (Singleton, 1994), the other dealing with the scale-up of bioremediation processes (Sturman *et al.*, 1995). More recent research has provided us with a wealth of information about metabolic pathways for pollutant degradation and some knowledge about the interaction between different microorganisms in complex microbial consortia exposed to pollutants (Anderson and O'Toole, 2008; Stewart and Franklin, 2008; Landini, 2009).

Biofilm resistance to environmental contaminants and stress

Microorganisms can partially or completely degrade a large variety of organic and inorganic pollutants, under specific conditions. This plastic adaptation to a variety of environmental conditions, organic carbon and energy source availability is the result of the fast growth and high mutation rates of microorganisms, together with their capability to transfer genetic information within the same and among different species (Cooper, 2007; Smith and Romesberg, 2007). In addition to these adaptive features, which have been widely investigated in detoxification and biodegradation studies, biofilm provides: (a) a protective microenvironment that reduces the toxicity of pollutants to microorganisms; (b) an organized scaffold that optimizes nutrient delivery and by-products removal; (c) a three-dimensional structure that brings in close proximity various microorganisms, including both aerobes and anaerobes, thereby increasing syntrophy and accelerating the rate of complex degradation processes. The following paragraphs further depict these three important provisions.

The biofilm microenvironment comprises, in addition to viable cells, a layer of extracellular polymeric substances (EPS) that retains a significant amount of dead cells. In fact, it plays a crucial role in biofilm development. The EPS buffers the biofilm microbial community from external toxic chemicals so that only a relatively small concentration reaches the viable cells (Flemming and Wingender, 2010). It has been shown that biofilms can withstand up to 1000 times the toxic concentration of a pollutant than the planktonic form (Buchholz *et al.*, 2010). Also, the EPS works as a biosorbent by removing a large fraction of the pollutant from the water phase. Other observed passive physicochemical effects are chelation and complexation (Pal and Paul, 2008). These passive mechanisms are particularly important in metal bioremediation, where the pollutant cannot be eliminated but only changed in its redox state or location (Marsili *et al.*, 2007; Cummings and Kovacic, 2009; Kazy *et al.*, 2009; Wang *et al.*, 2010). Also, apart from passive resistance mechanisms, the EPS can contribute to the degradation of organic pollutants and heavy metals into less dangerous forms for the inhabiting microorganisms, for instance, through the use of extracellular enzymes (Pal and Paul, 2008; Flemming and Wingender, 2010). A recent study showed that *Stenotrophomonas* biofilms grown with 100 mg/l Cr(VI) produce EPS with a higher Cr(VI) binding capability than the EPS observed in the control biofilm (Pal and Paul, 2008; Morel *et al.*, 2009).

Biofilm microstructure provides a solid scaffold that protects viable cells from the prevalent shear stresses present in large particulate soils (Vu *et al.*, 2009). The EPS improves shear stress resistance, thus increasing the amount of immobilized biomass (e.g. biobarriers) and reducing biofilm shedding. Increased stress resistance in biofilms allows a wider operational range for bioreactor parameters (osmolarity, pH, etc.), than their planktonic counterpart. Biofilm is also more resistant than its planktonic counterpart to UV damage, desiccation, predation, and pH gradients (Elasri and Miller, 1999; Jacques *et al.*, 2008; Monds and O'Toole, 2009; Morel *et al.*, 2009; Zhang *et al.*, 2010).

Early biofilm research suggests that biofilm microstructure aims to maximize organic carbon and nutrient delivery to all viable cells in the consortium. Quorum sensing (see Chapter 1) was often proposed as the main regulatory strategy for biofilm growth (Kjelleberg and Molin, 2002) and the observation of water channels, niches, and other local heterogeneity features were usually explained in terms of efficient circulation of nutrients and metabolites (Costerton *et al.*, 1995; Zhang *et al.*, 1995). However, environmental biofilms may be exposed also to strong shear stress and toxic contaminants. Therefore, the microstructure should satisfy the efficient delivery of nutrients to all cells, while increasing the toxicity resistance (Harrison *et al.*, 2007) and the mechanical resistance to shear stress (Beyenal and Lewandowski, 2002).

While the ultimate causes of biofilm microstructure are still controversial, the benefits of oxygen gradients in biofilm-based bioremediation processes are clearly documented. Typically, oxygen is consumed in the outer layers of biofilm, providing favourable conditions for growth of facultative and strict anaerobes in the deep layers of the biofilm. This ecological diversity favours the oxidation and removal of complex and often toxic contaminants by slow-growing strains. For example, oxygen consumption in the upper layers of multispecies biofilms triggers U(VI) reduction in the bottom layers, where U(IV) reoxidation is negligible (Krawczyk-Bärsch *et al.*, 2008). Similar results were obtained for other metal reduction processes, where oxygen causes redissolution of the immobilized contaminants. Other researchers reached a similar conclusion for Cr(VI) to Cr(III) bioremediation (Smith and Gadd, 2000). Species interactions in mixed aerobic/anaerobic biofilms favour hybrid degradation pathways of aromatic hydrocarbons (Cao *et al.*, 2009). A mixed-species biofilm composed of *Burkholderia* and *Pseudomonas aeruginosa* can degrade chlorobenzoate more efficiently than the single strain of *Burkholderia*. Moreover, the presence of *P. aeruginosa* resulted in higher biomass within the dual-species biofilm (Yoshida *et al.*, 2009). In some cases, the advantage of dual-species biofilm is that the strain with best degradative properties is not a biofilm-forming strain. The biofilm-forming microorganism provides a robust structure where the degradative microorganism can thrive. For example, in the degradation of 3,5-dinitrobenzoic acid, the degradative strain *Comamonas testosteroni* A3-I grows together with five other *Pseudomonas* spp. strains. The resulting biofilm provide both increased resistance to toxicity and a secure location for the *C. testosteroni* cells to thrive (Meng-Ying *et al.*, 2009).

The dynamics of genomic material transfer in the subsurface also affects biofilm structure and stability. It has been shown that biofilms facilitate horizontal gene transfer in the subsurface (Davey and O'Toole, 2000). This phenomenon results in syntrophic associations and facilitates plasmid transfer of important genes such as those associated with toluene degradation (Christensen *et al.*, 1998). Gene transfer also helps to maintain biofilm structure (Meng-Ying *et al.*, 2009). It has been further shown that bacteria release genomic DNA following cell lysis, which then has an essential role in intercellular adhesion and biofilm stability (Bayles, 2007).

Biofilm bioremediation methods

Biofilm bioremediation methods may be divided into *in situ* and *ex situ* approaches. *In situ* methods are applied to contaminated soil or groundwater systems in their original location. Conversely, *ex situ* methods are applied to contaminated soil or groundwater that has to be physically removed from its original location and then treated in a bioreactor. We provide here only a brief review of this complex topic. The reader is referred to Rittman and McCarty (2000) and Farhadian *et al.* (2008a,b) for further technical information.

In situ

These treatments take place in the original location, where contamination occurs. Depending on the degree of contamination and the size of area to be treated, engineering methodologies can also be included in the process. In general, *in situ* methods require a thorough characterization of the contaminant chemistry, site hydrology, and the composition of the microbial community in the subsurface system. The following are the most widely used *in situ* methods.

Natural attenuation

Natural attenuation relies on natural processes, within the context of a carefully controlled and monitored site clean-up approach, to achieve site-specific remediation objectives within a time frame that is reasonable compared with that offered by other more active methods. Here, we focus on the biological (biodegradation, biosorption, stabilization, etc.) processes that, under favourable conditions, reduce the mass, toxicity, mobility, volume, or concentration of contaminants in soil or groundwater (EPA and Office of Solid Waste and Emergency Response, 1999). The indigenous microbial community and the pollutant profile are monitored at defined times and locations. This treatment is not equivalent to a 'do-nothing' approach, instead, it provides crucial information that helps in the design of a more active bioremediation treatment, when needed (Brockmann and Stelzer, 2009). Natural attenuation is a viable option where contaminants do not pose immediate threat to groundwater, either because of their low concentration, or because of the large distance from populated areas (Bento *et al.*, 2005). Natural attenuation can also be used as the final step of an engineered bioremediation treatment, when the native microbial community can be maintained and continue the degradation process (Widdowson *et al.*, 2008).

Bioaugmentation

Bioaugmentation is the inoculation of a culture, selected using either conventional or genetic engineering methods, into a contaminated subsurface environment (Kulik *et al.*, 2006; Wang *et al.*, 2010). Microbial consortia that can degrade complex contaminant mixtures are often found in highly polluted subsurface environments and can be maintained in laboratory conditions similar to those encountered in contaminated environments. These consortia can be added to other subsurface environments with a similar contaminant composition and concentration, accelerating contaminant degradation and removal (Jacques *et al.*, 2008). However, numerous studies indicate that introduced microorganisms often do not survive in the environment and thus do not increase contaminant remediation. Therefore, bioremediation becomes limited also by the concentration of degrading microorganisms added in the subsurface (Kuhn *et al.*, 2009; N'Guessan *et al.*, 2010). Bioaugmentation is usually recommended for recently contaminated sites. In these sites, the indigenous microbially community may be seriously damaged and the residual microorganisms may require years to develop resistance to the novel contaminants and efficient degradation mechanisms. Because of the low relative abundance of electron donors/acceptors and organic carbon sources, the addition of microorganisms able to degrade the contaminant result in the rapid establishment of a suitable microbial community. For example, bioaugmentation with *Pseudomonas* can lead to a fast removal of dinitrobenzoic acid (Meng-Ying *et al.*, 2009) in contaminated soils and the addition of *Dehalococcoides ethenogenes* significantly increased the dehalogenation of chlorinated products during groundwater bioremediation (Conrad *et al.*, 2010).

Bioaugmentation efficiency may be improved by implementing engineering strategies into the process to increase the concentration and availability of limiting nutrients, electron acceptors, electron donors, and biomass as per the bioremediation process requirements. Simple engineering strategies include air venting and addition of appropriate organic carbon sources to the contaminated subsurface, as described below.

Air venting

In general, aerobic degradation proceeds at faster rates than anaerobic degradation, especially for hydrocarbon compounds and aromatics, where the breaking of stable chemical bonds requires high energy levels. In order to establish aerobic conditions in the subsurface, air is delivered in the soil by injection into aeration wells through centrifugal pumps. The number, the depth, and the location of the aeration wells depends on the subsurface morphology (Rittman and McCarty, 2000). The increase in oxygen concentration causes the enrichment of the aerobic microbial community, and an increase in biofilm microbial activity. It may be costly, depending on the morphology of the subsurface environment, therefore this technique is only suitable for small to medium-size contaminated areas where fast treatment is required (Bopp *et al.*, 2010; Chen *et al.*, 2010).

Addition of carbon source/electron donor/electron acceptor

This method supplies additional nutrients to the existing subsurface microbial community, thus enhancing bacterial activity and remediation rates. Addition of chemicals is needed where the contaminated environment is highly depleted in organic carbon (Pandey *et al.*, 2009) or other limiting nutrients. Contrary to bioaugmentation, additions of chemicals result in the enrichment of specific strains or microbial families derived from the existing microbial community. In a typical example, injection of the electron donor lactate in a trichloroethene (TCE) contaminated aquifer (i.e. using injection wells) stimulated microbial activity and promoted the biological de-halogenation of TCE by organisms native to the site, including *Dehalococcoides ethenogenes* (Conrad *et al.*, 2010). Other common chemicals added to contaminated subsurface include nutrient rich solutions such as molasses, acetate and ethanol. For example, the *in situ* bioreduction of U(VI) was achieved through the biostimulation of denitrifying, Fe(III)-reducing, and sulfate-reducing bacteria. The organic electron donor and carbon source ethanol was added to the subsurface through injection wells (Cardenas *et al.*, 2008). In another example, several gaseous electron donors, including hydrogen, 1-hexene, ethyl acetate and liquefied petroleum gas (LPG) were added to stimulate the biological reduction of perchlorate, under various operative conditions and soil moisture (Cai *et al.*, 2010).

Biobarriers

Biobarriers are subsurface structures created by biofilm-forming microorganisms, which can degrade groundwater contaminants or change their mobility, thus removing them from the groundwater flow. Biobarriers are based on enhancing the biomass and/or activity of bacterial communities in natural or artificially created bottlenecks in aquifers or other subsurface environments (Rittman and McCarty, 2000). The formation of a biobarrier can be promoted through the addition of high-surface support material, such as plant residues (Seo *et al.*, 2009). The addition of nutrients (stimulation) and/or suitable microorganisms (bioaugmentation) produce a dense biomass layer, which offer strong diffusion resistance to the pollutants, thus increasing its degradation or immobilization rate. There are two types of biobarrier: (a) Static biobarriers, where biofilms and biobarrier material adsorb the contaminants from the groundwater (Stroo *et al.*, 2010); and (b) Reactive biobarriers (Fig. 5.2), where the pollutant is degraded through a combination of biological and chemical reactions (Lennox and Ashe, 2009). Numerous laboratory and field studies have demonstrated that biobarriers are effective for bioremediation of various pollutants, such as crude oil, aromatic hydrocarbons (e.g. benzene, toluene), polycyclic aromatic hydrocarbons (e.g. naphthylene, phenanthrene), and chlorinated cyclic hydrocarbons (e.g. pentachlorophenol) (Tiehm *et al.*, 2008). Biobarriers are also able to immobilize metals, such as mercury, zinc, manganese and selenium (Hunter, 2009; Michalsen *et al.*, 2009), and to remove inorganic pollutants such as nitrate from groundwater (Hunter, 2009).

In order to clarify how a static biobarrier works, we report a study examining a vynil chloride-contaminated subsurface (Lennox and Ashe, 2009). The authors describe a fictitious scenario, where vinyl chloride was accidentally released in the proximity of a river. *Klebsiella oxytoca*, a facultative and biofilm-forming bacterium that has no pathogenic effects, was used to create a biobarrier. Briefly, the bacteria, previously depleted of organic carbon, were introduced to the contaminated soil through injection wells. The starvation process reduces the bacterial size, thereby improving their ability to penetrate into the soil matrix. The addition of molasses and nitrate stimulated the facultative metabolism of *K. oxytoca*. As a result, biofilm with a high EPS content grew rapidly and was able to significantly reduce the flow of vinyl chloride towards the subsurface water (Lennox and Ashe, 2009).

Ex situ

Biofilms also offer advantages in *ex situ* bioremediation. In fact, in a typical immobilized biomass reactor, the presence of biofilms increases the efficiency of contaminant removal and reduces

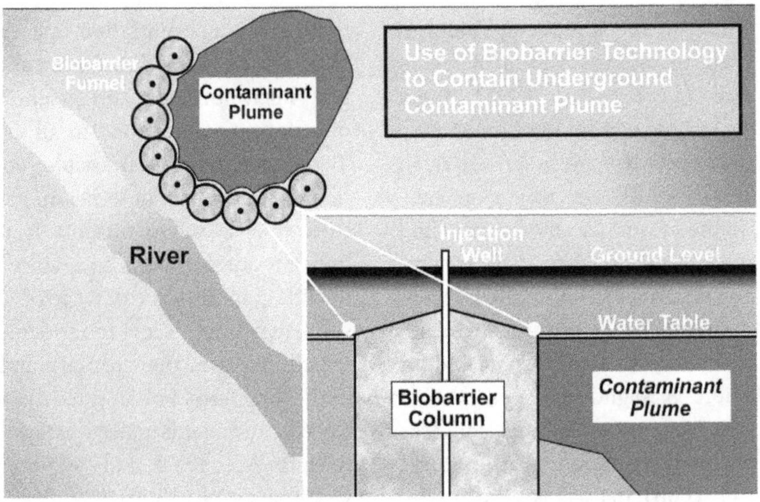

Figure 5.2 Schematic of a dynamic biobarrier for containment of contaminant plume. Reproduced with permission from the Center for Biofilm Engineering, MSU-Bozeman.

the time needed for the bioremediation process (Gusmão *et al.*, 2007). Bioreactors such as silicon tube membrane, granular activated-carbon biofilm, fluidized bed biofilm, and hollow-fibre biofilters have been used in *ex situ* treatment of numerous priority contaminants. For example, a recent study described the removal of pyrene from simulated contaminated soil in bio-slurry phase reactors operated in periodic discontinuous batch mode under an anoxic–aerobic–anoxic–anoxic microenvironment (Venkata Mohan *et al.*, 2008b). The simple aromatic hydrocarbon phenol was degraded in a polymer bed two-phase reactor (Prpich *et al.*, 2006). In other works, the *ex situ* bioremediation of a petroleum-contaminated soil was investigated in a laboratory scale aerated reactor, in which an indigenous bacterium isolated from a contaminated site close to a refinery, *Stenotrophomonas maltophilia*, were grown. A first order bioremediation reaction rate (Abbassi and Shquirat, 2008) was observed, while pyrene bioremediation rates in the bioslurry phase bioreactor showed a retention time dependent efficiency (max 90%) (Venkata Mohan *et al.*, 2008a). *Ex situ* treatments are usually very efficient and can be optimized through established bioreactor practices. The reader is referred to some of the numerous reviews on this topic (Farhadian *et al.*, 2008a,b; Robles-Gonzalez *et al.*, 2008).

Monitoring biofilm bioremediation

Monitoring bioremediation processes is of paramount importance to assess the efficacy of field applications. Subsurface environments are not easily accessible, and contaminated areas are usually far from analytical and research laboratories. Researchers have developed several strategies to monitor microbial processes, specifically in biofilm (Maila and Cloete, 2005). Small wells (e.g. injection wells) excavated in the area under examination are the easiest way to access the subsurface and to analyse soil and groundwater samples. Long-term sensors capable of wireless transmission to an integrated data analysis system are preferable, in which they provide a continuous data flow and reduce the need of expert presence on the bioremediation site. Battery replacement does constitute a cost and an environmental burden in bioremediation processes. Innovative low-power sensors and short-term pulse operation is needed to minimize the costs and impacts of wireless distributed sensors in the environment (Mathúna *et al.*, 2008). An efficient sensor measures the pollutant concentration but, most importantly, it should also measure microbial activity, to identify changes in the microbial consortia as a consequence of variations in pollutant concentration, or the hydrodynamic regime of the soil. Low-cost sensors based on bacterial biofilms have been proposed (Williams *et al.*, 2010). For example, dissimilatory metal-reducing biofilms (DMRB) are used for bioremediation of metal-contaminated soils (Lovley *et al.*, 1991). These biofilms have the unique capability of interacting with electrode surfaces that are polarized at an oxidative potential. DMRB are grown on electrodes inserted in the subsurface environment and produce a current proportional to their activity, thereby providing information about the concentration of electron donor in the site and the health of the microbial biofilms that are responsible for the bioremediation. A slightly different design proposes the use of a microbial fuel cell (see Chapter 10) to monitor bioremediation treatments (Tront *et al.*, 2008). In this case, the biofilm grown on the electrode uses only a negligible part of the electron donor that is delivered to the soil, generating the energy needed by the sensor. These and other sensors provide an early warning about the status of the bioremediation process, thus enabling further and in-depth control by trained operators.

Review of bioremediation treatment by pollutants

Since its beginnings, bioremediation has been applied successfully to remove aromatic hydrocarbons, chlorinated compounds, pesticides, metals, and radionuclides. Here, we will describe in detail only a few pollutants, in terms of biofilm bioremediation potential. Further details on specific pollutants may be found in Martin and Ruby (2004), who provide a complete review for *in situ* remediation approaches; and in Gadd (2010) for a clear explanation of the complex relationships between geomicrobiology and bioremediation.

Metals and metalloids

Metals and metalloids are among the most dangerous soil and groundwater contaminants. Most national regulations set thresholds for drinking water, wastewater and agricultural soils. The US Environmental Protection Agency (EPA) establishes thresholds about 5–50 μg/l for most metals and metalloids (EPA, 2009). Metals enter the organism mostly through ingestion, and can accumulate in tissues and organs, like the kidney and liver. Whilst some metals are known to be highly toxic at low concentration (e.g. As, Hg), several metals, such as Mn, Co, Zn, Cu and Fe are important micronutrients and essential components of redox enzymes found in all living organisms. Their maximum permissible concentration is therefore higher. For other metals, such as Cr, their role as micronutrients is not completely understood. Actinides and, among them, radionuclides, are a particularly toxic class of metals, since they possess low to moderate radioactivity, which increases their toxicity and health effects in the long term. For an extensive review of metal toxicity and metal toxicity assays the reader may refer to Wolterbeek and Verburg (2001) and to Goldfrank *et al.* (2006).

Differently from organic contaminants, metals cannot be degraded, only transformed. Their presence and toxicity are generally related to their solubility in water since soluble metals are typically more mobile in the environment. In general, metals are soluble in water at low pH. However, their distribution between liquid and solid phase depends largely on the presence of other inorganic ions, such as carbonates, phosphates, etc. Changes in metal mobility increases the risk of contamination for key drinking water sources and agricultural soils. Metal contamination originates from metallurgy plants, electroplating industries, etc. Mining and mineral processing also generate large amounts of water that contain metal levels beyond the acceptable threshold. A number of heavily contaminated sites have been reported. One of the most studied, because of its large size, is the Berkeley Pit in Butte, MT, USA. This large lake was formed by groundwater and surface water in a previous open-pit copper mine, following cessation of pumping and mining in 1982. The water is sulfate-rich, acidic (pH 3.0) and has a moderate oxidizing potential. Berkeley Pit is among the most contaminated water body in the USA. The main metals are Zn (414 mg/l), Al (mg/l), Cu (125 mg/l), Cd (2 mg/l) and As (3 mg/l). The level of the water is growing a few metres per year and poses a serious risk for contamination of surrounding drinking water sources. (Shokes and Möller, 1999).

If the metal concentration is very high, a chemical pre-treatment (i.e. neutralization, chelation, or precipitation) is needed before bioremediation.

It is often the case that metal-contaminated sites are treated years after the initial contamination occurred. Consequently, a large volume of water and soil needs to be remediated, thus becoming a costly enterprise.

Contamination by actinides have a different source. In the past, uranium (U) extraction and processing were often managed by military personnel, and little or no care was taken to avoid soil and water contamination from uranium waste heaps. Forty years of 'cold war' have left a heavy burden of contaminated sites across the USA and former Soviet Union. Consequently, a number of high-profile actinide and radionuclide-related bioremediation studies are carried out in the USA, where large-scale research programmes and facilities are financed for this specific purpose (http://www.epa.gov/superfund/).

Bioremediation is a viable option for the treatment of metal and actinide contaminated areas, particularly for the long-term management of this type of pollution. Here we provide an updated list of metal bioremediation treatments, with particular emphasis in biofilm studies and the few field applications available.

Arsenic (As)

Arsenic (As) is a metalloid used in mining, wood processing, pigment production, drugs and as a component in numerous pesticides (E-PRTR, 2008). Arsenic is a significant by-product of copper, zinc and lead smeltering activities. Other important sources are mine drainage, leaching from abandoned mines and coal combustion. The inorganic compounds of arsenic, in which As is combined with sulfur, chlorine and oxygen, are highly toxic to most living organisms. In the subsurface environment, arsenic mostly exists in inorganic forms as oxyanions of arsenite As(III) or arsenate As(V). Under oxidizing conditions in the surface soil, the predominant form of arsenic is As(V). In humans, exposure to inorganic arsenic has both acute and chronic toxicity effects and it is a direct cause of skin, lung and bladder cancer. Arsenic is also associated with the insurgence of cardiovascular disease. However, the organic compounds of arsenic pose in general lower risk (Sinicropi *et al.*, 2010). Owing to its high toxicity, As is classified as priority pollutant. The US Environmental Protection Agency enforces a maximum concentration of 10 µg/l (http://water.epa.gov/drink/contaminants/index.cfm). The last edition of drinking water guidelines published by the World Health Organization (WHO) also established 10 µg/l as a provisional guideline value for arsenic in drinking water, because available information on health effects at lower concentration is limited (World Health Organization, 2008). As any other metal, arsenic cannot be degraded and therefore must be physically removed from contaminated water and soils during the remediation process. This is usually achieved in two steps: (i) oxidation of the highly toxic As(III) to the less-toxic and less mobile As(V), and (ii) adsorption of As(V) on inert material such as cement or pozzolana, or precipitation with ferric chloride ($FeCl_3$). Aeration alone is ineffective at oxidizing As(III) to As(V). Numerous chemicals, such as chlorine, chlorine dioxide, ozone and potassium permanganate can be used to this purpose (Castro de Esparza, 2006). However, chemical oxidation is expensive and not easy to apply in field where the abundance of reduced organic compounds (e.g. humic acids) and the presence of redox mediators may affect the oxidation rate (Haffert *et al.*, 2010). The recent discovery of As(III) oxidizing bacteria has provided an alternative to physicochemical treatments. As(III) oxidizing consortia, such as the Chéni Arsenic Oxidizing 1 (CAsO1) consortium, were found in As(III) contaminated sediments recovered from the disused Chéni gold mine site (Limousin, France), where 600,000 tonnes of a fine arsenic-containing silt were dumped (Battaglia-Brunet *et al.*, 2002). Various studies have demonstrated the effectiveness of such As(III)-oxidizing biofilms in As bioremediation. For example, Battaglia-Brunet *et al.* (2006) grew naturally occurring As(III) oxidizing consortia as biofilm on high porosity column reactors in the laboratory (e.g. filled with pozzolana). When As-contaminated mine drainage water was pumped through the biofilm column reactors, As(III) oxidation rate was double than that observed on site. However, biofilms rich in extracellular polymeric substances may provide a barrier against As diffusion and the induction of As(III) oxidative metabolism, thus decreasing the rate of As(III) oxidation (Michel *et al.*, 2007). Arsenic biological oxidation occurs also under anaerobic conditions, as demonstrated in a long-term experiment (18 months) with an upflow anaerobic sludge bed (UASB) reactor, inoculated with anaerobically digested sewage sludge. The biofilm coupled As(III) oxidation to ClO_3– reduction, with >98% efficiency of As(III) oxidation. The main strains responsible for As(III) oxidation were *Dechloromonas* and *Stenotrophomonas* as well as genera within the family *Comamonadaceae* (Sun *et al.*, 2011).

A thorough knowledge of As(III) oxidizing biofilm adhesion and activity is needed to obtain quantitative results and to study the effect of the main process parameters, such as the growth medium composition and the support for biofilm growth. However, most methods for biofilm quantification were developed for young biofilms grown on flat and smooth surfaces, which are quite different from the high surface area materials used for biofilm reactors in bioremediation. Challan Belval *et al.* (2009) carried out a systematic study of As(III)-oxidizing biofilms grown on pozzolana. They used a simple sonication method and an adapted crystal violet staining assay to quantify young and mature biofilms, respectively. The standard deviation was comparable to that obtained in the investigation of medical biofilms. Also, they assessed the importance of numerous parameters, such as the operation mode (fed-batch vs. continuous), the temperature, and the influent As(III) concentration. The authors found that fed-batch mode enables the maintenance of high As(III) oxidation rates, even at the low temperature typical of drinking water sources.

While As(III) oxidizing microorganisms allow the removal of soluble As from water, As(V) reducing bacteria can be used for soil bioremediation. In fact, As(V) is less mobile than As(III) and is strongly associated to soil particles. The contact between As(V) contaminated soil and As(V) reducing bacteria release As(III) that can be subsequently removed from water through further reoxidation (either biological or chemical) and precipitation. Soda *et al.* (2009) demonstrated As(V) desorption from forest and paddy field soils, with lactate as the electron donor and carbon source. They used *Bacillus selenatarsenatis* SF-1 under anoxic conditions and with high concentrations of As (V) in the soils (>2000 mg/kg). The authors used a adsorption-reaction predictive model, coupled to known speciation data of As, to predict the dynamic of As(III) and As(V) in the soil and water phases.

Cadmium (Cd)

Cadmium (Cd) is released in the environment from the processing of non-ferrous metals, such as zinc and aluminium, and from iron and steel production. Cd is also released through phosphate fertilizers.

Cadmium can enter the body of humans and animals by inhalation and skin contact, and may accumulate in the kidneys, potentially causing organ failure. The US Environmental Protection Agency enforces a maximum concentration of 5 µg/l (http://water.epa.gov/drink/contaminants/index.cfm). The last edition of drinking water guidelines published by the World Health Organization (WHO) established 3 µg/l as a provisional guideline value for cadmium in drinking water (World Health Organization, 2008). Owing to its low mobility, Cd remains a contaminant of concern for many years after its release into the environment and has a strong negative effect on aquatic life and on crop plants (Agency for Toxic Substances and Disease, 2008a). Cd removal from contaminated soil is typically achieved by solidification/stabilization, vitrification and soil flushing. All these techniques have high costs and their application is limited to those cases where Cd concentration is very high (Martin and Ruby, 2004). Cd bioremediation is still in its infancy. Recent studies have shown that in the presence of sufficient organic matter, the subsurface biofilms may produce high EPS concentrations. Cd may be removed through biosorption in such cases (Fernandes *et al.*, 2009; Jiajun *et al.*, 2010).

In a recent laboratory investigation, Patel *et al.* (2010) engineered *Caulobacter crescentus* JS4022/p723–6H by inserting a hexahistidine peptide to a permissive site of the host surface layer (S-layer) protein RsaA. Planktonic cultures of this engineered strain are highly effective in Cd biosorption. The researchers observed showed a very high Cd bioimmobilization rate by *C. crescentus* biofilms grown in Cd-contaminated water.

Chromium (Cr)

Chromium (Cr) has three main valence forms, Cr(0), Cr(III), and Cr(VI). Cr(III) occurs naturally in the environment (e.g. in chromite), while Cr(0) and Cr(VI) are industrial by-products. Chromium is an essential alloy in stainless steel and is used in furnaces, bricks, dyes and pigments, chrome plating, chemical catalysts, leather tanning and wood preserving. Chromium is discharged in the environment from mining, chemical manufacturing, fossil fuels, combustion, waste incineration and steel making.

Cr(III) is an essential element in animal nutrition, functioning mainly in glucose metabolism and also possibly fat metabolism (Di Bona *et al.*, 2011). It is considered to be non-essential for plant growth, although it is essential for some microbes, possibly as a cofactor for specific enzyme systems (Stearns, 2000). Cr(VI) compounds have no known biological role. Owing to its strong oxidizing properties, high mobility and bioavailability, Cr(VI) is included in the priority pollutants list, while the less soluble and poorly bioavailable Cr(III) pose much lower health and environmental risks (Hamilton and Wetterhahn, 1988; Agency for Toxic Substances and Disease, 2008b). In humans and animals, short-term exposure to Cr(VI) causes skin irritation or allergic reaction while long term exposure, especially by inhalation, has been linked to the development of lung tumours. For this reason Cr (VI) and its compound are now listed as carcinogenic elements (Rowbotham *et al.*, 2000; Agency for Toxic Substances and Disease, 2008b). Cr(VI) is also known to alter plant germination and growth, photosynthetic process, water relations and mineral nutrition (Rowbotham *et al.*, 2000; Shanker *et al.*, 2005).

Cr(VI) bioremediation occurs through the biosorption of Cr(VI) in the biofilm EPS or the bioreduction of the toxic form Cr(VI) to the less toxic form Cr(III). In the first case, biofilms need to be periodically removed when saturated in Cr(VI). This strategy is therefore less fit for *in situ* bioremediation but more efficient for *ex situ* processes (Gadd, 2009). Bioreduction to Cr(III) provides a long-term solution to Cr(VI) contamination. Although there is no known strain capable of growing with Cr(VI) as sole electron acceptor, numerous facultative and strict anaerobic microorganisms can reduce Cr(VI) to Cr(III) via their co-metabolism. Dissimilatory metal reducing bacteria (DMRB) predominate in natural environments as biofilms and have shown marked Cr(VI) bioreduction capability (Gadd, 2010). For example, *Cellulomonas*, a biofilm-forming, fermentative subsurface isolate can reduce Cr(VI) to Cr(III) directly or through Fe(III) reduction (Viamajala *et al.*, 2008). A well-known methanogenic strain, *Methylococcus capsulatus* was found to have Cr(VI) reductase activity. Cr(III) was accumulated both on cell surfaces and in the growth medium (Al Hasin *et al.*, 2010). In another laboratory study, Chai *et al.* (2009) demonstrated the Cr(VI) bioreduction capability of a native soil microbial consortium recovered from a contaminated subsurface in close proximity to a steel alloy factory. A novel (facultative/anaerobic) microorganism tolerant to high Cr(VI) concentrations, *Pannonibacter phragmitetus* was isolated from the soil and characterized (Borsodi *et al.*, 2003). Other experiments with soil microcosms contaminated with Cr(VI) showed the presence of Cr(VI) tolerant *Bacillus* and *Enterobacteria* (Molokwane and Nkhalambayausi-Chirwa, 2009).

Cr(III) is usually immobilized in extracellular aggregates and in the EPS. This may be also due to the anionic nature of many EPS compounds, which assist in the metal immobilization process and determine the morphology of metal precipitates in the biofilms (Gadd, 2009). There is evidence that Cr(VI) reduction occurs not only in close proximity to the cells, but also at large distances, indicating indirect biochemical reduction. It is not yet clear if these effects are related to biofilm growth mode. Energy calculations suggest that biofilms should have higher Cr(VI) reduction power, since they can immobilize precious Cr(VI) reductases, rather than dispersing them into the environment. The chemical form of Cr(III) in biofilms is little known.

Cr(VI) reduction in the field has been also attempted. In order to avoid the periodic injection of electron donor in the subsurface, Faybishenko *et al.* (2008) used a slow hydrogen releasing compound to provide constant electron donor concentration in a Cr(VI) contaminated groundwater. Hydrogen served as the electron donor and bacteria removed Cr(VI) below the background concentration.

Copper (Cu)

Cu is found in numerous minerals including cuprite, malachite, azurite, chalcopyrite and bornite. The main anthropogenic sources of copper are coal-fired power stations, iron and steel industries, non-ferrous metal industries and waste incinerators. Cu release in the subsurface also occurs from the large-scale application of agricultural chemicals and from the application of sewage sludge to land (E-PRTR, 2008).

When copper is released into soil, it forms stable complexes with organic material and other soil components, such as clay and sand. Because of these strong chemical associations, Cu tends to remain in the top soil layers. When Cu is released into water, it can occur either as free Cu or, more probably, as complexation products with other compounds present in the water (e.g. humic acids). Even though Cu binds strongly to suspended particles and sediments, there is evidence to suggest that some water-soluble copper compounds do enter groundwater (Agency for Toxic Substances and Disease, 2008c). Cu may enter the body of humans and animals by inhalation, drinking water, food, and by skin contact with soil, water and other copper-containing materials (Agency for Toxic Substances and Disease, 2008c). Long-term exposure to Cu dust causes irritation and inflammation of tissues. High Cu intakes can cause serious liver and kidney damage. Currently, the US-EPA does not list Cu as a human carcinogen because of lack of convincing evidence (Agency for Toxic Substances and Disease, 2008c).

Cu removal is achieved through bioreduction to Cu(0) or through biosorption on suitable biomass (Gadd, 2009). Typical examples include permeable reactive barriers with zero-valent iron, where Fe(0) reduces copper to the metallic state. Depending on the biobarrier design, the Fe(III) resulting from this process may be reduced by dissimilatory metal reducing bacteria or the biobarrier may be removed at scheduled times to regenerate the iron. Cu bioreduction may be also achieved through indigenous microbial consortia. Jackson *et al.* (2009) assessed the Cu removal property of the autochthon microbial community from the water of a South African river, heavily contaminated with metals, including Cu. They set up a biobarrier with a material termed Bioball™. Bioballs™ are composed of acrylonitrile, butadiene and styrene. These polymers are a well-known support for biofilm growth in trickling filters for aquaculture. Each Bioball™ has a surface area of 20 cm^2 to support biofilm growth. The high density biofilm grown on the biobarrier reduced more than 90% of the influent Cu into Cu(0). Analysis of the composition of the Cu-reducing microbial consortia revealed that more than 50 bacterial strains were involved in the bioreduction process. As expected, facultative metal-tolerant strains, such as *Pseudomonas* sp. and *Sphingomonas* sp., were found in the mature biofilms. However, the microbial ecology of the biofilms changed with biofilm age. Since Cu is sequestered through a variety of strategies, such as chelation, precipitation, and sorption, it is probable that naturally occurring Cu-tolerant consortia are far more complex than the corresponding laboratory biofilms.

Zinc (Zn)

Zinc occurs in numerous minerals, sediments, and freshwater. Man-made Zn releases are primarily due to iron and steel production, non-ferrous metal production and road transport, as it is present in tyre rubber.

Zinc enters the chain food especially through fish. In fact, fish retain Zn from the filtration of water needed for their respiration and from their food. High Zn intake has adverse health effects. However, the toxic concentration of Zn is much higher than the recommended daily dose. Toxic effects range from stomach cramps to anaemia and damage to the pancreas (Agency for Toxic Substances and Disease, 2008d).

Zinc contamination affects the trace metal transport system in plants and is a major concern for agricultural production. *Ex situ* solidification/stabilization (S/S) is the most common approach used to immobilize zinc from soils and wastes. The process involves the addition of binding agents to the soil matrix. Consequently, the metal contaminant is immobilized within the soil matrix through a combination of chemical reactions, encapsulation, and a reduced permeability/

surface area. Cement-based and pozzolana materials (e.g. fly ash) are commonly used as S/S binders for metal remediation with the predominant immobilization mechanism for metals being precipitation of hydroxides, carbonates, and silicates (Martin and Ruby, 2004). This process dramatically lowers the Zn bioavailability and prevents metal leaching. Similar to other metals, Zn bioremediation frequently occurs through precipitation within biofilms. For example, *Desulfotomaculum nigrificans* biofilms removed Zn from contaminated water (Radhika *et al.*, 2006). Ancion *et al.* (2010) investigated the absorption rates of zinc in a freshwater biofilm and assessed whether biofilm bacterial populations are affected by exposure to environmentally relevant concentrations of these metals in a flow chamber microcosms. Metals were rapidly accumulated by the biofilm and then retained for at least 14 days after transfer to uncontaminated water. These results strongly support the possibility of using bioaccumulation as a remediation procedure for these metals.

Actinides

The group of actinides include elements such as uranium (U), technetium (Tc), plutonium (Pu) and neptunium (Np). Actinides occur in particular geological formations around the world and in gas and crude oil reservoirs. Their concentration is usually very low, but it may arise as a consequence of oil processing and natural gas burning. Actinides have no known biological function; therefore microorganisms have not developed specific transporter systems to reduce their toxicity. However, isolates from actinide-contaminated soils show high resistance to actinide toxicity. Many microorganisms can immobilize actinides on the outer cell membrane (Gadd, 2009), reduce them in the extracellular medium or in the EPS matrix via microbially produced redox shuttles (Borch *et al.*, 2010), or change their oxidation state through the oxido-reductases and outer membrane cytochromes used in extracellular respiration (Lovley *et al.*, 1991). The latter phenomenon occur among DMRB and sulfate-reducing bacteria (SRB), which are the best-known microorganisms capable of actinide bioremediation, both in the laboratory set-up and in field tests. In the bioremediation process, the water-soluble, toxic oxidized U (VI), Tc (VII), Pu (VI) and Np (V) are reduced by viable microorganisms to the insoluble, less toxic U (IV), Tc (IV), Pu (IV) and Np (IV). Because of their very low solubility, the reduced actinides are removed from the groundwater and accumulated on the outer cell membranes or in close vicinity to the bacteria. In DMRB and SRB biofilms, EPS provide an excellent defence for microorganisms against the chemical and radioactive toxicity of actinides. Furthermore, biofilm form a barrier against the reoxidation of actinides from oxidized compounds in subsurface environments, such as Fe(III), nitrates and traces of oxygen.

Uranium

^{238}U is the most common radionuclide in soils, sediments, and groundwater. Following its release in the environment, U(VI) leaches into the groundwater and drinking water. The groundwater flow transports the soluble U(VI) beyond their original boundaries, so the U(VI) contamination spreads far from the pollution source. ^{238}U has been discharged into the environment for more than 60 years as a result of nuclear energy and weapons production and incorrect mine management. The United States Environmental Protection Agency (US-EPA) remediation programme lists 21 sites in the US in which uranium contaminates groundwater and soil (Cao *et al.*, 2009). *In situ* 238U bioremediation is the most successful method for subsurface environment decontamination, leading to the bioreduction of U(VI) to U(IV) and its immobilization. Early studies showed that sulfate-reducing bacteria (Lovley *et al.*, 1993) and dissimilatory Fe(III) reducing bacteria (Lovley *et al.*, 1991) were involved in uranium cycles in anaerobic marine sediments. In the following years, it was observed how the environmental biofilms immobilized uranium and other minerals through a combination of biosorption and energy metabolism (Brown *et al.*, 1994). In the years 1996–2005, the US Department of Energy funded the Natural and Accelerated Bioremediation Research (NABIR) programme, with the goal of developing novel uranium bioremediation process for field applications. The interactions between uranium and biofilms were investigated in depth, using advanced methods such as X-ray spectroscopy, electron microscopy, microelectrodes, and nuclear magnetic resonance.

Both non-biological and biological reactions contribute to the microbially stimulated reduction of uranium. Non-biologically, uranium can be reduced by microbially produced sulfides in sulfate-reducing bacteria (Marsili *et al.*, 2007). Microbially produced redox mediators or electron shuttles available in the soil enhance the bioremediation process. For example, the radiation-tolerant microorganism *Deinococcus radiodurans* use the humic acid analogue anthraquinone-2,6-disulfonate (AQDS) to increase uranium reduction rates (Fredrickson *et al.*, 2000). Biologically, uranium U(VI) is reduced by electrons derived directly from organics. Reduced uranium deposits are usually associated with microbial membranes but found also in the intracellular space (Suzuki *et al.*, 2005). This enzymatic reduction of uranium can be carried out by numerous microorganisms, through their outer membrane cytochromes or oxidoreductases. For an extensive review of uranium reduction by DMRB we refer the reader to Mohapatra, 2010.

The accumulation of U(IV) in SRB biofilms of *Desulfovibrio desulfuricans* was studied in laboratory experiments. Marsili *et al.* (2005) showed that U(IV) accumulates within the EPS up to several mg/g in concentration. Similar biofilm experiments were repeated on fixed bed reactors, with mineral matrices that simulate subsurface conditions. The concentration of carbonates and other ions affect the bioavailability of U(VI) to DMRB and reduce drastically the amount of uranium removed from the contaminated water as U(IV). The stability of reduced uranium is a critical point for field applications of bioremediation. Several studies demonstrated that crystallized U(IV) is more stable and less amenable to reoxidation by Fe(III) or nitrates (Suzuki *et al.*, 2005; Marsili *et al.*, 2007).

Owing to their intrinsic heterogeneity, biofilms pose challenges in predicting uranium mobility. Although recent studies have improved our understanding of uranium immobilization processes in biofilms, further research about the biofilm-U(VI)-minerals system is needed before bioremediation can become a safe and reliable option for long-term treatment of U(VI) contaminated sites.

Even though hydrated U(VI) ions in pure water are amenable for reduction by microbially generated H_2S, U(VI) in natural waters forms complexes with carbonates, where the non-biological U(VI) reduction becomes very difficult. On the other hand, the efficiency of the enzymatic reduction of uranium does not seem to decrease in the presence of uranium when complexed with other minerals, which, hypothetically, makes possible uranium removal from contaminated waters using enzymatic reduction. However, it is not clear whether the enzymatic reduction of uranium is efficient enough to remove considerable concentrations of uranium in the contaminated groundwaters and whether the possible toxic effects of uranium to microorganisms make the process sustainable for long periods of time.

Despite the large number of laboratory studies, only a few pilot-scale field test systems have been reported. Field studies are needed to assess the feasibility of uranium remediation under natural conditions and the interaction of oxidized and reduced uranium with other minerals in the system. N'Guessan *et al.* (2010) showed that the function of a microbial community cannot be inferred based on laboratory experiments alone. The effect of Fe(III) addition to the microbial reduction of uranium in sediments were studied, but there is no agreement about the reoxidation of reduced uranium in contaminated sediments under iron-reducing and sulfate-reducing conditions (Sani *et al.*, 2004). A similar study assessed the role of nitrate in U(IV) reoxidation and shows that careful monitoring and continuous biostimulation is needed to avoid remobilization of reduced uranium (Wu *et al.*, 2010). Subsurface heterogeneities and biofilm formation may alter bioremediation monitoring. Two-dimensional modelling is needed to account for biofilm abundance along the carbon source down-gradient in the vicinity of the injection wells (Li *et al.*, 2010). Fe(0) loaded in biobarriers serves as electron donor to DMRB such as *Geobacter* spp. and increases U(VI) bioremediation rates. A pilot-scale field test system was set up on a previous military installation for uranium enrichment at Oak Ridge National Laboratory, TN, USA. They added ethanol in the subsurface environment for two years in order to biostimulate uranium reduction in the native microbial consortia. During that period, the U(VI) concentration dropped below thresholds for drinking water (30 µg/l). The structure of the stimulated microbial community was determined using a functional gene array and compared with a community from the same environment that did not receive the biostimulation. Results showed an increase in activity of the community and a marked enrichment in well-known anaerobes and DMRB, such as *Desulfovibrio, Geobacter, Anaeromyxobacter* and *Shewanella* (Xu *et al.*, 2010). Isotope ratio determination provided an accurate assessment of bioremediation progress in this field study (Bopp *et al.*, 2010).

Other actinides

Plutonium

Although tiny amounts of plutonium naturally occur in the environment, most of the plutonium in the environment results from uranium enrichment waste, spent fissile fuel and fallout from previous nuclear energy and weapon experiments. Its chemical toxicity is low, but the radioactive isotopes are considered a dangerous pollutant. New (2006) reported the toxic effect of uranium and plutonium on aerobic *Pseudomonas putida* biofilms. Biofilm growth was inhibited at sub-mM concentrations of U(VI) and Pu(IV). Experiments at sublethal concentrations of U(VI) and Pu(IV) showed that EPS production was not significantly altered. Pu(IV) was found associated with cell membranes, while U(VI) was probably complexed by siderophores or other chelating agents, and therefore found also in the EPS (New, 2006). The effect on Fe(III)-reducing *G. sulfurreducens* and *S. oneidensis* on Pu(IV) reduction was studied by Renshaw *et al.* (2009). Their results show that Pu(IV) was initially adsorbed on the outer cell membrane, followed by slow reduction to Pu(III).

Neptunium

Early bioremediation methods were based on chemisorption and biosorption in *Citrobacter* cells immobilized on a high surface area support, such as polyacrylamide gels and Raschig rings. *Citrobacter* are known to produce phosphate ligands that bioprecipitate actinides (Macaskie *et al.*, 1996; Basnakova *et al.*, 1998). Planktonic suspensions of the well-known U(VI) reducing bacteria *G. metallireducens* and *S. oneidensis* show a low resistance to Np(V) toxicity, less than 2.0 mM. *S. oneidensis* cell suspension reduced Np(V) to Np(IV), which then precipitated. The presence of Fe(III) as the main electron acceptor (reduced to Fe (II)) was needed to sustain bacterial growth. However, *S. oneidensis* was not able to grow with Np(V) as the only electron acceptor. *G. metallireducens* cell suspensions could not reduce Np(V) to Np(IV). In the presence of the Np complexing agent, citrate, Np(V) reduction rates increased in both strains, but they were not able to grow with Np(V) as the sole electron acceptor. The difference in reduction rates may be explained by the fact that *S. oneidensis* produce the soluble electron mediator riboflavin (Marsili *et al.*, 2008) that enable extracellular reduction of Np(V), thereby reducing the toxic effect. *G. metallireducens* can transfer electrons only through the outer membrane cytochrome complex, so it is more sensitive to Np(V) toxicity (Icopini *et al.*, 2007). A microbiological characterization of actinides waste from a Siberian Chemical Complex (Russia) was recently carried out. The results show that metal-reducing and sulfate-reducing activities increased in contaminated groundwater. A *Shewanella* strain capable of Np(V) reduction was isolated (Nazina *et al.*, 2010). To our best knowledge, no biofilm study has been performed on the bioreduction of Np(V).

Bio-electroremediation

This innovative process rests on the unique physiology of DMRB. These microorganisms conserve energy through extracellular electron transfer (EET) to solid conductive surfaces. In the natural environment, DMRB reduce metals such as Fe(III) and Mn(IV) to Fe(II) and Mn(0), respectively. They are crucial to metal biogeochemistry and are used also in the bioremediation of metals, as described in previous section of this chapter. In the presence of an electrode polarized at positive potential and at very low concentrations of soluble electron acceptors, DMRB may transfer electrons from an initial electron donor, which gets reduced, to the electrode (anodic process or electrode reduction). In the inverse process, in the presence of an electrode polarized at negative potential and low concentration of electron donors, DMRB may receive electrons from the electrode (cathodic process or electrode oxidation) and transfer them to a final electron acceptor, thus reducing it.

This EET process can be used to remediate organic pollutants. For example, nitrate in absence of organic compounds was reduced by mixed microbial consortia that used the electrode as an electron donor (Park *et al.*, 2005). A subsequent study from the same authors revealed that the majority of the electroactive community consisted of proteobacteria (Park *et al.*, 2006). This is not a surprise, since numerous electroactive strains have been reported among alpha-, beta-, gamma-, and delta-proteobacteria classes (Logan, 2009). Uranium bioreduction at the laboratory scale in a bioelectrochemical process was described by Gregory and Lovley (2005). *Geobacter sulfurreducens* biofilms were grown on a solid-state electrode polarized at −300 mV versus a standard hydrogen electrode (SHE) as the sole electron donor. Almost 90% of initial U(VI) was reduced to the less toxic species U(III) in one step. In a more recent study, the cathodic process was used to degrade trichloroethene (TCE) to ethene in two steps, with cis-dichloroethene (cis-DCE) as an intermediate. For the first step (TCE to cis-DCE), the authors used a culture highly enriched in *Desulfitobacterium* spp., while for the second step (cis-DCE to ethene) they used a culture highly enriched in *Dehalococcoides* spp. (Aulenta *et al.*, 2008). Biologically mediated de-halogenation of chlorophenol to phenol was achieved in a laboratory set-up with an *Anaeromyxobacter dehalogenans* biofilm, where the electrode was the sole electron donor (Strycharz *et al.*, 2010). The inverse process where the pollutant works as an electron donor and the polarized electrode as an electron acceptor was also studied. For example, providing graphite electrodes as an electron acceptor in hydrocarbon-contaminated sediments significantly stimulated the removal of added toluene and benzene (Farhadian *et al.*, 2008b; Zhang *et al.*, 2010).

Because these bioelectrochemical processes for bioremediation are only at an early stage of study, they have several disadvantages, such as the high energy cost needed for electrode polarization and the relative low columbic efficiency (i.e. the percentage of charge that goes directly into the reduction of the organic pollutant). However, bioelectrochemical remediation is attractive for its easy implementation in contaminated environments and the full control of the bioremediation rates achievable through the adjustment of electrode potential, both in anodic and cathodic mode. Provided the correct microbial community is established and maintained throughout the process, bio-electroremediation show great promise for fast clean-up of localized pollution.

Techniques to improve bioremediation processes

Genetic methods

Current bioremediation processes use native microorganisms, environmental isolates from contaminated soil and sediments, and strains selected through conventional subculturing procedures. Although existing microorganisms are effective for bioremediation, numerous laboratory studies demonstrated the usefulness of genetic engineering methods for the creation of strains with (a) higher toxicity resistances, (b) aggregated degradation pathways, and (c) faster bioremediation rates. These genetic strain improvement methods mimic natural processes of gene flow and DNA evolution that occur in subsurface biofilms (Massoudieh *et al.*, 2007).

Increased resistance to toxicity

The low resistance of microorganisms used in bioremediation to the toxicity of soil contaminants limits the field application of bioremediation. Heavy metals, novel chemicals, solvents and pesticides are of particular concern since microorganisms may not have developed metabolic pathways to degrade these pollutants. Several works report methods and strategies to improve microbial resistance. The most common strategies are based on genetic engineering. For example, *Stenotrophomonas* sp. strain YC-1, a native soil bacterium that produces methyl parathion hydrolase was genetically engineered by incorporating

Pseudomonas syringae genes possessing a broader substrate range for organophosphates (OPs), allowing the strain to degrade a mixture of six OP pesticides (0.2 mM each) completely within 5 h. The broader substrate specificity in combination with the rapid degradation rate made the engineered strain a promising candidate for *in situ* remediation of OP-contaminated sites (Yang *et al.*, 2010). In another study, an engineered *Deinococcus radiodurans* expressing toluene dioxygenase was able to oxidize toluene, chlorobenzene, 3,4-dichloro-1-butene, and indole, and was also tolerant to the solvent effects of toluene and trichloroethylene at levels exceeding those of many radioactive waste sites in the presence of radionuclide irradiation (Lange *et al.*, 1998; Zhang *et al.*, 2010). Also, eight DNA-shuffled toluene ortho-monooxygenase genes expressed by engineered *Escherichia coli* showed an improved degradation capability against chlorinated compounds (Rui *et al.*, 2004). Toxicity resistance is complex, and may be the result of several site-specific genetic modifications. Genome shuffling generates a large amount of mutants, among which the ones that present the desired characteristics can be isolated with a well-defined screening method.

Aggregated degradation pathways

Existing degradation pathways from different microorganisms can be combined in a single strain through genetic engineering methods. The insertion of foreign degradation cluster(s) in the chromosome or in plasmids result in stable strains, even in field conditions (Lajoie *et al.*, 1993). The catabolic genes can be stably maintained also in groundwater aquifer material without specific selective pressure for the introduced genotypes (Jain *et al.*, 1987). The resulting strain can degrade completely a target pollutant, thereby simplifying the bioremediation treatment in the laboratory and in the field. For example, complete mineralization of polychlorinated biphenyl (PCB) was achieved through the stable incorporation of *Burkholderia xenovorans* LB400 genes into the chlorobenzoate-degrading bacterium *Cupriavidus necator* JMP134-X3 through the mini-TN5 transposon system and biparental mating (Matias Saavedra *et al.*, 2010). The genetically modified strain was added to non-sterile, PCB contaminated soil and a strong increase in PCB mineralization was observed. Genes encoding the 2,4-dinitrotoluene (DNT) degradation pathway enzymes were taken from another *Burkholderia* strain and incorporated into the genome of the psychrotolerant rhizobacterium *Pseudomonas fluorescens*. The engineered strain was stable and capable of growing on 2,4-DNT as sole nitrogen source, even at low temperatures (Monti *et al.*, 2005). In another study, the genes encoding an alcohol dehydrogenase and an esterase from *Pseudomonas fluorescens* were cloned into a vector and functionally expressed in *E. coli*, demonstrating that three recombinant enzymes can be applied in alkane biodegradation (Kirschner *et al.*, 2007).

Accelerated bioremediation rate

A quite different approach was adopted to increase metal reduction rates in the dissimilatory metal reducing bacterium (DMRB) *Geobacter sulfurreducens*. Following *in silico* model prediction, the authors engineered the strain in order to increase ATP consumption. This was achieved through the creation of ATP-consuming futile cycles and through the reduction of internal ATP reserves. Genome-wide analysis of gene transcript levels showed that the modified strain produced a lower amount of biomass per equivalent of electron transfer and had a higher rate of respiration than the control strain. Respiration rates in DMRB are strictly correlated with metal reduction rates, therefore the engineered strain may increase metal bioimmobilization rates (Izallalen *et al.*, 2008).

As with any other application of genetically engineered microorganism, risks and ethical concern may arise from the open-field application and should be discussed within the current regulatory regime (Snow *et al.*, 2005). Restrictions on the use of genetically modified microorganisms can be avoided by other using genetic modification methods instead, such as genome shuffling. In fact, an organism treated by genome shuffling may be not considered a genetically modified organism (Ahmed, 2003).

Genome shuffling is a powerful tool to improve bacterial phenotypes producing highly recombined and evolved progeny in a short time. In brief, a large number of mutants, generated through random mutagenesis are screened for improved phenotype characteristics (e.g. higher degradation rate of a pollutant) and recombined by multiple rounds of protoplast fusion. Genome shuffling saves work and time with respect to conventional mutagenesis methods. However, few applications of genome shuffling to bioremediation have been published. For example, Dai and Copley (2004) used genome shuffling to improve the degradation rate of PCP by *Sphingobium chlorophenolicum*. After three rounds of shuffling, several faster and more PCP tolerant strains were obtained.

The authors of this chapter are currently working on the improvement of the Cr(VI) reduction capability and tolerance of the DMRB *Geobacter metallireducens*, which conserve energy through extracellular electron transfer to Fe(III) and heavy metals. *G. metallireducens* use the same respiratory chain for Fe(III) and Cr(VI) bioreduction. In the case of Cr(VI), the reduction to Cr(III) serves as protection against the toxicity of Cr(VI). Genome shuffling will enhance the Cr(VI) tolerance and Cr(VI) bioremediation rate of *G. metallireducens.*

Engineering improvement

The limiting factors in bioremediation process design are the bioavailability of the pollutant to degrading microorganisms and its solubility. This is particularly important for polyaromatic hydrocarbons (PAH) and PCBs. The addition of chemical surfactants improves PAH bioavailability (Rodriguez and Bishop, 2008) and increases degradation rates. The addition of Tween 90, a non-ionic surfactant, increases bioremediation rates of chlorinated pesticide contaminated soil in methanogenic granular sludge (Baczynski and Pleissner, 2010). In another example, coconut fatty alcohol sulfate (CFAS) was added to soil in order to increase the bioremediation rate of chlorpyrifos, a widely spread organophosphate insecticide. CFAS addition increased chlorpyrifos bioavailability and resulted in higher biomass concentration chlorpyrifos removal rates with respect to unamended soil (Bobiles *et al.*, 2010). However, the addition of synthetic surfactants or surfactants derived from agricultural products increases costs, particularly for inaccessible soils and groundwater, and may raise environmental concerns. A possible solution to this problem is the use of environmentally friendly, microbially produced surfactants, which can be generated *in situ* during the bioremediation treatment, thereby reducing costs and simplifying the bioremediation treatment. Many microorganisms produce biosurfactants when they grow on water-immiscible substrates. The most common biosurfactants are glycolipids, where the carbohydrates are attached to a long-chain aliphatic acid. However, other more complex surfactants, such as lipopeptides, lipoproteins, and heteropolysaccharides, were reported (Desai and Banat, 1997). Also, bioemulsifiers, such as glycolipids and polymeric surfactants (e.g. lipoprotein and lipopeptides) increase bioremediation rates of hydrocarbon polluted soils, because they increase the displacement of hydrocarbons from soil particles (Calvo *et al.*, 2009; Al Hasin *et al.*, 2010). The most studied biosurfactants are the rhamnolipids, produced by *Pseudomonas aeruginosa*. They are probably the oldest biosurfactants identified (Zhang and Miller, 1992) and one of the first that entered industrial production (Mukherjee *et al.*, 2009). Many other microbially produced surfactants have been identified and it is expected that they will contribute to faster oil product biodegradation (Okoh and Trejo-Hernandez, 2006). While crude oil is a common substrate for many aerobic and anaerobic microorganisms, it has became difficult to degrade it in a cost-effective manner when spread over a large surface, as in the case of the recent (April, 2010) oil spill in the offshore oil rig off the Louisiana coast (USA). Microbially produced biosurfactants will enable faster and more sustainable bioremediation treatment after these rare but catastrophic events occur.

Conclusions

After 20 years of laboratory research, bioremediation is now coming out of its infancy, thus giving rise to field applications. Large oil spills and highly contaminated land around the world have provided an excellent 'playground' for bioremediation scientists. At the same time, thanks to consistent advances in the understanding of microbial consortia, virtually every pollutant can be completely degraded today into less harmful compounds, at least in the laboratory. Biofilms are the main agents in bioremediation, both at pilot scale studies and in the field. Not only are biofilms the predominant mode of growth in the environment, they are favoured over planktonic suspensions because of their higher toxicity tolerance, and resistance to mechanical and environmental stress. Bioremediation studies designed to test biofilm effectiveness under conditions similar to those encountered in natural environments still remain few. Many issues remain unclear, such as (1) the correlation between the biofilm microstructure and the biodegradation process; (2) the long-term behaviour of biofilms exposed to fluctuations in pollutant concentration; and (3) the details of the correlation between soil composition and morphology and biofilm behaviour.

Solutions to these issues will provide a predictive and quantitative model for biofilm-based bioremediation processes, so to improve large-scale application of this environmentally sustainable technology.

References

Abbassi, B.E., and Shquirat, W.D. (2008). Kinetics of indigenous isolated bacteria used for *ex situ* bioremediation of petroleum contaminated soil. Water Air Soil Pollut. *192*, 221–226.

Agency for Toxic Substances and Disease. (2008a). Public Health Statement for Cadmium. *CAS# 1306-19-0.*

Agency for Toxic Substances and Disease. (2008b). Public Health Statement for Chromium. *CAS# 95-57-8.*

Agency for Toxic Substances and Disease. (2008c). Public Health Statement for Copper. *CAS# 7440-50-8.*

Agency for Toxic Substances and Disease. (2008d). Public Health Statement for Zinc. *CAS# 7440-66-6.*

Ahmed, F.E. (2003). Genetically modified probiotics in foods. Trends Biotechnol. *21*, 491–497.

Akinson, B. (1981). Immobilized biomass-a basis for process development in wastewater treatment. In Biological Fluidized Bed Treatment of Water and Wastewater, Cooper P.E., and Akinson, B., eds. (New York, Ellis Horwood), pp.22–34.

Al Hasin, A., Gurman, S.J., Murphy, L.M., Perry, A., Smith, T.J., and Gardiner, P.H.E. (2010). Remediation of chromium(VI) by a methane-oxidizing bacterium. Environ. Sci. Technol. *44*, 400–405.

Ancion, P., Lear, G., and Lewis, G.D. (2010). Three common metal contaminants of urban runoff (Zn, Cu & Pb) accumulate in freshwater biofilm and modify embedded bacterial communities. Environ. Pollut. *158*, 2738–2745.

Anderson, G.G., and O'Toole, G.A. (2008). Innate and induced resistance mechanisms of bacterial biofilms. In Bacterial Biofilms, Romeo, T., ed. (Springer-Verlag, Berlin, Germany), pp. 85–105.

Antizar-Ladislao, B., and Galil, N.I. (2010). Biofilm and colloidal biomass dynamics in a shallow sandy contaminated aquifer under *in situ* remediation conditions. Int. Biodeterior. Biodegrad. *64*, 331–338.

Aulenta, F., Fuoco, M., Canosa, A., Papini, M.P., and Majone, M. (2008). Use of poly beta-hydroxy-butyrate as a slow-release electron donor for the microbial reductive dechlorination of TCE. Water Sci. Technol. *57*, 921–925.

Baczynski, T.P., and Pleissner, D. (2010). Bioremediation of chlorinated pesticide-contaminated soil using anaerobic sludges and surfactant addition. J. Environ. Sci. Heal. B. *45*, 82–88.

Basnakova, G., Finlay, J.A., and Macaskie, L.E. (1998). Nickel accumulation by immobilized biofilm of Cirobacter sp. containing cell-bound polycrystalline hydrogen uranyl phosphate. Biotechnol. Lett. *20*, 949-952.

Battaglia-Brunet, F., Dictor, M.C., Garrido, F., Crouzet, C., Morin, D., Dekeyser, K., Clarens, M., and Baranger, P. (2002). An arsenic(III)-oxidizing bacterial population: selection, characterization, and performance in reactors. J. Appl. Microbiol. 93, 656–667.

Battaglia-Brunet, F., Itard, Y., Garrido, F., Delorme, F., Crouzet, C., Greffie, C., and Joulian, C. (2006). A simple biogeochemical process removing arsenic from a mine drainage water. Geomicrobiol. J. *23*, 201–211.

Bayles, K.W. (2007). The biological role of death and lysis in biofilm development. Nat. Rev. Microbiol. *5*, 721–726.

Beyenal, H., and Lewandowski, Z. (2002). Internal and external mass transfer in biofilms grown at various flow velocities. Biotechnol. Prog. *18*, 55–61.

Bento, F.M., Camargo, F.A.O., Okeke, B.C., and Frankenberger, W.T. (2005). Comparative bioremediation of soils contaminated with diesel oil by natural attenuation, biostimulation and bioaugmentation. Bioresour. Technol. *96*, 1049–1055.

Bobiles, S.C., Librojo-Basilio, N.T., and Varca, L.M. (2010). Surfactant-aided bioremediation of chlorpyrifos in the soil. Asia Life Sci. *19*, 207–214.

Bopp, C.J., Lundstrom, C.C., Johnson, T.M., Sanford, R.A., Long, P.E., and Williams, K.H. (2010). Uranium 238U/235U isotope ratios as indicators of reduction: results from an *in situ* biostimulation experiment at Rifle, Colorado, U.S.A. Environ. Sci. Technol. *44*, 5927–5933.

Borch, T., Kretzschmar, R., Kappler, A., Van Cappellen, P., Ginder-Vogel, M., Voegelin, A., and Campbell, K. (2010). Biogeochemical redox processes and their impact on contaminant dynamics. Environ. Sci. Technol. *44*, 15–23.

Borsodi, A.K., Micsinai, A., Kovacs, G., Toth, E., Schumann, P., Kovacs, A.L., Boddi, B., and Marialigeti, K. (2003). *Pannonibacter phragmitetus* gen. nov., sp. nov., a novel alkalitolerant bacterium isolated from decomposing reed rhizomes in a Hungarian soda lake. Int. J. Syst. Evol. Microbiol. 53, 555–561.

Bragg, J.R., Prince, R.C., Harner, E.J., and Atlas, R.M. (1994). Effectiveness of bioremediation for the Exxon Valdez oil spill. Nature *368*, 413–418.

Brockmann, C., and Stelzer, K. (2009). Monitoring of water quality in the coastal zone using optical remote sensing. Oceans 2009. Europe *1,2*, 1160–1164.

Brown, D.A., Choudari Kamineni, D. Sawicki, J.A., and Beveridge, T.J. (1994). Minerals associated with biofilms occuring on exposed rock in a granitic underground research laboratory. Appl. Environ. Microbiol. *60*, 3182–3191.

Buchholz, F., Wolf, A., Lerchner, J., Mertens, F., Harms, H., and Maskow, T. (2010). Chip calorimetry for fast and reliable evaluation of bactericidal and bacteriostatic treatments of biofilms. Antimicrob. Agents. Ch. *54*, 312–319.

Cai, H., Eramo, A.G., Evans, P.J., Fricke, R., and Brennan, R.A. (2010). *In situ* bioremediation of perchlorate in vadose zone soil using gaseous electron donors: Microcosm treatability study. Water Environ. Res. *82*, 409–417.

Calvo, C., Manzanera, M., Silva-Castro, G.A., Uad, I., and González-López, J. (2009). Application of bioemulsifiers in soil oil bioremediation processes. Future prospects. Sci. Total Environ. *407*, 3634–3640.

Cao, B., Nagarajan, K., and Loh, K. (2009). Biodegradation of aromatic compounds: current status and opportunities for biomolecular approaches. Appl. Microbiol. Biotechnol. *85*, 207–228.

Cardenas, E., Wu, W., Leigh, M.B., Carley, J., Carroll, S., Gentry, T., Luo, J., Watson, D., Gu, B., Ginder-Vogel, M., *et al.* (2008). Microbial communities in contaminated sediments, associated with bioremediation of uranium to submicromolar levels. Appl. Environ. Microbiol. *74*, 3718–3729.

Castro de Esparza, M.L. (2006). Removal of arsenic from drinking water and soil bioremediation. paper presented at International Congress on Natural Arsenic in Groundwaters of Latin America, Mexico City, 20–24 June 2006.

Chai, L., Huang, S., Yang, Z., Peng, B., Huang, Y., and Chen, Y. (2009). Cr (VI) remediation by indigenous bacteria in soils contaminated by chromium-containing slag. J. Hazard. Mater. *167*, 516–522.

Challan Belval, S., Garnier, F., Michel, C., Chautard, S., Breeze, D., and Garrido, F. (2009). Enhancing pozzolana colonization by As(III)-oxidizing bacteria for bioremediation purposes. Appl. Microbiol. Biotechnol. *84*, 565–573.

Chen, K., Kao, C., Chen, C., Surampalli, R.Y., and Lee, M. (2010). Control of petroleum-hydrocarbon contaminated groundwater by intrinsic and enhanced bioremediation. J. Environ. Sci. *22*, 864–871.

Christensen, B.B., Sternberg, C., Andersen, J.B., Eberl, L., Moller, S., Givskov, M., and Molin, S. (1998). Establishment of new genetic traits in a microbial biofilm community. Appl. Environ. Microbiol. *64*, 2247–2255.

Conn, K.E., Siegrist, R.L., Barber, L.B., and Meyer, M.T. (2010). Fate of trace organic compounds during vadose zone soil treatment in an onsite wastewater system. Environ. Toxicol. Chem. *29*, 285–293.

Conrad, M.E., Brodie, E.L., Radtke, C.W., Bill, M., Delwiche, M.E., Lee, M.H., Swift, D.L., and Colwell, F.S. (2010). Field evidence for co-metabolism of trichloroethene stimulated by addition of electron donor to groundwater. Environ. Sci. Technol. *44*, 4697–4704.

Cooper, T.F. (2007). Recombination speeds adaptation by reducing competition between beneficial mutations in populations of *Escherichia coli*. Plos Biol. *5*, 1899–1905.

Costerton, J.W., Lewandowski, Z., and Cladwell, D.E. (1995). Microbial biofilms. Ann. Rev. Microbiol. *49*, 711–745.

Cummings, J.E., and Kovacic, J.P. (2009). The ubiquitous role of zinc in health and disease. J. Vet. Emerg. Crit. Car. *19*, 215–240.

Cunningham, A.B., Characklis, W.G., Abedeen, F., and Crawford, D. (1991). Influence of biofilm accumulation on porous media hydrodynamics. Environ. Sci. Technol. *25*, 1305–1311.

Dai, M., and Copley, S. (2004). Genome shuffling improves degradation of the anthropogenic pesticide pentachlorophenol by *Sphingobium chlorophenolicum* ATCC 39723. Appl. Environ. Microbiol. *70*, 2391–2397.

Davey, M.E., and O'Toole, G.A. (2000). Microbial biofilms: from ecology to molecular genetics. Microbiol. Mol. Biol. Rev. *64*, 847–867.

Desai, J., and Banat, I. (1997). Microbial production of surfactants and their commercial potential. Microbiol. Mol. Biol. R. *61*, 47–64.

Di Bona, K.R., Love, S., Rhodes, N.R., McAdory, D.A., Halder Sinha, S., Kern, N., Kent, J. Strickland, J. Wilson, A., Beaird, J., *et al.* (2011). Chromium is not an essential trace element for mammals: effects of a 'low-chromium' diet. J. Biol. Inorg. Chem. *16*, 381–390.

Elasri, M.O., and Miller, R.V. (1999). Study of the response of a biofilm bacterial community to UV Radiation. Appl. Environ. Microbiol. *65*, 2025–2031.

EPA, U.S. (2009). Drinking Water Contaminants. *816-F-09-0004*,

EPA, U.S., and Office of Solid Waste and Emergency Response. (1999). Use of monitored natural attenuation at Superfund, RCRA Corrective Action, and underground storage tank sites. *9200.4–17P.*

E-PRTR. (2008). The European Pollutant Release and Transfer Register PER. Web-based database. Available at: http://prtr.ec.europa.eu/.

Erable, B., Duţeanu, N.M., Ghangrekar, M.M., Dumas, C., and Scott, K. (2010). Application of electro-active biofilms. Biofouling *26*, 57.

Farhadian, M., Duchez, D., Vachelard, C., and Larroche, C. (2008a). Monoaromatics removal from polluted water through bioreactors – A review. Water Res. *42*, 1325–1341.

Farhadian, M., Vachelard, C., Duchez, D., and Larroche, C. (2008b). *In situ* bioremediation of monoaromatic pollutants in groundwater: A review. Bioresour. Technol. *99*, 5296–5308.

Faybishenko, B., Hazen, T.C., Long, P.E., Brodie, E.L., Conrad, M.E., Hubbard, S.S., Christensen, J.N., Joyner, D., Borglin, S.E., Chakraborty, R., *et al.* (2008). *In situ* long-term reductive bioimmobilization of Cr(VI)

in groundwater using hydrogen release compound. Environ. Sci. Technol. *42*, 8478–8485.

Fernandes, V.C., Albergaria, J.T., Oliva-Teles, T., Delerue-Matos, C., and De Marco, P. (2009). Dual augmentation for aerobic bioremediation of MTBE and TCE pollution in heavy metal-contaminated soil. Biodegradation *20*, 375–382.

Flemming, H., and Wingender, J. (2010). The biofilm matrix. Nat. Rev. Microbiol. *8*, 623–633.

Fredrickson, J.K., Kostandarithes, H.M., Li, S.W., Plymale, and A.E., Daly, M. J. (2000). Reduction of Fe(III), Cr(VI), U(VI), and Tc(VII) by *Deinococcus radiodurans* R1. Appl. Environ. Microbiol. *66*, 2006–2011.

Gadd, G.M. (2009). Biosorption: critical review of scientific rationale, environmental importance and significance for pollution treatment. J. Chem. Technol. Biot. *84*, 13–28.

Gadd, G.M. (2010). Metals, minerals and microbes: geomicrobiology and bioremediation. Microbiology *156*, 609–643.

Ghigo, J. (2001). Natural conjugative plasmids induce bacterial biofilm development. Nature *412*, 442–445.

Goldfrank, L.R., Nelson, L.S., Howland, M.A., Lewin, N.A., Flomenbaum, N.E., and Hoffman, R.S. (2006). Goldfrank's Toxicologic Emergencies, 8th ed. (The McGraw-Hill Companies, Inc., New York).

Gregory, K., and Lovley, D. (2005). Remediation and recovery of uranium from contaminated subsurface environments with electrodes. Environ. Sci. Technol. *39*, 8943–8947.

Gusmão, V.R., Chinalia, F.A., Sakamoto, I.K., and Varesche, M.B.A. (2007). Performance of a reactor containing denitrifying immobilized biomass in removing ethanol and aromatic hydrocarbons (BTEX) in a short operating period. J. Hazard. Mater. *139*, 301–309.

Hamilton, J.W., and Wetterhahn, K.E. (1988). Differential effect of chromium(vi) on expression of inducible and constitutive genes *in vivo*. Abstr. Pap. Am. Chem. S. *195*, 98.

Harrison, J.J., Ceri, H., and Turner, R.J. (2007). Multimetal resistance and tolerance in microbial biofilms. Nat. Rev. Microbiol. *5*, 928–938.

Hunter, W. (2009). Vadose zone microbial biobarriers remove nitrate from percolating groundwater. Curr. Microbiol. *58*, 622–627; 627.

Icopini, G.A., Boukhalfa, H., and Neu, M.P. (2007). Biological reduction of Np(V) and Np(V) citrate by metal-reducing bacteria. Environ. Sci. Technol. *41*, 2764–2769.

Izallalen, M., Mahadevan, R., Burgard, A., Postier, B., Didonato, J.,Raymond, Sun, J., Schilling, C.H., and Lovley, D.R. (2008). *Geobacter sulfurreducens* strain engineered for increased rates of respiration. Metab. Eng. *10*, 26–275.

Jackson, V.A., Paulse, A.N., Bester, A.A., Neethling, J.H., Khan, S., and Khan, W. (2009). Bioremediation of metal contamination in the Plankenburg River, Western Cape, South Africa. Int. Biodeterior. Biodegrad. *63*, 559–568.

Jacques, R.J.S., Okeke, B.C., Bento, F.M., Teixeira, A.S., Peralba, M.C.R., and Camargo, F.A.O. (2008). Microbial consortium bioaugmentation of a polycyclic aromatic hydrocarbons contaminated soil. Biores. Technol. *99*, 2637–2643.

Jain, R.K., Sayler, G.S., Wilson, J.T., Houston, L., and Pacia, D. (1987). Maintenance and stability of introduced genotypes in groundwater aquifer material. Appl. Environ. Microbiol. *53*, 996–1002.

Jiajun, X., Linjun, G., Shipeng, W., and Yitong, L. (2010). Comparative impact of cadmium on two phenanthrene-degrading bacteria isolated from cadmium and phenanthrene co-contaminated soil in China. J. Hazard. Mater. *174*, 818–823.

Joannis-Cassan, C., Delia, M.-., and Riba, J. (2007). Biofilm growth kinetics on hydrocarbon in a porous medium under biostimulation conditions. Environ. Prog. *26*, 140–148.

Kazy, S.K., D'Souza, S.F., and Sar, P. (2009). Uranium and thorium sequestration by a Pseudomonas sp.: Mechanism and chemical characterization. J. Hazard. Mater. *163*, 65–72.

Kjelleberg, S., and Molin, S. (2002). Is there a role for quorum sensing signals in bacterial biofilms? Curr. Opin. Microbiol. *5*, 254–258.

Kirschner, A., Altenbuchner, J., and Bornscheuer, U.T. (2007). Design of a secondary alcohol degradation pathway from *Pseudomonas fluorescens* DSM 50106 in an engineered *Escherichia coli*. Appl. Microbiol. Biot. *75*, 1095–1101.

Krawczyk-Bärsch, E., Grossmann, K., Arnold, T., Hofmann, S., and Wobus, A. (2008). Influence of uranium (VI) on the metabolic activity of stable multispecies biofilms studied by oxygen microsensors and fluorescence microscopy. Geochim. Cosmochim. Acta 72, 5251–5265.

Kuhn, T.K., Hamonts, K., Dijk, J.A., Kalka, H., Stichler, W., Springael, D., Dejonghe, W., and Meckenstock, R.U. (2009). Assessment of the intrinsic bioremediation capacity of an eutrophic river sediment polluted by discharging chlorinated aliphatic hydrocarbons: A compound-specific isotope approach. Environ. Sci. Technol. *43*, 5263–5269.

Kulik, N., Goi, A., Trapido, M., and Tuhkanen, T. (2006). Degradation of polycyclic aromatic hydrocarbons by combined chemical pre-oxidation and bioremediation in creosote contaminated soil. J. Environ. Manag. *78*, 382–391.

Lajoie, C.A., Zylstra, G.J., Deflaun, M.F., and Strom, P.F. (1993). Development of field application vectors for bioremediation of soils contaminated with polychlorinated-biphenyls. Appl. Environ. Microbiol. *59*, 1735–1741.

Landini, P. (2009). Cross-talk mechanisms in biofilm formation and responses to environmental and physiological stress in *Escherichia coli*. Res. Microbiol. *160*, 259–266.

Lange, C.C., Wackett, L.P., Minton, K.W., and Daly, M.J. (1998). Engineering a recombinant *Deinococcus radiodurans* for organopollutant degradation in radioactive mixed waste environments. Nat. Biotechnol. *16*, 929–933.

Lennox, J., and Ashe, J. (2009). Biofilms as biobarriers. Am. Biol. Teach. *71*, 20–26.

Li, L., Steefel, C.I., Kowalsky, M.B., Englert, A., and Hubbard, S.S. (2010). Effects of physical and geochemical heterogeneities on mineral transformation and biomass accumulation during biostimulation experiments at Rifle, Colorado. J. Contam. Hydrol. *112*, 45–63.

Logan, B.E. (2009). Exoelectrogenic bacteria that power microbial fuel cells. Nat. Rev. Microbiol. 7, 375–381

Lovley, D.R., Phillips, E.J.P., Gorby, Y.A., and Landa, E.R. (1991). Microbial reduction of uranium. Nature *350*, 413–416.

Lovley, D.R., Roden, E.E., Phillips, E.J.P, Woodward, J.C. (1993). Enzymatic iron and uranium reduction by sulfate-reducing bacteria. Mar. Geol. *113*, 41–53.

Macaskie, L.E., Lloyd, J.R., Thomas, R.A.P., and Tolley, M.R. (1996). The use of micro-organisms for the remediation of solutions contaminated with actinide elements, other radionuclides, and organic contaminants generated by nuclear fuel cycle activities. Nucl. Energ. J. Br. Nucl. *35*, 257–271.

Maila, M.P., and Cloete, T.E. (2005). The use of biological activities to monitor the removal of fuel contaminants – perspective for monitoring hydrocarbon contamination: A review. Int. Biodeterior. Biodegrad. *55*, 1–8.

Marsili, E., Beyenal, H., Di Palma, L., Merli, C., Dohnalkova, A., Amonette, J.E., and Lewandowski, Z. (2005). Uranium removal by sulfate reducing biofilms in the presence of carbonates. Water Sci. Technol. *52*, 49–55.

Marsili, E., Beyenal, H., Palma, L.D., Merli, C., Dohnalkova, A., Amonette, J., E., and Lewandowski, Z. (2007). Uranium immobilization by sulfate-reducing biofilms grown on hematite, dolomite, and calcite. Environ. Sci. Technol. *41*, 8349–8354.

Marsili, E., Baron, D.B., Shikhare, I.D., Coursolle, D., Gralnick, J.A., and Bond, D.R. (2008). Shewanella Secretes flavins that mediate extracellular electron transfer. P. Natl. Acad. Sci. U.S.A. *105*, 3968–3973.

Martin, T.A., and Ruby, M.V. (2004). Review of *in situ* remediation technologies for lead, zinc, and cadmium in soil. Remediation *14*, 35–53.

Martínková, L., Uhnáková, B., Pátek, M., Nešvera, J., and Křen, V. (2009). Biodegradation potential of the genus *Rhodococcus*. Environ. Int. *35*, 162–177.

Massoudieh, A., Mathew, A., Lambertini, E., Nelson, K.E., and Ginn, T.R. (2007). Horizontal gene transfer on surfaces in natural porous media: Conjugation and kinetics. Vadose Zone J. *6*, 306–315.

Mathúna, C.Ó., O'Donnell, T., Martinez-Catala, R.V., Rohan, J., and O'Flynn, B. (2008). Energy scavenging for long-term deployable wireless sensor networks. Talanta 75, 613–623.

Matias Saavedra, J., Acevedo, F., Gonzalez, M., and Seeger, M. (2010). Mineralization of PCBs by the genetically modified strain Cupriavidus necator JMS34 and its application for bioremediation of PCBs in soil. Appl. Microbiol. Biot. *87*, 1543–1554.

Meng-Ying, L., Ji, Z., Peng, L., Jing-Liang, X., and Shun-Peng, L. (2009). Evaluation of biological characteristics of bacteria contributing to biofilm formation. Pedosphere *19*, 554–561.

Michalsen, M.M., Peacock, A.D., Smithgal, A.N., White, D.C., Spain, A.M., Sanchez-Rosario, Y., Krumholz, L.R., Kelly, S.D., Kemner, K.M., McKinley, J., *et al.* (2009). Treatment of nitric acid-, U(VI)-, and Tc(VII)-contaminated groundwater in intermediate-scale physical models of an *in situ* Biobarrier. Environ. Sci. Technol. *43*, 1952–1961.

Michel, C., Jean, M., Coulon, S., Dictor, M.-., Delorme, F., Morin, D., and Garrido, F. (2007). Biofilms of As(III)-oxidising bacteria: Formation and activity studies for bioremediation process development. Appl. Microbiol. Biotechnol. 77, 457–467.

Mohapatra, B.R. Dinardo O., Douglas Goulda, W., and Koren, D.W. (2010). Biochemical and genomic facets on the dissimilatory reduction of radionuclides by microorganisms-a review,' Miner. Eng. 23, 591–599.

Molokwane, P.E., and Nkhalambayausi-Chirwa, E.M. (2009). Microbial culture dynamics and chromium (VI) removal in packed-column microcosm reactors. Water Sci. Technol. *60*, 381–388.

Monds, R.D., and O'Toole, G.A. (2009). The developmental model of microbial biofilms: ten years of a paradigm up for review. Trends Microbiol. *17*, 73–87.

Monroe, D. (2007). Looking for chinks in the armor of bacterial biofilms. PLOS Biol. *5*, 2458–2461.

Monti, M., Smania, A., Fabro, G., Alvarez, M., and Argarana, C. (2005). Engineering *Pseudomonas fluorescens* for biodegradation of 2,4-dinitrotoluene. Appl. Environ. Microb. *71*, 8864–8872.

Morel, M.A., Ubalde, M.C., Olivera-Bravo, S., Callejas, C., Gill, P.R., and Castro-Sowinski, S. (2009). Cellular and biochemical response to Cr(VI) in *Stenotrophomonas* sp. FEMS Microbiol. Lett. *291*, 162–168.

Mukherjee, P.K., Chand, D.V., Chandra, J., Anderson, J.M., and Ghannoum, M.A. (2009). Shear stress modulates the thickness and architecture of *Candida albicans* biofilms in a phase-dependent manner. Mycoses *52*, 440–446.

Nazina, T.N., Luk'yanova, E.A., Zakharova, E.V., Konstantinova, L.I., Kalmykov, S.N., Poltaraus, A.B., and Zubkov, A.A. (2010). Microorganisms in a disposal site for liquid radioactive wastes and their influence on radionuclides. Geomicrobiol. J. *27*, 473–486.

New, M. (2006). Annual progress Report on research related to our research project 'Stabilization of Plutonium in Subsurface Environments via Microbial Reduction and Biofilm Formation' funded by the Environmental Remediation Sciences Division (ERSD). Technical Report ERSD-1024817–2006, Los Alamos National Laboratory (LANL), USA.

N'Guessan, A.L., Elifantz, H., Nevin, K.P., Mouser, P.J., Methe, B., LWoodard, T., Manley, K., Williams, K.H., Wilkins, M.J., Larsen, J.T., Long, P.E., and Lovley, D.R.

(2010). Molecular analysis of phosphate limitation in Geobacteraceae during the bioremediation of a uranium-contaminated aquifer. ISME J. *4*, 253–266.

Okoh, A.I., and Trejo-Hernandez, M.R. (2006). Remediation of petroleum hydrocarbon polluted systems: Exploiting the bioremediation strategies. Afr. J. Biotechnol. *5*, 2520–2525.

Pal, A., and Paul, A. (2008). Microbial extracellular polymeric substances: central elements in heavy metal bioremediation. Indian J. Microbiol. *48*, 49–64.

Pandey, J., Chauhan, A., and Jain, R.K. (2009). Integrative approaches for assessing the ecological sustainability of *in situ* bioremediation. FEMS Microbiol. Rev. *33*, 324–375.

Park, H.I., Kim, D.K., Choi, Y.J., and Pak, D. (2005). Nitrate reduction using an electrode as direct electron donor in a biofilm-electrode reactor. Process Biochem. *40*, 3383–3388.

Park, H.I., Kim, J.S., Kim, D.K., Choi, Y.J., and Pak, D. (2006). Nitrate-reducing bacterial community in a biofilm-electrode reactor. Enz. Microb. Tech. *39*, 453–458.

Patel, J., Zhang, Q., McKay, R.M.L., Vincent, R., and Xu, Z. (2010). Genetic engineering of *Caulobacter crescentus* for removal of cadmium from water. Appl. Biochem. Biotech. *160*, 232–243.

Prpich, G., Adams, R., and Daugulis, A. (2006). *Ex situ* bioremediation of phenol contaminated soil using polymer beads. Biotechnol. Lett. *28*, 2027–2031.

Radhika, V., Subramanian, S., and Natarajan, K.A. (2006). Bioremediation of zinc using *Desulfotomaculum nigrificans*: bioprecipitation and characterization studies. Water Res. *40*, 3628–3636.

Renshaw, J., Law, N., Geissler, A., Livens, F., and Lloyd, J. (2009). Impact of the Fe(III)-reducing bacteria *Geobacter sulfurreducens* and *Shewanella oneidensis* on the speciation of plutonium. Biogeochemistry *94*, 191–196.

Rittman, B.E., and McCarty, P.L. (2000). Environmental biotechnology: principles & applications. (New York, McGraw-Hill Science Engineering).

Robles-Gonzalez, I.V., Fava, F., and Poggi-Varaldo, H.M. (2008). A review on slurry bioreactors for bioremediation of soils and sediments. Microbial Cell Factories *7*, 5.

Rodriguez, S.J., and Bishop, P.L. (2007). Three-dimensional quantification of soil biofilms using image analysis. Environ. Eng. Sci. *24*, 96–103.

Rodriguez, S.J., and Bishop, P.L. (2008). Enhancing the biodegradation of polycyclic aromatic hydrocarbons: effects of nonionic surfactant addition on biofilm function and structure. J. Envir. Eng. *134*, 505–512.

Rowbotham, A., Levy, L., and Shuker, L. (2000). Chromium in the environment: an evaluation of exposure of the UK general population and possible adverse health. J. Toxicol. Env. Heal. B. 3, *145*–178.

Rui, L.Y., Kwon, Y.M., Reardon, K.F., and Wood, T.K. (2004). Metabolic pathway engineering to enhance aerobic degradation of chlorinated ethenes and to reduce their toxicity by cloning a novel glutathione S-transferase, an evolved toluene o-monooxygenase, and gamma-glutamylcysteine synthetase. Environ. Microbiol. *6*, 491–500.

Sabater, S., Guasch, H., Ricart, M., Romaní, A., Vidal, G., Klünder, C., and Schmitt-Jansen, M. (2007). Monitoring the effect of chemicals on biological communities. The biofilm as an interface. Anal. Bioanal. Chem. *387*, 1425–1434.

Salinero, K., Keller, K., Feil, W., Feil, H., Trong, S., Di Bartolo, G., and Lapidus, A. (2009). Metabolic analysis of the soil microbe Dechloromonas aromatica str. RCB: indications of a surprisingly complex life-style and cryptic anaerobic pathways for aromatic degradation. BMC Genomics *10*, 351.

Sani, R.K., Peyton, B.M., Amonette, J.E., and Geesey, G.G. (2004). Reduction of uranium(VI) under sulfate-reducing conditions in the presence of Fe(III)-(hydr) oxides. Geochim. Cosmochim. Acta *68*, 2639–2648.

Seo, Y., Lee, W., Sorial, G., and Bishop, P.L. (2009). The application of a mulch biofilm barrier for surfactant enhanced polycyclic aromatic hydrocarbon bioremediation. Environ. Poll. *157*, 95–101.

Shafahi, M., and Vafai, K. (2009). Biofilm affected characteristics of porous structures. Int. J. Heat Mass Transfer *52*, 574–581.

Shanker, A., Cervantes, C., Loza-Tavera, H., and Avudainayagam, S. (2005). Chromium toxicity in plants. Environ. Int. *31*, 739–753.

Shokes, T.E., and Möller, G. (1999). Removal of dissolved heavy metals from acid rock drainage using iron metal. Environ. Sci. Technol. 33, 282–287.

Singleton, I. (1994). Microbial metabolism of xenobiotics: Fundamental and applied research. J. Chem. Technol. Biotechnol. *59*, 9–23.

Sinicropi, M., Amantea, D., Caruso, A., and Saturnino, C. (2010). Chemical and Biological Properties of Toxic Metals and Use of Chelating Agents for the Pharmacological Treatment of Metal Poisoning. (Springer, Berlin/Heidelberg).

Smith, P.A., and Romesberg, F.E. (2007). Combating bacteria and drug resistance by inhibiting mechanisms of persistence and adaptation. Nat. Chem. Biol. 3, 549–556.

Smith, W.L., and Gadd, G.M. (2000). Reduction and precipitation of chromate by mixed culture sulfate-reducing bacterial biofilms. J. Appl. Microbiol. *88*, 983–991.

Snow, A., Andow, D., Gepts, P., Hallerman, E., Power, A., Tiedje, J., and Wolfenbarger, L. (2005). Genetically engineered organisms and the environment: Current status and recommendations. Ecol. Appl. *15*, 377–404.

Soda, S., Kanzaki, M., Yarnamuara, S., Kashiwa, M., Fujita, M., and Ike, M. (2009). Slurry bioreactor modeling using a dissimilatory arsenate-reducing bacterium for remediation of arsenic-contaminated soil. J. Biosci. Bioeng. *107*, 130–137.

Stearns, D.M. (2000). Is chromium a trace essential metal? Biofactors *11*, 149–162.

Stewart, P.S., and Franklin, M.J. (2008). Physiological heterogeneity in biofilms. Nat. Rev. Microbiol. *6*, 199–210.

Stroo, H.F., Major, D.W., and Gossett, J.M. (2010). Bioaugmentation for anaerobic bioremediation of chlorinated solvents. In *In Situ* Remediation of Chlorinated Solvent Plumes, Ward, C.H., ed. (Springer, New York), pp. 425–454.

Strycharz, S.M., Gannon, S.M., Boles, A.R., Franks, A.E., Nevin, K.P., and Lovley, D.R. (2010). Reductive dechlorination of 2-chlorophenol by *Anaeromyxobacter dehalogenans* with an electrode serving as the electron donor. Environ. Microbiol. Rep. *2*, 289–294.

Sturman, P.J., Stewart, P.S., Cunningham, A.B., Bouwerd, E.J., and Wolfram, J.H. (1995). Engineering scale-up of *in situ* bioremediation processes: a review. J. Cont. Hydr. *19*, 171–203.

Sun, W., Sierra-Alvarez, R., Field, J.A. (2011). Long term performance of an arsenite-oxidizing-chlorate-reducing microbial consortium in an upflow anaerobic sludge bed (UASB) bioreactor. Bioresour. Technol. *102*, 5010–5016.

Suzuki, Y., Kelly, S.D., Kemner, K.M., and Banfield, J.F. (2005). Direct microbial reduction and subsequent preservation of uranium in natural near-surface sediment. Appl. Environ. Microbiol. *71*, 1790–1797.

Tiehm, A., Mueller, A., Alt, S., Jacob, H., Schad, H., and Weingran, C. (2008). Development of a groundwater biobarrier for the removal of polycyclic aromatic hydrocarbons, BTEX, and heterocyclic hydrocarbons. Water Sci. Technol. *58*, 1349–1355.

Tront, J.M., Fortner, J.D., Ploetze, M., Hughes, J.B., and Puzrin, A.M. (2008). Microbial fuel cell biosensor for *in situ* assessment of microbial activity. Biosens. Bioelectron. *24*, 586–590.

Venkata Mohan, S., Prasanna, D., Purushotham Reddy, B., and Sarma, P.N. (2008). *Ex situ* bioremediation of pyrene contaminated soil in bio-slurry phase reactor operated in periodic discontinuous batch mode: Influence of bioaugmentation. Int. Biodeterior. Biodegrad. *62*, 162–169.

Viamajala, S., Peyton, B.M., Gerlach, R., Sivaswamy, V., Apel, W.A., and Petersen, J.N. (2008). Permeable reactive biobarriers for *in situ* Cr(VI) reduction: Bench scale tests using *Cellulomonas* sp. strain ES6. Biotechnol. Bioeng. *101*, 1150–1162.

Vu, B., Chen, M., Crawford, R.J., and Ivanova, E.P. (2009). Bacterial extracellular polysaccharides involved in biofilm formation. Molecules *14*, 2535–2554.

Wang, F., Dorfler, U., Schmid, M., Fischer, D., Kinzel, L., Scherb, H., Munch, J.C., Jiang, X., and Schroll, R. (2010). Homogeneous inoculation vs. microbial hot spots of isolated strain and microbial community: What is the most promising approach in remediating 1,2,4-TCB contaminated soils? Soil. Biol. Biochem. *42*, 331–336.

Widdowson, M.A., El-Sayed, A., and Landmeyer, J.E. (2008). Numerical model for the uptake of groundwater contaminants by phreatophytes. Paper presented at: Water Pollution IX. (Ashurst Lodge, Southampton, England).

Williams, K.H., Nevin, K.P., Franks, A., Englert, A., Long, P.E., and Lovley, D.R. (2010). Electrode-based approach for monitoring *in situ* microbial activity during subsurface bioremediation. Environ. Sci. Technol. *44*, 47–54.

Wolterbeek, H.T., and Verburg, T.G. (2001). Predicting metal toxicity revisited: general properties vs. specific effects. Sci. Total Environ. *279*, 87–115.

World Health Organization. (2008). Guidelines for drinking-water quality (3rd Ed.): incorporating 1st and 2nd addenda, Vol.1. Recommendations, 186.

Wu, W., Carley, J., Green, S.J., Luo, J., Kelly, S.D., Nostrand, J.V., Lowe, K., Mehlhorn, T., Carroll, S., Boonchayanant, B., *et al.* (2010). Effects of Nitrate on the Stability of Uranium in a Bioreduced Region of the Subsurface. Environ. Sci. Technol. *44*, 5104–5111.

Xu, M., Wu, W., Wu, L., He, Z., Van Nostrand, J.D., Deng, Y., Luo, J., Carley, J., Ginder-Vogel, M., Gentry, T.J., Gu B., Watson D., Jardine P.M., Marsh T.L., Tiedje J.M., Hazen T., Criddle C.S., Zhou J.. (2010). Responses of microbial community functional structures to pilot-scale uranium *in situ* bioremediation. ISME J. *4*, 1060–1070.

Yang, C., Song, C., Mulchandani, A., and Qiao, C. (2010). Genetic engineering of *Stenotrophomonas* Strain YC-1 to possess a broader substrate range for organophosphates. J. Agric. Food Chem. *58*, 6762–6766.

Yoshida, S., Ogawa, N., Fuji, T., and Tsushima, S. (2009). Enhanced biofilm formation and 3-chlorobenzoate degrading activity by the bacterial consortium of *Burkholderia sp.* NK8 and *Pseudomonas aeruginosa* PAO1. J. Appl. Microbiol. *106*, 790–800.

Zhang, T., Gannon, S.M., Nevin, K.P., Franks, A.E., and Lovley, D.R. (2010). Stimulating the anaerobic degradation of aromatic hydrocarbons in contaminated sediments by providing an electrode as the electron acceptor. Environ. Microbiol. *12*, 1011–1020.

Zhang, Y., and Miller, R. (1992). Enhanced octadecane dispersion and biodegradation by a pseudomonas rhamnolipid surfactant (biosurfactant). Appl. Environ. Microbiol. *58*, 3276–3282.

Biofilms in Wastewater Treatment Systems

6

G.A. Clark Ehlers and Susan J. Turner

Abstract

Biofilms occur frequently inside various engineered systems for wastewater treatment. These include traditional trickling filter systems, modified lagoons, and specialized supplementary systems for nutrient removal or treatment of specialized wastes. The major advantages of biofilm systems over suspension treatment is the high microbial density that can be achieved, leading to smaller treatment system footprints, and the inherent development of aerobic, anoxic and anaerobic zones which enable simultaneous biological nutrient removal. The intrinsic resistance of biofilm communities to changing environmental conditions creates the added advantage that biofilm-based treatment systems are more resilient to influent variation in toxicity and nutrient concentrations. In contrast to biofilms of environmental or biomedical relevance comparatively little is known about development and stability in waste treatment systems. The advent of tools that enable the study of biofilms in reactor systems on a molecular level has enabled greater insight into the physiologically and biochemically relevant pathways that may facilitate optimized processes. In this chapter, the current literature on biofilms in wastewater treatment systems is reviewed and opportunities for further development in this field are identified.

Introduction

Almost all modern wastewater treatment systems include some form of biological treatment that capitalizes on the microbial assimilation and mineralization processes that occur in natural environments. Two factors are critical to effective and reliable treatment performance; the retention of an active microbial biomass within the system, and the ability to separate biomass from effluents in the process of clarification. In the activated sludge process, which is typically used in municipal treatment plants, the biomass is present as a suspension of single cells, microcolonies and larger aggregates known as flocs. Clarification of effluents is achieved through gravity-settling of flocs or filtration of effluents, whereupon a portion of the settled sludge is returned to the reactor (return activated sludge) to maintain the levels of active biomass (Seviour and Nielsen, 2010). Biofilm-based systems, as the name suggests, are engineered to retain a component of the microbial biomass as surface-associated communities on living or inert substrates. The key functional differences between the two types of system are that transformational processes are limited by mass transfer in biofilm systems while suspension-based systems tend to be limited by the amount of biomass (Boltz and Daigger, 2009). A broad range of biofilm-based systems have been described and are now used in the wastewater treatment industry. These include natural and engineered wetlands (Vymazal, 2009), trickling filters, fluidized bed (Sharma *et al.*, 2008) or moving bed (Ødegaard, 2006) bioreactors, rotating biological contactors (Cortez *et al.*, 2008) and membrane bioreactors (Syron and Casey, 2008). The fundamental differences between these systems lie in the nature of the substrate on which the biofilm is formed; the mechanism by which the biofilm is exposed to the wastewater substrate; and the degree of engineering or operational control that is exerted over the system.

As with all forms of biological treatment, a major challenge in the design and operation of biofilm systems is to maintain functional stability within a complex biological system that is inherently responsive to nutritional and physicochemical cues. The control of microbial processes in bioreactor systems is therefore reliant on fundamental understanding of biofilm dynamics and how the desired functionality can be optimized and maintained against a backdrop of changing environmental conditions.

This chapter reports on the application of biofilms in the wastewater treatment industry and the research that underpins future developments in this field. It describes the application of biofilms in wastewater treatment systems from a functional perspective, examining the utilization of biofilm communities to accomplish a desired end-result. In this context, biofilm functionality is discussed in terms of (i) biological nutrient removal; and (ii) pollutant degradation and mineralization. We conclude by providing an outlook on the opportunities for technical advancements in this field and questions that need to be addressed to achieve them.

New tools for understanding biofilm systems

The evolution of biofilm-based reactor systems has, and will continue to be, reliant on our increasing ability to undertake an integrated analysis of both the macroenvironment of the bulk liquid phase and the biofilm microenvironment. A suite of both online and offline analytical methods are currently available, and are routinely employed, for monitoring the chemical and physical parameters of bulk fluids. Typical analyses include determinations of organic carbon in the form of BOD or COD, pH, dissolved oxygen, and nutrients such as nitrogen and phosphorus (Eaton *et al.*, 2005). Until comparatively recently, wastewater treatment processes have been defined on the basis of results from these analyses and comparatively few studies have attempted to monitor the biofilm microenvironment. However, this area of research has benefitted from the development of microsensor and molecular approaches that have been applied to biofilm research in other fields and which are now finding increasing use in wastewater research (Rittmann, 2006; Rittmann *et al.*, 2006).

To-date most studies have focused on systems that are tailored towards specific treatment applications, particularly biological nutrient removal, where the use of biofilm processes can offer significant treatment efficiencies over more conventional floc-based systems. For example, the performance of a biofilm reactor configured to remove nitrogen species is dependent on the physical constraints of mass transfer and the interactions between nitrifying and denitrifying populations alongside other competing species. Greater insight into how these divergent groups of microorganisms interact in shaping both the physical structure of the biofilm, and the food webs within it, may prove advantageous in defining treatment system design and operation for greater nutrient removal efficiency.

New insights into microbial processes in wastewater treatment have been accomplished through the development and application of new molecular microbiology and microscopy tools. These have been reviewed in a number of recent articles (Daims *et al.*, 2006; Talbot *et al.*, 2008; Wojnowska-Baryła *et al.*, 2010). Tools such as fluorescence *in situ* hybridization (FISH) and quantitative real-time PCR (q-PCR) enable tracking and quantification of specific bacterial groups, while techniques such as microautoradiography combined with FISH (FISH-MAR) and stable isotope probing (SIP) facilitate evaluation of metabolic functions within microbial communities. The application of microelectrodes that enable *in situ* measurement of physical parameters across steep gradients, is also enabling further exploration of metabolic parameters and diffusivity of electroactive species as a function of depth within biofilms. Collectively these tools now create the opportunity for studies that will elucidate the linkages between microbial community structure and metabolic function.

While there is much progress to be made from *in situ* analysis of mixed communities, considerable insight into the metabolic capability of these systems can also be obtained from the genome sequencing and functional analysis of individual members of the wastewater microbial community (Siezen and Galardini, 2008; Heijstra *et al.*, 2009). The increasing accessibility of high-throughput next generation sequencing technology and advanced bioinformatics has also enabled more extensive culture-independent characterization of the microbiome (Bibby *et al.*, 2010; McLellan *et al.*, 2010) and metagenomes (Sanapareddy *et al.*, 2009) of waste treatment consortia. The latter approach has enabled genomic analysis of uncultured organisms involved in nutrient removal from floc-based systems, particularly phosphorus (Martin *et al.*, 2006) and nitrogen (Lücker *et al.*, 2010). Such studies give new insight into the metabolic complexity of these systems and provide the first step towards the development of models for systems-based prediction of functionality within multispecies assemblages.

Biofilm systems – establishment, stability and operation

As already stated, a wide range of wastewater treatment technologies have been developed in which some component of the microbial biomass is retained as a biofilm community. A detailed technical review of these is beyond the scope of this chapter and the reader is referred to specialist texts for further information on such systems (Lazarova and Manem, 1995; Tchobanoglous *et al.*, 2002). This section draws on specific examples to discuss the factors influencing biofilm development and structure and how these factors might influence the performance of wastewater treatment systems.

Crucial issues that determine the performance of wastewater systems are the establishment and stability of the microbial community, and the functionality of the community with respect to biotransformation or remediation services. Biofilm systems are reported to offer a number of advantages over conventional suspension-based waste water treatment processes in this regard. It has been postulated that the higher biomass density and longer sludge retention times achieved in biofilm systems leads to increased levels of mineralization and less sludge in the bulk phase reducing costs for downstream treatment and disposal of waste sludge (Van Loosdrecht and Heijnen, 1993).

Another advantage is that biofilm systems enable long biomass retention times supporting development and retention of active populations of slow-growing organisms. The highly structured biofilm environment also provides for development of oxygen gradients enabling simultaneous nitrification and denitrification (Van Loosdrecht and Heijnen, 1993). A more detailed discussion of biofilm processes in the context of nutrient removal is presented in the next section. The combination of oxygen gradients and support of slow-growing organisms has also made biofilm reactor systems the method of choice for treatment of xenobiotics (Syron and Casey, 2008).

The protective nature of the biofilm environment, which makes them a problem in biomedical contexts (see Chapter 2) is potentially an advantage in wastewater treatment as it is presumed to confer resilience to transient levels of toxicants that might otherwise disrupt suspended biomass (Hall-Stoodley *et al.*, 2004). The effect of antibiotics on suspension-based systems has been demonstrated by a recent study in which intermittent dosing of a cocktail of antibiotics impaired phosphorus removal in a lab-scale granular reactor configured for enhanced biological phosphorus removal (EBPR) (Slater *et al.*, 2011). Whether biofilm systems do indeed offer greater resistance than floc or granular systems is yet to be confirmed experimentally.

Another presumed advantage of biofilm systems, which has been brought into question, is that retention of the biomass on a solid matrix overcomes some of the problems associated with solid-liquid separation in floc-based systems. However, biofilm attachment is never absolute and recent studies suggest that these systems never achieve a steady state owing to periodic events that result in removal of large chunks of biomass. Though shear forces may contribute to this, a more significant cause is the controlled microbial detachment processes that are part of the normal successional development, or which occur in response to environmental cues, such as nutritional stress. Continual loss of biomass from the biofilm is not necessarily detrimental to system performance in terms of nutrient transformation as sufficient levels of functional redundancy are retained within the biomass present in the system (Lewandowski and Boltz, 2011). However, biofilm detachment can generate appreciable amounts of suspended particulates in effluents. Fine particulates may also derive from microbial cells that are either biofilm in-competent or, under the conditions present within the reactor, favour growth in a planktonic state. Clarification by settling or membrane filtration is therefore necessary in systems that are required to achieve a high level of solids removal.

A comparatively recent advance in wastewater treatment technology has been the development of hybrid systems that combine elements of both floc-based and fixed film reactors. One such configuration is the moving bed biofilm reactor (MBBR) in which the majority of the biomass is attached to solid support media that are held in suspension through active mixing and aeration (Ødegaard *et al.*, 1994). A variant of this is the integrated fixed-film activated sludge (IFAS) system in which a high density of floccular biomass is also maintained by the return of activated sludge (RAS) that has settled out in the clarifier (Mannina and Viviani, 2009). The latter is also referred to as a hybrid MBBR system (HMBBR). Both systems rely on effective biofilm formation and a range of support media have been developed offering varying surface area, buoyancy, and shear factors to suit different applications. The main focus of studies on these systems has been on physical aspects of biofilm density and mass transfer (Mannina *et al.*, 2007) and few studies have compared the effects of substrate on microbial community composition and dynamics.

One of the main advantages of HMBBR-type systems is the high throughput within a relatively compact system, owing to reduced hydraulic retention times required for effective biological transformation. Though previous studies suggest that the improved treatment efficiency is due to the higher biomass density in such systems, respirometric studies suggest that biological activity is comparable between activated sludge (AS) and HMBBR systems (Di Trapani *et al.*, 2008). The factors contributing to more rapid transformation in this context remain unclear but it is possible that the combination of biofilm and suspended communities provides for more efficient cometabolism and trophic interactions within these systems.

Although it is presumed that the biofilm and planktonic communities in MBBR systems differ substantially in composition, there are only limited data available on full-scale systems to validate this assertion. A pilot study of microbial communities in two MBBR treatment systems for municipal waste in central New Zealand revealed distinct differences between the dominant bacterial groups in 16S clone libraries prepared from biofilm and suspended particulates. Biofilm communities were dominated by anaerobes and included members of the *Firmicutes* and sulfate-reducing bacteria from the *Deltaproteobacteria*, while suspended particulates were dominated by members of the *Gammaproteobacteria*. In contrast, conventional activated sludge is typically dominated by members of the *Betaproteobacteria* and the *Cytophaga Bacteriodetes* group. One explanation for this difference is that MBBR systems typically operate at a very high hydraulic loading, selecting for organisms that are either retained on surfaces, or have a growth-rate that is high-enough to withstand washout (K. Biswas and S.J. Turner, unpublished).

The observation that neither the biofilm nor the suspended communities resembled conventional AS raises the question of whether improved performance in carbon/nutrient removal is simply due to the increased biomass achieved in MBBR systems. An alternative proposition is that transformational processes are more efficient owing to either the symbiotic associations within biofilms or the different community composition within the reactor as a whole, or a combination of these factors. At the present time there are limited data to address these questions. While a number of mathematical models have been developed which describe biofilm processes in the context of wastewater treatment, the interactions between suspended and attached biomass remain poorly understood (Lewandowski and Boltz, 2011). A further question of intense interest is whether biofilm systems can be manipulated to improve stability and reduce the fine particulate population. As noted above, resolution of this awaits more validated data on the community composition of biofilms and suspension particulates from full scale systems. A small number of studies on the bacterial populations in full-scale treatment reactors have been conducted and a summary of these studies is presented in Table 6.1.

Another area of biofilm research concerning wastewater treatment is addressing process initiation which can take extended periods of time, e.g. between 4 and 9 months in the case of anaerobic biofilm reactors (Annachhatre and Bhamidimarri, 1992; Cresson *et al.*, 2006). Insight into how biofilms form and at what rates functionality can be obtained in a bioreactor have been signalled as important areas for future research. The influence of solid media properties and the operation of bioreactors on attachment and growth of microbes will enable the acceleration of process initiation into process stabilization. For example, low biomass yield and biomass retention have been cued as an important component in managing the anammox pathway. Researchers have shown that zeolite particles and non-woven materials provide superior support for anammox biomass owing to large specific surface area, high porosity and diffusivity of substrates (Gong *et al.*, 2007; Fernández *et al.*, 2008). The effects of carrier structure and substrate on biofilm development and stability have been investigated in a number of studies. Oxygen consumption rates were used to investigate biofilm development and stability on a range of carrier substrates designed for use in fluidised bed reactors (Bolton *et al.*, 2006). Distinct differences were found between the fourteen different carriers with biofilm accumulation correlating most strongly with the surface roughness and surface area of the carrier. Garcia *et al.* (2008) used cell morphology and denaturing gradient gel electrophoresis (DGGE) to investigate the biofilm community structure and stability on four different carrier materials (polyurethane foam, vegetal carbon, synthetic pumice, low-density polyethylene) in an anaerobic biofilm reactor. The carrier materials showed different biofilm densities and community structure diversity (Garcia *et al.*, 2008).

Biofilm functionality in wastewater treatment applications

Bioreactor systems are tailored to perform a particular function with respect to the type and constituents of the wastewater it has to treat. The function prescribes reactor configuration and mode of operation which, in turn, is driven by its end use. An understanding of the physical and chemical processes that drive these metabolic functions has permitted researchers to engineer process that should technically enable fully functional systems. This has been the output in a considerable number of studies performed in the laboratory and in the field (Daims *et al.*, 2001; Rowan *et al.*, 2003; Lydmark *et al.*, 2006; Chen *et al.*, 2009; Koschorreck *et al.*, 2010). Knowledge of how nutrients, whether they are in the form of BOD, N, P or SO_4, are cycled and mineralized, or how complex chemical structures in detoxification processes are destroyed, has shaped the advancement of bioreactor system configurations. The emergence of tools that facilitate the study of microbial communities and their performance in wastewater treatment processes on a microscale has provided important insight into potentially significant processes occurring in biofilms. Biofilms are structurally and compositionally heterogeneous structures which impact uptake and diffusion rates of nutrients and contaminants. The ability to map and study the nature of biofilms on various scales and heterogeneities aids the modelling of biofilm performance (Phoenix and Holmes, 2008). This has also facilitated the discovery of novel microbial communities and beneficial metabolic pathways. As a consequence it has enabled researchers to identify bottlenecks in respiratory chains that could impair essential metabolic reactions in treatment processes (Singh *et al.*, 2006; Kindaichi *et al.*, 2007). Biofilm functionality will be discussed in relation to two areas that serve as models to exemplify the role of biofilms in wastewater treatment processes as crucial agents to drive advantageous biological processes. The areas are nutrient removal and pollutant detoxification.

Table 6.1 Summary of community profile studies conducted in full-scale wastewater treatment systems

	Bioreactor system	Microorganisms in biofilm samples	References
N-removal (Anammox)	Anammox reactor	*Brocadia anammoxidans*, *Kuenenia stuttgartiensis*	Roest *et al*. (2005), Tsushima *et al*. (2007)
N-removal (nitrification)	Nitrifying trickling filter reactor	*Nitrosomonas oligotropha*, *Nitrosomonas europaea*, *Nitrosomonas communis*	Lydmark *et al*. (2006)
N-removal (ammonia oxidizing)	Trickling filter	*Nitrosococcus mobilis*	Rowan *et al*. (2003)
Detoxification of organic pollutants (phenolic)	Granular activated carbon – anaerobic fluidized bed	*Methanosaeta*, *Syntrophorhabdaceae*	Chen *et al*. (2009)
Detoxification of organic pollutants (hydrocarbons)	Anaerobic fluidized bed	*Methanomethylovorans thermophila*, *Methanolinea tarda*	Liu *et al*. (2010)
Acid mine drainage treatment, sulfate removal	Submerged carrier media (polyethylene)	*Desulfomonile*, *Desulfobacterium* sp.	Koschorreck *et al*. (2010)

Nutrient removal

Various treatment systems are available for the effective cycling and biological removal of nutrients in wastewater. The systems vary from simple treatment systems to more complex and hybrid configurations that capitalize on the inherent ability of specific groups of microbes to take up and convert specific nutrient species. Nitrogen removal, for example, is driven largely through the operation of nitrification and denitrification. This is accomplished by two physiologically distinct groups of microorganisms, i.e. autotrophic aerobic nitrifiers and heterotrophic anaerobic denitrifiers, with distinct metabolic differences (Clifford *et al.*, 2010). This translates into synthesizing a treatment process which has nitrogen removal as its primary aim, incorporating both aerobic and anaerobic stages. This can also be accomplished by configuring reactor systems that support simultaneous nitrification and denitrification by favouring the establishment of both groups of bacteria in different regions in a reactor.

A large body of research work documents biological nutrient removal processes as increasing demands are placed on finding efficient and cost effective treatment technologies that will optimize the operation of existing wastewater treatment facilities or laboratory-based systems. The larger percentage of studies record laboratory investigations on synthetic but also real wastewater streams. These studies employ traditional and emerging techniques to study biological processes and are providing insight into how these processes can be optimized (Garzón-Zúñiga and Gonzlez-Martínez, 1996; Gieseke *et al.*, 2001; Ito *et al.*, 2002; Persson *et al.*, 2002; Koschorreck *et al.*, 2010). It is imperative to develop more efficient systems and given the fact that biofilms are and will continue to be crucial components in reactor treatment technology, knowledge of factors that impact metabolic processes will galvanize progress in this area.

The utilization of molecular tools to study biofilm operation in relation to its functionality, whether it be nitrogen or carbon removal, is documented in a number of recent articles (Daims *et al.*, 2001, 2006; Wagner *et al.*, 2002; Terada *et al.*, 2003. 2006; Clifford *et al.*, 2010). These authors have applied different reactor systems to power nutrient removal whilst employing techniques such as microsensors or q-PCR to gain a greater understanding of processes in biofilms. Clifford *et al.* (2010) for example studied N transformation processes in a horizontal flow biofilm reactor (HFBR). It operated on the premise that the two distinct groups of bacteria in the traditional aerobic-anaerobic N removal processes will occupy spatially different regions in a single unit reactor system that uses horizontal stacks of polystyrene sheets. This, the authors postulated, allowed examination of the biotransformation processes linked with the establishment and distribution of microbial communities in different regions in the reactor. Terada *et al.* (2003) showed by microsensor monitoring and FISH analysis that nitrifying and denitrifying bacterial communities occupy distinct physical areas in a biofilm that facilitated simultaneous nitrification and denitrification. The outcome of this investigation established the construction of an efficient membrane-aerated bioreactor configured to remove N in high-strength wastewaters.

A large number of papers document the synthesis of mathematical models that describe nutrient removal processes which should also advance reactor operation strategies (Persson *et al.*, 2002; Tchobanoglous *et al.*, 2002; Villaverde, 2004; Wang *et al.*, 2006; Lücker *et al.*, 2010). Through the employment of emerging tools that make information available which is fundamental in piecing together biological processes operational in a biofilm, improved models are generated. The validity of reactor models can also be tested by using molecular techniques such as FISH in conjunction with confocal laser scanning microscopy (CLSM). Detailed analysis of nitrification processes have provided complex multispecies models which can be used to evaluate operating strategies of treatment systems (Persson *et al.*, 2002). *In situ* hybridization studies have shown different degrees of associations and spatial segregation between ammonia-oxidizing bacteria (AOB) and nitrite oxidizing bacteria (NOB) in nitrifying trickling filter (NTF) reactors and sequencing batch biofilm reactors (Daims *et al.*, 2001; Pär *et al.*, 2006). This has contributed to our understanding of factors that may influence nitrifying biofilm structures and identify the impact of environmental conditions on the ecology of nitrifying bacteria. Persson *et al.* (2002), for example, have shown that proportions of AOB at different depths in NTF reactors can be compared with predictions that were made in multispecies biofilm reactor models. Clear zones of stratification have been seen in biofilms where populations of AOB and NOB occupy different depths within NTF reactor biofilms (Pär *et al.*, 2006). Levels of oxygen penetration into biofilm layers play a central role in the establishment of *Nitrosomonas* species (AOB) and NOB populations of which *Nitrospira* is the most common genus. Recent investigations however have given further insight into an array of processes that govern biofilm formation and stratification in NTF reactors. The appearance of rough ridge-like biofilm structures was linked with extracellular polysaccharides (EPS) production and low biomass at particular depths inside NTF reactors (Pär *et al.*, 2006). The presence of *Nitrosomonas oligotropha*, the most common AOB found in their NTF reactor, has previously been shown to produce large quantities of EPS and essentially provide protection against environmental factors such as toxic substances and grazing by protozoa and metazoa. This generates localized microenvironments within biofilms which stimulates the proliferation of specific species of bacteria relevant to nitrogen transformation processes. Knowledge about parameters present in microenvironments generated by microsensor measurements may offer recognition of the spatial arrangements that groups of bacteria important in nutrient removal process may occupy.

Molecular tools have contributed to advancing the application of novel and highly efficient nitrogen removal pathways, such as the anaerobic ammonium oxidation (anammox) and completely autotrophic nitrogen removal over nitrite (CANON) processes, through to pilot investigations more economically. Real-time quantitative PCR and FISH analyses have shown that aerobic ammonia oxidizing bacteria and anammox bacteria co-existed mainly in surface biofilm regions which the authors postulated is an indication of metabolic dependency of the anammox population on the AOB through the creation of microenvironments (Gieseke *et al.*, 2001; Gong *et al.*, 2007; Kindaichi *et al.*, 2007; Tsushima *et al.*, 2007). This leads to the development of community structures and spatial organization of populations in biofilms that can function optimally to obtain the desired output of the treatment system. Slow-growing anammox bacteria translate into long start-up periods in bioreactors. The immobilization of an anammox population is an efficient means of retaining low levels of biomass during the early stages of operation. For example, the immobilization of anammox bacteria on zeolite promoted the establishment of an anaerobic ammonia oxidizing population and prevented washout in a SBR during the start-up period (Fernández *et al.*, 2008). The co-existence of these two groups of functionally and metabolically diverse bacteria makes it possible to provide a platform for the establishment of a single stage autotrophic nitrogen removal process during which O_2 is rapidly consumed and favourable niches are created for accelerated nitrite conversion by anammox bacteria. Anammox bacteria were detected in different regions and different depths in biofilms grown on non-woven polyester sheets in a fixed-bed column reactor using FISH (Kindaichi *et al.*, 2007). This study showed the putative sensitivity of anammox bacteria to organic matter decay and detritus liberated from deeper sections in the biofilm that may have supported the growth of other heterotrophic bacteria. These competing populations impaired the function of the anammox populations since reduced anammox activity was measured downstream in the reactor where higher levels of dissolved organic carbon (DOC) was present even though high numbers of anammox bacteria were monitored. Their sensitivity to DOC presents a potential bottleneck in the application of this beneficial nitrogen removal pathway. A greater understanding of these restrictions and potential measures to alleviate metabolic impairment will support its use as a manageable and efficient process.

Concomitant removal of C, P and N can be obtained in biofilms since they provide a spatially heterogeneous environment for aerobic and anaerobic processes. The co-existing relationship between nitrogen removing bacteria and phosphate accumulating bacteria was revealed in studies that employed online and microscale investigations and enabled engineers to propose reactor modes of operation (Garzón-Zúñiga and Gonzlez-Martínez, 1996; Gieseke *et al.*, 2002; Terada *et al.*, 2006). This made possible the proliferation of reactor systems tailored to take advantage of the interdependent metabolic relationship and incorporate knowledge of competing metabolic requirements. Both nitrification and the uptake of phosphorous require aerobic conditions. However, an impediment to their concurrent activity is the different affinity for O_2 that nitrifying bacteria and polyphosphate accumulating organisms (PAO) have (Gieseke *et al.*, 2002). Awareness of the mechanisms that govern coexistence between functionally important groups of bacteria is essential to create optimum nutrient removal processes. Gieseke *et al.* (2002) used on-line and microsensor monitoring to establish profiles of O_2, PO_4^{3-} and nitrogen species in various sections of biofilms during treatment cycles in a sequencing batch biofilm reactor. The authors proposed that nitrification occurs after the uptake of phosphate owing to oxygen limitation. This develops because of the competition between PAO and nitrifiers for O_2 and the problem may be circumvented by extending the aeration phase in the reaction phase of SBR operation. In an earlier study, the authors found spatial and temporal separation of different populations of AOB, that are related to *N. europaea*, *N. oligotropha* and *N. communis* lineages, in a phosphate-removing biofilm in an SBR. This suggests that segregation is required for management of the different activities of these populations (Gieseke *et al.*, 2001). It is speculated that in the case of *N. europaea*-like AOB which were found in deeper regions of the biofilm, temporal coexistence with the other AOB is favoured as a result of O_2-limiting conditions that develop after heterotrophic PAO have taken up most of the O_2. Therefore, the appearance of anoxic conditions may support continuation of ammonia oxidization owing to *N. europaea* metabolic activity that can utilize alternative electron acceptors.

The simultaneous removal of nitrogen and phosphorous in a membrane-aerated biofilm reactor was shown to occur in different regions of a biofilm using online monitoring and FISH analysis (Terada *et al.*, 2006). The appearance of denitrifying polyphosphate-accumulating bacteria was shown to operate in deeper anoxic regions in the biofilm that are capable of utilizing nitrate as a terminal electron acceptor. This enabled engineers to propose a reactor configuration in which the immobilization of bacterial cells on gas permeable membrane tubes allowed for the physical separation of nitrifying and denitrifying phosphate accumulators. This configuration aided sustainable removal efficiency. Molecular and microsensor tools have also been applied in the characterization and activity measurements of sulfate-reducing bacterial populations in biofilms (Ito *et al.*, 2002; Koschorreck *et al.*, 2010).

Detoxification of pollutants

The major advantages of biofilm-based reactors over suspended biomass systems for the biodegradation of pollutants have been documented across numerous publications. These advantages include greater biomass retention, greater BOD removal, resistance to toxic load concentrations, and presence of anaerobic and aerobic zones in a single system. Biofilms are essentially communities of various microbial species and this arrangement provides for unparalleled functionality for the biodegradation of xenobiotics and other complex molecules. The biodegradative properties of the communities within the biofilm is greater that the properties of individual cells and can best be illustrated in co-metabolic processes (Sutherland, 2001). Current research topics that are garnering increased interest within the area of organic or inorganic solute removal in biofilm reactors include mathematical modelling to simulate and predict the biodegradation of pollutants in fixed-film biofilm reactors (de Mello *et al.*, 2010), the removal of volatile contaminants into the gaseous phase due to stripping in aerobic bioreactor operation (Farhadian *et al.*, 2008), and sorption sites in biofilms where contaminants are preferentially retained and remobilized (Späth *et al.*, 1998; Wicke *et al.*, 2007).

Important insights into the functionality of biofilms that detoxify wastewaters have been made. Xenobiotic compounds are removed through the processes of biosorption (passive uptake by interactions with biomass), biotransformation (alteration of molecules into less or non-hazardous compounds) and biomineralization (formation of inorganic break-down end products generated through microbial metabolic pathways). Heavy metal bioremediation can be achieved by immobilization, precipitation and partitioning to environmental compartments (biosorption) (Costley and Wallis, 2001; Singh *et al.*, 2006; Srivastava and Majumder, 2008). Modelling studies and laboratory investigations have confirmed that (i) increasing levels of toxic compound concentrations in wastewater necessitates an increase in the contact surface area inside biofilm reactors (Van Loosdrecht and Heijnen, 1993); (ii) a minimum residence time is required for biodegradation of organic pollutants including monoaromatic compounds even with a higher biomass retention inside a reactor (de Mello *et al.*, 2010); and (iii) sorption routes of hydrophobic organic compounds (HOC) such as polycyclic aromatic hydrocarbons (PAH) and heavy metals into biofilms comprised a two-stage process: an initial rapid sorption process followed by a second phase of slower diffusion (Wicke *et al.*, 2007; Phoenix and Holmes, 2008). The latter provides predictive data to establish kinetic parameters in pollutant uptake processes.

Sorption sites within biofilms and sorption mechanisms are important parameters that influence the distribution of pollutions within treatment systems. Possible sorption sites include cell walls, cytoplasmatic membranes, the cytoplasm and EPS substances (Flemming, 1995). A study into the partitioning of heavy metals (Cd, Zn) and BTX (benzene, toluene and *m*-xylene) compounds into biofilm fractions revealed that more than 80% of the metals were sorbed into the cellular fractions whereas more than 60% of the BTX compounds were found in the EPS, signalling important consideration when removal efficiency into different biomass fractions are investigated (Späth *et al.*, 1998). The probability exists that desorption and remobilization of heavy metal contaminants may occur upon cell lysis. The remobilization of heavy metals can be capitalized on by implementing desorption cycles in treatment systems. Costley and Wallis (2001) demonstrated effective recovery of biosorbed Cu, Cd and Zn in a rotating biological contactor. Sorption studies can also give valuable data with respect to the kinetics involved in removal of HOC from wastewater streams in biofilm reactors. It has been shown that biofilms are preferential sinks for PAH but also for more hydrophilic compounds such as glycerol and that diffusion is slower through the network of EPS layers than into biofilm pores (Vogt *et al.*, 2000; Wicke *et al.*, 2007). Prolonged periods to reach maximum sorption capacity and states of equilibrium between soluble and immobilized HOC concentrations have been shown to occur in suspended bacterial sorption studies. This contrasts with more rapid sorption maxima and equilibrium periods when the same bacterial cells used in the suspended study (*Kluvera cryocrescens*) were immobilized onto kaolin (Wang *et al.*, 2002). This indicates that processes of pollutant uptake are accelerated in biofilm-based reactors compared with suspended cell systems. The diffusion of nutrients and contaminants may nevertheless be one of the rate-limiting steps in biofilm operation. The use of analytical tools such as magnetic resonance imaging (MRI) has enabled the mapping of heavy metal diffusion and sequestration in biofilms (Phoenix and Holmes, 2008).

A potential bottleneck in the mineralization of HOC compounds in *in situ* remediation efforts has been identified as bioavailability restrictions where these chemicals are adsorbed to, for example, soil particles. Chemotaxis is one mechanism that can be exploited for bacteria to migrate towards these compounds to form biofilms and produce surfactants that will lead to an increase in availability (Singh *et al.*, 2006).

Conclusions and future trends

The design and operation of biofilm reactors for improved nutrient removal and detoxification of wastewater is ultimately dependent on our ability to gain a clearer understanding of both bulk-phase and biofilm processes. A crucial step towards this will be the development of research initiatives that integrate chemical, microbiological and molecular analyses with new mathematical modelling and engineering design approaches. Progress in this field will benefit considerably from the continued advancement of molecular tools that provide insight into metabolic processes within complex and dynamic microbial consortia. There is great scope for these tools to aid in the development of novel bioreactor configurations that will operate optimally in line with a particular application, as well as diagnostic tools for process monitoring.

The wastewater treatment sector also stands to benefit from the intense focus on biofilm research in the context of biomedicine which will generate new insight into fundamental processes in biofilm establishment and stability. However, there will still be a need for fundamental research into the establishment and maintenance of functional bacterial groups in reactor systems.

In summary, it has long been recognized that insight into the structure and metabolic characteristics of immobilized microbial communities is crucial to drive biofilm system design in wastewater treatment (Van Loosdrecht and Heijnen, 1993). The increasing suite of new molecular and modelling tools that enable microscale investigations is certainly accelerating our understanding of these processes. However, the challenge for the future will be the development of molecular tools that are sufficiently robust and streamlined to be accessible to wastewater treatment plant operators for routine management and trouble-shooting of process performance.

References

Annachhatre, A.P., and Bhamidimarri, S.M.R. (1992). Microbial attachment and growth in fixed-film reactors: Process startup considerations. Biotech. Adv. *10*, 69–91.

Bibby, K., Viau, E., and Peccia, J. (2010). Pyrosequencing of the 16S rRNA gene to reveal bacterial pathogen diversity in biosolids. Water Res. *44*, 4252–4260.

Bolton, J., Tummala, A., Kapadia, C., Dandamudi, M., and Belovich, J.M. (2006). Procedure to quantify biofilm activity on carriers used in wastewater treatment systems. J. Environ. Eng.*132*, 1422–1430.

Boltz, J.P., and Daigger, G.T. (2009). Modeling integrated fixed-film activated sludge and moving-bed biofilm reactor systems I: Mathematical treatment and model development. Water Environ. Res. *81*, 555–575.

Chen, C.L., Wu, J.H., Tseng, I.C., Liang, T.M., and Liu, W.T. (2009). Characterization of active microbes in a full-scale anaerobic fluidized bed reactor treating phenolic wastewater. Microbe Environ. *24*, 144–153.

Clifford, E., Nielsen, M., Sørensen, K., and Rodgers, M. (2010). Nitrogen dynamics and removal in a horizontal flow biofilm reactor for wastewater treatment. Water Res. *44*, 3819–3828.

Cortez, S., Teixeira, P., Oliveira, R., and Mota, M. (2008). Rotating biological contactors: A review on main factors affecting performance. Rev. Environ. Sci. Biotechnol. 7, 155–172.

Costley, S.C., and Wallis, F.M. (2001). Bioremediation of heavy metals in a synthetic wastewater using a rotating biological contactor. Water Res. 35, 3715–3723.

Cresson, R., Carrère, H., Delgenès, J.P., and Bernet, N. (2006). Biofilm formation during the start-up period of an anaerobic biofilm reactor-Impact of nutrient complementation. Biochem. Eng. J. *30*, 55–62.

Daims, H., Nielsen, J.L., Nielsen, P.H., Schleifer, K.-H., and Wagner, M. (2001). *In situ* characterization of Nitrospira-like nitrite-oxidizing bacteria active in wastewater treatment plants. Appl. Environ. Microbiol. *67*, 5273–5284.

Daims, H., Taylor, M.W., and Wagner, M. (2006). Wastewater treatment: a model system for microbial ecology. Trends Biotechnol. *24*, 483–489.

Di Trapani, D., Mannina, G., Torregrossa, M., and Viviani, G. (2008). Hybrid moving bed biofilm reactors: a pilot plant experiment. Water Sci. Technol. *57*, 1539–1545.

Eaton, A.D., Clesceri, L.S., Rice, E.W., Greenberg, A.E., and Franson, M.A.H. (2005). Standard Methods for the Examination of Water and Wastewater, APHA (American Waterworks Association, Water Environment Federation, Baltimore, MD).

Farhadian, M., Duchez, D., Vachelard, C., and Larroche, C. (2008). Monoaromatics removal from polluted water through bioreactors – a review. Water Res. *42*, 1325–1341.

Fernández, I., Vázquez-Padín, J.R., Mosquera-Corral, A., Campos, J.L., and Méndez, R. (2008). Biofilm and granular systems to improve Anammox biomass retention. Biochem. Eng. J. *42*, 308–313.

Flemming, H.-C. (1995). Sorption sites in biofilms. Water Sci. Technol. *32*, 27–33.

Garcia, M.L., Lapa, K.R., Foresti, E., and Zaiat, M. (2008). Effects of bed materials on the performance of an anaerobic sequencing batch biofilm reactor treating domestic sewage. J. Environ. Manag. *88*, 1471–1477.

Garzón-Zúñiga, M.A., and Gonzlez-Martínez, S. (1996). Biological phosphate and nitrogen removal in a biofilm sequencing batch reactor. Water Sci. Technol. *34*, 293–301.

Gieseke, A., Purkhold, U., Wagner, M., Amann, R., and Schramm, A. (2001). Community Structure and activity dynamics of nitrifying bacteria in a phosphate-removing biofilm. Appl Environ. Microbiol. *67*, 1351–1362.

Gieseke, A., Arnz, P., Amann, R., and Schramm, A. (2002). Simultaneous P and N removal in a sequencing batch biofilm reactor: insights from reactor- and microscale investigations. Water Res. *36*, 501–509.

Gong, Z., Yang, F., Liu, S., Bao, H., Hu, S., and Furukawa, K. (2007). Feasibility of a membrane-aerated biofilm reactor to achieve single-stage autotrophic nitrogen removal based on Anammox. Chemosphere *69*, 776–784.

Hall-Stoodley, L., Costerton, J.W., and Stoodley, P. (2004). Bacterial biofilms: from the natural environment to infectious diseases. Nat. Rev. Microbiol. *2*, 95–108.

Heijstra, B.D., Pichler, F.B., Liang, Q., Blaza, R.G., and Turner, S.J. (2009). Extracellular DNA and Type IV pili mediate surface attachment by *Acidovorax temperans*. Anton. Leeuwen. Int. J. Gen *95*, 343–349.

Ito, T., Okabe, S., Satoh, H., and Watanabe, Y. (2002). Successional development of sulfate-reducing bacterial populations and their activities in a wastewater biofilm growing under microaerophilic conditions. Appl. Environ. Microbiol. *68*, 1392–1402.

Kindaichi, T., Tsushima, I., Ogasawara, Y., Shimokawa, M., Ozaki, N., Satoh, H., and Okabe, S. (2007). *In situ* activity and spatial organization of anaerobic ammonium-oxidizing (anammox) bacteria in biofilms. Appl. Environ. Microbiol. *73*, 4931–4939.

Koschorreck, M., Geller, W., Neu, T., Kleinsteuber, S., Kunze, T., Trosiener, A., and Wendt-Potthoff, K. (2010). Structure and function of the microbial community in an *in situ* reactor to treat an acidic mine pit lake. FEMS Microbiol. Ecol. *73*, 385–395.

Lazarova, V., and Manem, J. (1995). Biofilm characterization and activity analysis in water and wastewater treatment. Water Res. *29*, 2227–2245.

Lewandowski, Z., and Boltz, J.P. (2011). Biofilms in water and wastewater treatment. In Treatise on Water Science, Peter, W., ed. (Elsevier, Oxford), pp. 529–570.

Liu, R., Li, D., Gao, Y., Zhang, Y., Wu, S., Ding, R., Hesham, A.E.-L., and Yang, M. (2010). Microbial diversity in the anaerobic tank of a full-scale produced water treatment plant. Process Biochem. *45*, 744–751.

Lücker, S., Wagner, M., Maixner, F., Pelletier, E., Koch, H., Vacherie, B., Rattei, T., Damsté, J.S.S., Spieck, E., Le Paslier, D., *et al.* (2010). A Nitrospira metagenome illuminates the physiology and evolution of globally important nitrite-oxidizing bacteria. P. Natl. Acad. Sci. U.S.A. *107*, 13479–13484.

Lydmark, P., Lind, M., Sörensson, F., and Hermansson, M. (2006). Vertical distribution of nitrifying populations in bacterial biofilms from a full-scale nitrifying trickling filter. Environ. Microbiol. *8*, 2036–2049.

McLellan, S.L., Huse, S.M., Mueller-Spitz, S.R., Andreishcheva, E.N., and Sogin, M.L. (2010). Diversity and population structure of sewage-derived microorganisms in wastewater treatment plant influent. Environ. Microbiol. *12*, 378–392.

Mannina, G., and Viviani, G. (2009). Hybrid moving bed biofilm reactors: An effective solution for upgrading a large wastewater treatment plant. Water Sci. Technol. *60*, 1103–1116.

Mannina, G., Di Trapani, D., Torregrossa, M., and Viviani, G. (2007). Modelling of hybrid moving bed biofilm reactors: A pilot plant experiment. Water Sci. Technol. *55*, 237–246.

Martin, H.G., Ivanova, N., Kunin, V., Warnecke, F., Barry, K.W., McHardy, A.C., Yeates, C., He, S., Salamov, A.A., Szeto, E., *et al.* (2006). Metagenomic analysis of two enhanced biological phosphorus removal (EBPR) sludge communities. Nat. Biotech. *24*, 1263–1269.

de Mello, J.M.M.D., de Lima Brandão, H., de Souza, A.A.U., da Silva, A., and de Souza, S.M.D.A.G.U. (2010). Biodegradation of BTEX compounds in a biofilm reactor--Modeling and simulation. J. Petrol. Sci. Eng. *70*, 131–139.

Ødegaard, H. (2006). Innovations in wastewater treatment: The moving bed biofilm process. Water Sci. Technol. *53*, 17–33.

Ødegaard, H., Rusten, B., and Westrum, T. (1994). A new moving bed biofilm reactor: Applications and results. Water Sci. Technol. *29*, 157–165.

Pär, L., Magnus, L., Fred, S., and Malte, H. (2006). Vertical distribution of nitrifying populations in bacterial biofilms from a full-scale nitrifying trickling filter. Environ. Microbiol. *8*, 2036–2049.

Persson, F., Wik, T., Sörensson, F., and Hermansson, M. (2002). Distribution and activity of ammonia oxidizing bacteria in a large full-scale trickling filter. Water Res. *36*, 1439–1448.

Phoenix, V.R., and Holmes, W.M. (2008). Magnetic resonance imaging of structure, diffusivity, and copper immobilization in a phototrophic biofilm. Appl. Environ. Microbiol. *74*, 4934–4943.

Rittmann, B.E. (2006). Microbial ecology to manage processes in environmental biotechnology. Trends Biotechnol. *24*, 261–266.

Rittmann, B.E., Hausner, M., Loffler, F., Love, N.G., Muyzer, G., Okabe, S., Oerther, D.B., Peccia, J., Raskin, L., and Wagner, M. (2006). A vista for microbial ecology and environmental biotechnology. Environ. Sci. Technol. *40*, 1096–1103.

Roest, K., Heilig, H.G.H.J., Smidt, H., de Vos, W.M., Stams, A.J.M., and Akkermans, A.D.L. (2005). Community analysis of a full-scale anaerobic bioreactor treating paper mill wastewater. System. Appl. Microbiol. *28*, 175–185.

Rowan, A.K., Snape, J.R., Fearnside, D., Barer, M.R., Curtis, T.P., and Head, I.M. (2003). Composition and diversity of ammonia-oxidising bacterial communities in wastewater treatment reactors of different design treating identical wastewater. FEMS Microbiol. Ecol. *43*, 195–206.

Sanapareddy, N., Hamp, T.J., Gonzalez, L.C., Hilger, H.A., Fodor, A.A., and Clinton, S.M. (2009). Molecular diversity of a north Carolina wastewater treatment plant as revealed by pyrosequencing. Appl. Environ. Microbiol. *75*, 1688–1696.

Seviour, R.J., and Nielsen, P.H. (2010). Microbiology of Activated Sludge (IWA Publishing, London).

Sharma, K., Zhang, X., Liu, Y., Lin, W., and Yu, T. (2008). Biological fixed film systems. Water Environ. Res. *80*, 1078–1112.

Siezen, R.J., and Galardini, M. (2008). Genomics of biological wastewater treatment. Microb. Biotechnol. *1*, 333–340.

Singh, R., Paul, D., and Jain, R.K. (2006). Biofilms: implications in bioremediation. Trends Microbiol. *14*, 389–397.

Slater, F.R., Singer, A.C., Turner, S., Barr, J.J., and Bond, P.L. (2011). Pandemic pharmaceutical dosing effects on wastewater treatment: no adaptation of activated sludge bacteria to degrade the antiviral drug Oseltamivir (Tamiflu®) and loss of nutrient removal performance. FEMS Microbiol. Lett. *315*, 17–22.

Späth, R., Flemming, H.C., and Wuertz, S. (1998). Sorption properties of biofilms. Water Sci. Technol. *37*, 207–210.

Srivastava, N.K., and Majumder, C.B. (2008). Novel biofiltration methods for the treatment of heavy metals from industrial wastewater. J. Hazard. Mater. *151*, 1–8.

Sutherland, I.W. (2001). The biofilm matrix – an immobilized but dynamic microbial environment. Trends Microbiol. *9*, 222–227.

Syron, E., and Casey, E. (2008). Membrane-aerated biofilms for high rate biotreatment: Performance appraisal, engineering principles, scale-up, and development requirements. Environ. Sci.Technol. *42*, 1833–1844.

Talbot, G., Topp, E., Palin, M.F., and Massé, D.I. (2008). Evaluation of molecular methods used for establishing the interactions and functions of microorganisms in anaerobic bioreactors. Water Res. *42*, 513–537.

Tchobanoglous, G., Burton, F.L., and Stensel, H.D. (2002). Wastewater Engineering: Treatment and Reuse (Metcalf & Eddy Inc., New York).

Terada, A., Hibiya, K., Nagai, J., Tsuneda, S., and Hirata, A. (2003). Nitrogen removal characteristics and biofilm analysis of a membrane-aerated biofilm reactor applicable to high-strength nitrogenous wastewater treatment. J. Biosci. Bioeng. *95*, 170–178.

Terada, A., Yamamoto, T., Tsuneda, S., and Hirata, A. (2006). Sequencing batch membrane biofilm reactor for simultaneous nitrogen and phosphorus removal: Novel application of membrane-aerated biofilm. Biotechnol. Bioeng. *94*, 730–739.

Tsushima, I., Ogasawara, Y., Kindaichi, T., Satoh, H., and Okabe, S. (2007). Development of high-rate anaerobic ammonium-oxidizing (anammox) biofilm reactors. Water Res. *41*, 1623–1634.

Van Loosdrecht, M.C.M., and Heijnen, S.J. (1993). Biofilm bioreactors for waste-water treatment. Trends Biotechnol. *11*, 117–121.

Villaverde, S. (2004). Recent developments on biological nutrient removal processes for wastewater treatment. Rev. Environ. Sci. Biotechnol. *3*, 171–183.

Vogt, M., Flemming, H.C., and Veeman, W.S. (2000). Diffusion in *Pseudomonas aeruginosa* biofilms: a pulsed field gradient NMR study. J. Biotechnol. *77*, 137–146.

Vymazal, J. (2009). The use constructed wetlands with horizontal sub-surface flow for various types of wastewater. Ecol. Eng. *35*, 1–17.

Wagner, M., Loy, A., Nogueira, R., Purkhold, U., Lee, N., and Daims, H. (2002). Microbial community composition and function in wastewater treatment plants. Antonie van Leeuwenhoek *81*, 665–680.

Wang, C., Li, J., Wang, B., and Zhang, G. (2006). Development of an empirical model for domestic wastewater treatment by biological aerated filter. Process Biochem. *41*, 778–782.

Wang, W., Wang, W., Zhang, X., and Wang, D. (2002). Adsorption of *p*-chlorophenol by biofilm components. Water Res. *36*, 551–560.

Wicke, D., Böckelmann, U., and Reemtsma, T. (2007). Experimental and modeling approach to study sorption of dissolved hydrophobic organic contaminants to microbial biofilms. Water Res. *41*, 2202–2210.

Wojnowska-Baryła, I., Cydzik-Kwiatkowska, A., and Zielińska, M. (2010). The application of molecular techniques to the study of wastewater treatment systems. In Bioremediation, Cummings, S.P., ed. (New York, Humana Press), pp. 157–183.

Corrosion and Fouling

7

Steve Flint and Gideon Wolfaardt

Abstract

Biofilms can directly or indirectly be attributed to deterioration of the underlying substratum. Corrosion may result, particularly if the surface comprises metal or metal alloy. This phenomenon, referred to as microbially influenced corrosion (MIC) affects many industries from food manufacture to medicine. The economic impact of corrosion is significant owing to the need for replacing corroded equipment, repairs and attempts to prevent corrosion. MIC is believed to be responsible for one third of all metallic corrosion. Although there have been many studies into the mechanisms of MIC, the process is relatively poorly understood. Most information relates to pure cultures, however biofilms are rarely composed of single species thus most models are a simplification of the real process. It is probable the MIC depends on the composition of the biofilm and the environment surrounding the biofilm. Prevention and control methods rely on mechanical cleaning of fouling and chemical removal and killing of biofilms. Future control measures are likely to focus on preventing biofilm formation.

Introduction

Corrosion is a process whereby the electrical potential between a metallic surface differs to that of a surrounding conductive medium, such that one is anodic and the other cathodic (Coetser and Cloete, 2005). As later described, the resulting electrochemical reactions of the metal with its surroundings lead to the disintegration of the metal. The effects of corrosion in industry are significant and costly, affecting most major sectors of the modern world from food processing to water supply to the oil industry (Koch *et al.*, 2001).

Efforts to prevent corrosion and plant shutdown by addressing corrosion issues, repairing corrosion related damage and researching innovative techniques for mitigating undesirable corrosive effects are expensive (Parsek and Fuqua, 2004). In the industrial situation, many of the surfaces (pipelines, vats, marine structures, heat exchanges, etc.) are made of metal and as such are susceptible to corrosion (Fig. 7.1). Metals exist in nature typically in a low-energy oxide form and when manmade industrial structures are fabricated, the metal is tempered into a high energy form. It therefore has a tendency to revert back to an oxide form if given the opportunity and corrosion is a process that allows this transition. Water is one of the key requirements for corrosion and is readily available in many industrial systems. Whilst corrosion is generally exhibited on metal surfaces, polymers and concrete are also prone to corrosion to a lesser extent.

Many different types of bacteria exhibit corrosive effects through the action of their metabolic end products. Foremost among these are the sulfate reducing bacteria (SRB) such as *Desulfovibrio* spp. These facultative anaerobes have been studied extensively and the mechanisms by which they cause corrosion are becoming better understood. However, costly effects are still observed, particularly in the petroleum industry. The metabolism of SRB results in corrosion of drilling and pumping machinery, storage tanks, oil spoilage by sulfide production, plugging of machinery and rock

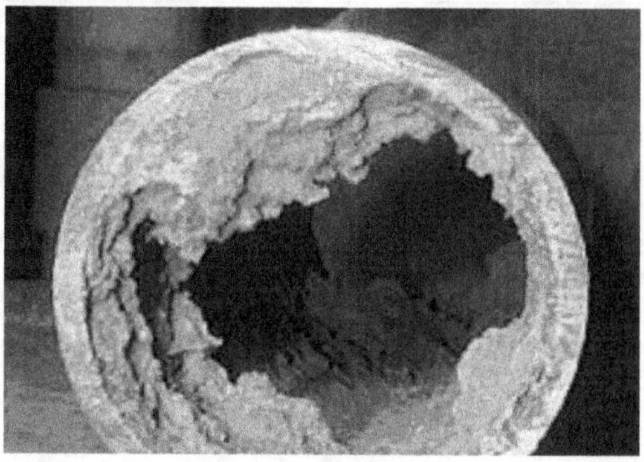

Figure 7.1 Corroding stainless steel pipe used for waste from a food manufacturing plant. Bacteria often play a pivotal role in the corrosion through the production of a large variety of extracellular polymeric substances that act as a matrix in which biotic and abiotic substances can accumulate. Note the corrosion of the metal surface can be seen in the cross section of the pipe.

pores with biomass and sulfide precipitates (biofouling) (Heidelberg *et al.*, 2004). In biofilms, the SRB have an advantage over other bacteria as they are capable of using sulfate as a terminal electron acceptor (Chan *et al.*, 2002). Acid producing bacteria such as *Thiobacillus* spp. are capable of producing sulfuric acid, an intrinsically corrosive metabolite. Hydrogen producing bacteria such as *Clostridium* spp. are capable of causing hydrogen embrittlement on metals as atomic hydrogen diffuses into the metal lattice. This can result in a loss of ductility and tensile strength, crack propagation or stress corrosion cracking and increase in the potential for catastrophic failure (Coetser and Cloete, 2005). Although this is not strictly corrosion, it does impact on the metal strength. Iron bacteria (*Gallionella*, *Crenothrix*, etc.) oxidize ferrous ions (Fe^{2+}) to ferric ions (Fe^{3+}). Ferric iron can bind free chloride ions in solution which react to form ferric chlorides that are deposited as corrosion products.

Microbial influenced corrosion (MIC) refers to the accelerated degradation of metals due to biofilm-associated organisms (Beech and Sunner, 2004). It has been estimated that at least one third of material loss that arises from corrosion can be attributed to MIC (Akid *et al.*, 2008). The presence of biofilm increases the likelihood of corrosion, owing to the creation of differential aeration cells on the metal surface. The ability of biofilms to withstand harsh conditions such as acid exposure, metal toxicity, dehydration, chemical sanitation and physical dislodgement, enable proliferation of biofilm in industrial environments and present problems in removing them once established (Lindsay and von Holy, 2006).

The first reports that speculated microorganisms were influencing the corrosion of metals, date back to the early twentieth century (Gaines, 1910; Videla, 1991).

It was not until 1934 that a reasonable attempt was made at explaining the processes/mechanisms responsible for the observed MIC (von wolzogen Kuhr and van der Flugt, 1934). The publication proposes a theory of cathodic depolarization. The basic principle of this theory was that the removal of hydrogen from the cathodic area on the metal surface by hydrogenases from the bacteria coupled with the reduction of sulfate to sulfide, would depolarize the cathodic reaction and consequently accelerate the corrosion process. Although not absolutely correct, this work provided the basis for further study of the mechanisms of MIC as reviewed in Videla and Herrera (2005).

The mechanism used by all microorganisms to produce and utilize energy from the environment, involves the flow of electrons from negative to more positive potentials. This process utilizes specialize electron carriers (e.g. ferredoxins, flavoproteins and cytrochromes). Different organisms produce different types of these components. This results in variation in the energy sources and terminal electron acceptors observed in the microbial world (Hamilton, 2003). With aerobic micro-organisms, oxygen is the terminal electron acceptor, whereas anaerobic microorganisms use various terminal electron acceptors (e.g. ferric ion, nitrate, sulfate, CO_2). These processes, often integrated with processes in microorganisms of different species, are the driving force behind MIC.

From a chemical perspective, MIC is relatively complex. The complete interactions governing MIC are inadequately understood. The corrosion process is facilitated by electrochemical reactions on the metal interface. The anodic and cathodic reactions and their relationship with one another, play an important role in determining the overall corrosion rate. This is determined by the slowest partial reaction, either reduction at the cathode or oxidation at the anode. As the two reactions are in equilibrium, any electron produced in the oxidation reaction at the anode, must be consumed in the reduction reaction at the cathode. Examples of anodic and cathodic reactions on ferrous metals in aerobic environments are detailed in the following equations.

$$Fe \rightarrow Fe^{2+} + 2e^{2-}$$

$$2H_2O + O_2 + e^- \rightarrow 4OH^-$$

Products of these reactions are converted into ferric hydroxides in the presence of oxygen and water.

$$Fe^{2+} + 2(OH) \rightarrow Fe(OH)_2$$

$$Fe(OH)_2 + \tfrac{1}{2}H_2O + \tfrac{1}{4}O_2 \rightarrow Fe(OH)_3$$

These ferric hydroxides are deposited and typically observed as rust. The simple reaction shown in the first reaction is sufficient to cause corrosion owing to physical dissolution of the metal. Biofilms can have a beneficial effect on MIC, minimizing aerobic corrosion by consuming oxygen and rendering it unusable as a terminal electron acceptor in aerobic reduction reactions.

The rate of corrosion can be quite astounding. Sheng *et al.* (2007) studied the corrosive effects that a SRB, *Desulfovibrio desulfuricans* exerted on stainless steel over a 44-day period. Using atomic force microscopy (AFM) they were able to determine the depth of pitting corrosion over the course of the experiments (Fig. 7.2). Pitting of up to 265 nm was evident after only 4 days, and this with a single species biofilm. It has also been reported that over 40 days, a SRB isolated from marine sediment, caused 1390 nm corrosion on mild steel (Xu *et al.*, 1999).

Recent research has focussed on extracellular polymeric substances (EPS) and their affect on electrochemical reactions at the metal surface. This is typically observed as an exacerbation of corrosion but can also be instrumental in inhibiting corrosion. One of the first studies to address this was by Roe *et al.* (1996). The authors showed that the presence of a cell free exopolysaccharide (alginate) on the surface of steel was sufficient to induce corrosion. Apart from the direct role the extracellular matrix plays in the mechanical inhibition of diffusion of reactants such as oxygen, the matrix may play a more direct role in corrosion. One of the key properties of the matrix is its potential ability to bind metal ions (Kinzler *et al.*, 2003; Rohwerder *et al.*, 2003). This ability to bind metal ions is dependent on interaction between the anionic functional groups (e.g. carboxyl, phosphate, sulfate, glycerate, pyruvate or succinate) or the polypeptide or polysacchraride components of the extracellular matrix (Beech and Sunner, 2004). The binding of metal ions allows them to act as electron 'shuttles', allowing for the formation of novel electron transfer pathways within the biofilm which can result in cathodic depolarization and subsequent corrosion. It has been recently demonstrated experimentally that biofilm-associated polysaccharides could bind ferric ions in the carboxyl groups of the polysaccharide resulting in iron oxyhydrate (Chan *et al.*, 2004). This, in theory, could result in cathodic depolarization. In another study, it was shown that the uronic acid residues of a polysaccharide produced by *Acidithiobacillus ferrooxidans* or *Leptospirillum ferrooxidans* can bind and concentrate ferric ions (Sand and Gehrke, 2006). In this study this process increased the rate of bioleaching (extracting metals from their ores through microbial activities), which is a very similar process to biocorrosion and utilizes many similar mechanisms. Thus, it is not unreasonable to assume a similar process may be occurring for biocorrosion.

The mode of corrosion inhibition by EPS is generally very intricate and not easily attributed to one mechanism. However, two widely accepted processes are thought to be central in this system (1) the formation of a protective film layer that prevents access of microorganisms to the metal and (2) the neutralization of corrosive substances, present in the biofilm (Videla and Herrera, 2009). Enzymatic reactions are also important in MIC. These can be associated with the microbial surface or secreted extracellularly in the biofilm matrix and beyond. The action and impact of hydrogenase in particular dominates the literature on this subject.

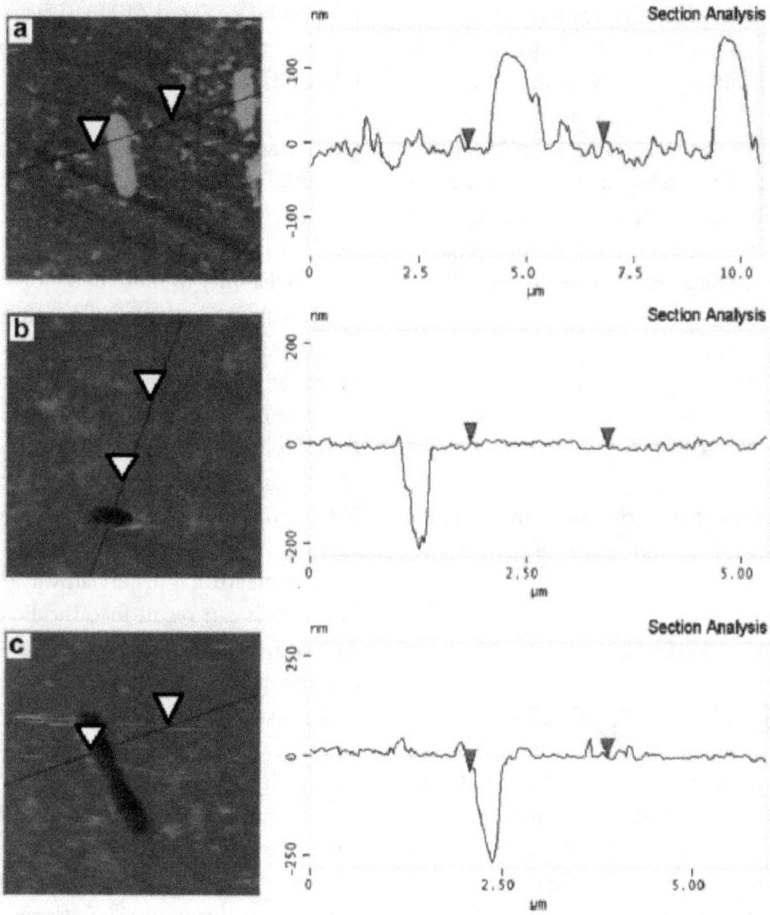

Figure 7.2 Pitting corrosion analysed by atomic force microscopy (Sheng *et al.*, 2007). (A) height profile of *Desulofibrio desulfuricans* cells (white patches). (B) Depth profile of a small pit (dark patch). (C) Depth profile of a large pit (dark patch). The y-axis is the measurement in nm. The triangles show relative positions on the graphs and the AFM image.

The role of microorganisms with known ability of causing biodeterioration of metals needs to be further understood before we can go some way to reducing their deleterious effects. The interactions of microbial consortia are complex and undoubtedly crucial to appreciating the bigger picture in biofilm ecology. Advances in molecular and genetic techniques provide an insight into the microbial ecology of biofilms. For example, the 3,570,858 bp genome sequence of the SRB *Desulfovibrio vulgaris* Hildenborough, has now been reported (Heidelberg *et al.* 2004).

Microbial induced corrosion is to a large degree the consequence of microbial colonization of surfaces as biofilm. Biofilms, widely recognized as the dominant form of bacterial existence, are typically communities of cells of differing species 'encased' in an extracellular matrix composed of polysaccharides, proteins and nucleic acids (Branda *et al.* 2005). Biofilms are highly structured and the concentration of various reactants differs across the biofilm. Microelectrode measurements and theoretical calculations have shown that the concentration of oxygen decreases with the thickness of the biofilm, with regions of the biofilm being completely anaerobic (Lewandowski *et al.*, 1989; Costerton *et al.*, 1995; Stewart, 2003). Bester *et al.* (2009) showed that in addition to the important role of biofilms to create a micro-habitat where cell function can be notably different from that of planktonic cells, biofilms may also play a key role as 'cell nurseries' and thus proliferation. The same authors (Bester *et al.*, 2010, 2011) showed that a small fraction (i.e. less than 5%) of carbon used by the biofilms is channelled into the yield of a notable number; up to 10^7 cells yielded per cm^2 per hour released from the biofilms to the planktonic state. This may be another factor that complicates efforts to control biofouling and MIC as the constant source for re-inoculation may be at an upstream location that is not treated.

Aerobic corrosion

In aerobic biofilms, oxygen is present in the liquid and penetrates throughout the entire biofilm. The best studied process contributing to MIC from aerobic biofilms is the formation of tubercles (Lee *et al.*, 1993a,b; Lee and Beer, 1995). These authors show that during the early stages of biofilm development with sulfate-reducing bacteria, scattered growth on the surface of metal results in a primary stage of oxygen-dependent corrosion and ultimately the formation of tubercles (or raised 'nodules') and localized regions of decreased pH. Oxygen reduction occurs at the outer border of the tubercle and iron dissociation occurs underneath/within the tubercle. Ferrous ions migrate out through the tubercle and this causes an influx of anions (such as Cl–) into the tubercle, which results in the gradual reduction of the pH within the tubercle.

Some bacteria can release metabolites such as organic or inorganic acids, that cause corrosion. The best characterized of these are thiobacilli. The thiobacilli are aerobic chemolithotrophic autotrophs that obtain energy from the fixation of CO_2 by the oxidation of sulfur or reduced sulfur components to sulfuric acid (Garrity, 2005). Two of these microorganisms, *Thiobacillus thiooxidans* and *Thiobacillus concretivorus*, can grow in acid conditions as low as pH 0.7 (Coetser and Cloete, 2005). These sulfur-oxidizing microorganisms produce sulfuric acid via the following reactions:

$$2H_2S + SO_2 \rightarrow H_2S_2O_2 + H_2O$$

$$5Na_2S_2O_3 + 4O_2 + H_2O \rightarrow 5Na_2SO_4 + H_2SO_4 + 4S$$

$$4S + 6O_2 + 4H_2O \rightarrow 4H_2SO_4$$

Iron-oxidizing bacteria (e.g. of the genera *Gallionella* and *Leptothrix*) can also contribute to the aerobic corrosion process. These bacteria oxidize ferrous ions (Fe^{2+}) to ferric ions (Fe^{3+}). Ferric ions can bind chloride ions to form ferric chloride. Ferric chloride is an aggressive corrosive and forms pits in stainless steel (Coester and Cloete, 2005).

Anaerobic corrosion

The most economically important and best studied (but still poorly understood) mechanisms for MIC, is the corrosion caused by the anaerobic sulfate-reducing bacteria (Hamilton, 1985, 1998; Hamilton and Lee, 1995; Lee *et al.*, 1995). The first mechanism sulfate-reducing bacteria use that promotes corrosion is the removal of hydrogen from a cathodic area with microbially produced hydrogenases (Iverson, 1987; Bryant *et al.* 1991; de Romero *et al.*, 2002, 2003). This is essentially the cathodic depolarization first described in 1934 (von Wolzogen Kuhr and van der Flugt, 1934). It has been shown that the presence of hydrogenase in the absence of viable bacteria is sufficient to induce this cathodic depolarization (Da Silva *et al.*, 2004). Removal of the hydrogen at the cathode drives the cathodic reaction and subsequently promotes further metal dissociation at the anode. The removal of hydrogen and subsequent reduction of sulfate (SO_4^{2-}) to S^{2-} results in the precipitation of ferrous sulfide corrosion product in the anaerobic region of the biofilm. A unifying electron transfer hypothesis was proposed to explain the various reactions occurring in a biofilm containing anaerobic sulfate reducing bacteria (Hamilton, 2003). The model proposes a reaction that takes into account the dynamic nature of the biofilm, in regards to the variable aerobic and anaerobic regions of the biofilm. The model proposed that the ferrous and sulfide ions can be oxidized in the more aerobic regions of the biofilm where oxygen can act as a terminal electron acceptor. This oxidation generates further corrosion products such as ferric oxide, ferric hydroxide and elemental sulfur. It has been noted that elemental sulfur can play a role in the following cathodic reaction (Lee *et al.*, 1993a,b; Neilsen *et al.*, 1993):

$$S + H_2O + 2e^- \rightarrow HS^- + OH^-$$

The formation of an iron sulfide corrosion product layer may first protect the metal surface from further corrosion, but these films are unstable and the breaking of this film results in a highly reactive site of corrosion between the iron sulfide (cathode) and the freshly exposed iron surface (anode). This model is almost an oversimplification but does provide a sound basis for the mechanisms of corrosion induced by sulfate-reducing bacteria. One of the key variables this model 'overlooks' is the fact that only very rarely are biofilms composed of a single species. It does not address the effects that other microbial metabolites may have on the reactions. The effect of mixed biofilms is best illustrated in two papers which demonstrated that the rates of corrosion induced by a mixed species biofilm is greater than those of biofilms composed of the respective single species (Valencia-Cantero *et al.*, 2003; Pitonzo *et al.*, 2004). In earlier work focussing on the effect of sulfate concentration on MIC rates by pure culture SRB in anaerobic laboratory reactors, Wolfaardt and Cloete (1992) observed that when the reactors were left aerobic and allowed aerobic bacteria to colonize the corrosion coupons, MIC rates in fact increased (data not shown).

The methanogenic bacteria are an important group that may play a role in mixed species biofilm induced corrosion. They can cause cathodic depolarization by mechanisms separate to those used by the sulfate reducing bacteria. Generally, they remove hydrogen from the cathodic area during the reduction of CO_2 to CH_4 although they can also use methanol or acetate (Boopathy and Daniels, 1991).

The iron reducing bacteria play an important role in microbially influenced corrosion. These microorganisms use iron ions as electron acceptors in their respiratory pathway (Gorby and Lovley, 1991; Lovley, 1991). Many bacterial species have been shown to be capable of using Fe^{3+} as an electron acceptor in their respiratory pathway (Lovley and Coates, 2000). It is still relatively poorly understood to what degree these microorganisms affect MIC (Lee and Newman, 2003). The reduction of iron ions is thought to result in an anodic depolarization and subsequent increase in the corrosion rate (Coester and Cloete, 2005). It has been suggested the combination of the reduction of iron ions and subsequent anodic depolarization and the cathodic depolarization caused by SRB may result in increased levels of corrosion (Ford and Mitchell, 1990). A different observation has been made in several papers where iron reducing microorganisms have been shown to inhibit corrosion (Potekhina *et al.*, 1999; Dubiel *et al.*, 2002). The theory behind this is that bacterial Fe^{3+} respiration protects from corrosion by using Fe^{2+} to reduce O_2 but this can only occur in aerobic biofilms. These studies used pure (single species) cultures. Thus it remains unclear how iron-reducing organisms affect iron corrosion in a natural dynamic mixed culture biofilm (Lee and Newman, 2003).

Extracellular polymeric substances and their role in corrosion

EPS is a complex macromolecular matrix composed of proteins, carbohydrates, lipids and nucleic acids (Beech and Sunner, 2004). A mature biofilm is only approximately 10% bacterial cells by weight and therefore EPS make up a considerable portion of the biofilm. EPS confers several interesting attributes to the mature biofilm including elasticity and therefore some resistance to shear forces (Hall-Stoodley *et al.*, 2004). EPS also plays an important role in facilitating biocorrosion of metal surfaces. Indirect effects of EPS are enhancing the ability to attach and persistence of biofilms providing a niche for the chemical reactions that promote corrosion.

Various components of the biofilm matrix are involved in metal corrosion. These include oxygen, chloride ions and metal ions (Chan *et al.*, 2002). Their presence and endurance in the biofilm matrix is stabilized by the presence of EPS (Johansson and Saastamoinen, 1999). Functional anionic groups on carbohydrates and proteins (carboxyl, succinate, pyruvate, phosphate, sulfate, glycerate, etc.) in the EPS have the capacity to bind metal ions, which can cause an increase in cathodic potential (Beech and Sunner, 2004).

The affinity of the EPS to bind metal ions in different oxidation states may result in diverse reduction potentials. Fe^{2+} and Fe^{3+} reduction potentials for example, fluctuate significantly with altered ligands (from +12 V to −0.4 V) (Beech and Sunner, 2004). This can create new electron pathways that lead to cathodic depolarization, and therefore increase the rate of corrosion. Fig. 7.3 shows ferrous ion–EPS complexes and their effect on oxidation and corrosion within an oxygenated biofilm.

Another example of EPS binding metal ions in biofilms was illustrated by Sand *et al.* (2006). They used two extremophilic bacteria *Acidithiobacillus ferrooxidans* and *Leptospirillum ferrooxidans,* in their experiments, looking into their role in bioleaching and biocorrosion. Sand and Gehrke (2006) found that in addition to the main components of EPS (lipids, sugars, etc.), glucuronic acid was also present, complexed with iron (III) ions. This gave the bacteria a net positive charge and their substrate, pyrite, has a net negative charge at low pH. This facilitated interactions between bacteria and substrate and helped in the initial phases of electrostatic attachment. It was also found that the ability of EPS to bind ion irons increased the iron concentration locally and this caused enhanced oxidation potential (Sand and Gehrke, 2006). Further research performed by the group found the composition of EPS could be altered according to the substrate being metabolized (Gehrke *et al.*, 1998).

Although Beech *et al.* (2005) and Sand and Gehrke (2006) showed the effects of EPS-bound metal ions, Chan *et al.* (2002) investigated the specific impact of purified EPS on corrosion. This was achieved by observing the effect that a 1% (w/w) EPS solution had on mild steel coupons over 60 days in the absence of the SRB that produced them. Pitting corrosion was evident at a 5-fold higher rate than that observed for the EPS free negative control as confirmed by atomic

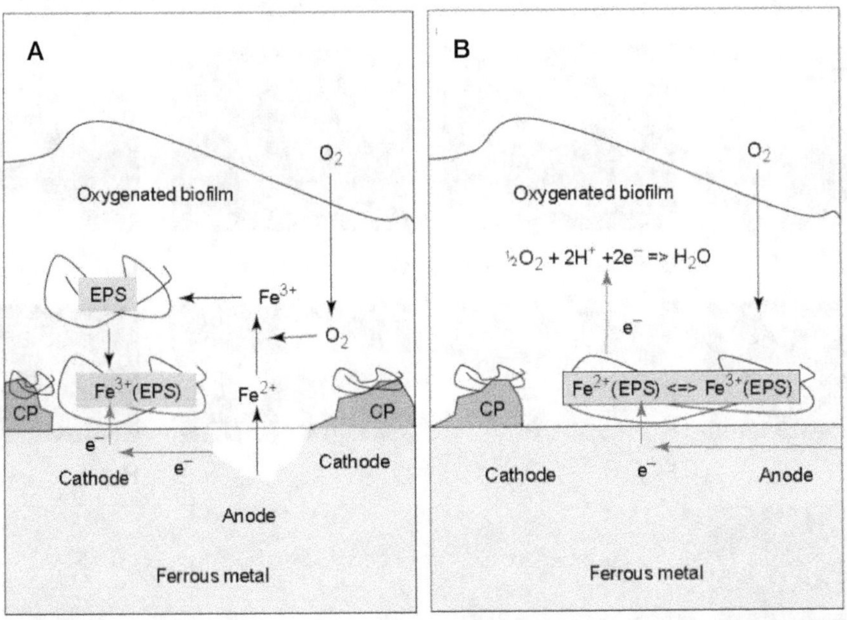

Figure 7.3 Schematic representation of the cathodic depolarization reaction of a ferrous material in the presence of an oxygenated biofilm, owing to Fe^{3+} binding by EPS. (A) Fe^{3+}, obtained as a result of, for example, oxidation of anodically produced Fe^{2+}, is bound by EPS and the Fe^{3+}–EPS complex is deposited on the metal surface. (B) Electrons are transferred directly from the zero valent Fe to EPS-bound Fe^{3+}, reducing it to Fe^{2+}. In the presence of oxygen, acting as terminal electron acceptor, Fe^{2+} in EPS is reoxidized to Fe^{3+}. Note that a similar type of reaction can take place on the surfaces of corrosion products (CP) such as oxides, hydroxides and sulfides, which contain divalent iron (Beech and Sunner, 2004).

force microscopy (AFM). This inherent property of EPS to exhibit corrosive effects could be due to the lower pH of the EPS solution although when this was adjusted towards a neutral pH, no difference in corrosiveness was observed (Chan *et al.*, 2002).

The structural role of EPS in biofilms is widely acknowledged, including the establishment of a matrix in which oxygen, nutrient and metabolite gradients exist. These gradients facilitate co-existence and cooperation of multiple groups of microorganisms. In the case of anaerobic SRB, their co-existence with other microorganisms extends their habitat range. In fact, such cooperative relationships have long been known, e.g. Dowling *et al.* (1988) described metabolic versatility amongst community members of complex biofilm consortia, and that many of the bacteria known to be involved in MIC grow best in the presence of metabolites of other bacteria. Saunders and Hamilton (1986) discussed the close interdependence between SRB and other functional groups that provide reduced microenvironments in otherwise aerobic conditions. Very few environments are dominated by pure cultures. This can be easily demonstrated by profiling the microbial community from MIC tubercles and other deposits. Fig. 7.4 shows different colony morphotypes of isolates from a corrosion pit in domestic copper tubing, illustrating the EPS producing ability of these strains and their diversity.

Interestingly, when the isolates were individually exposed to copper, four of the eight showed high sensitivity (more sensitive than the minimum inhibitory concentration for *E. coli*; Nies, 1999) to the metal (Table 7.1), demonstrating the beneficial result of the microbial communal lifestyle.

The ability of EPS to inhibit biocorrosion

In 2008, Sadler *et al.* (2008) hypothesized that EPS could inhibit the corrosion activities of microorganisms. EPS was isolated from SRB and *Pseudomonas* spp. and tested for its ability to inhibit corrosion. EPS from *Desulfovibrio vulgaris* and, *Desulfovibrio indonesiensis* was found to be inherently corrosive yet the EPS of *Desulfovibrio*

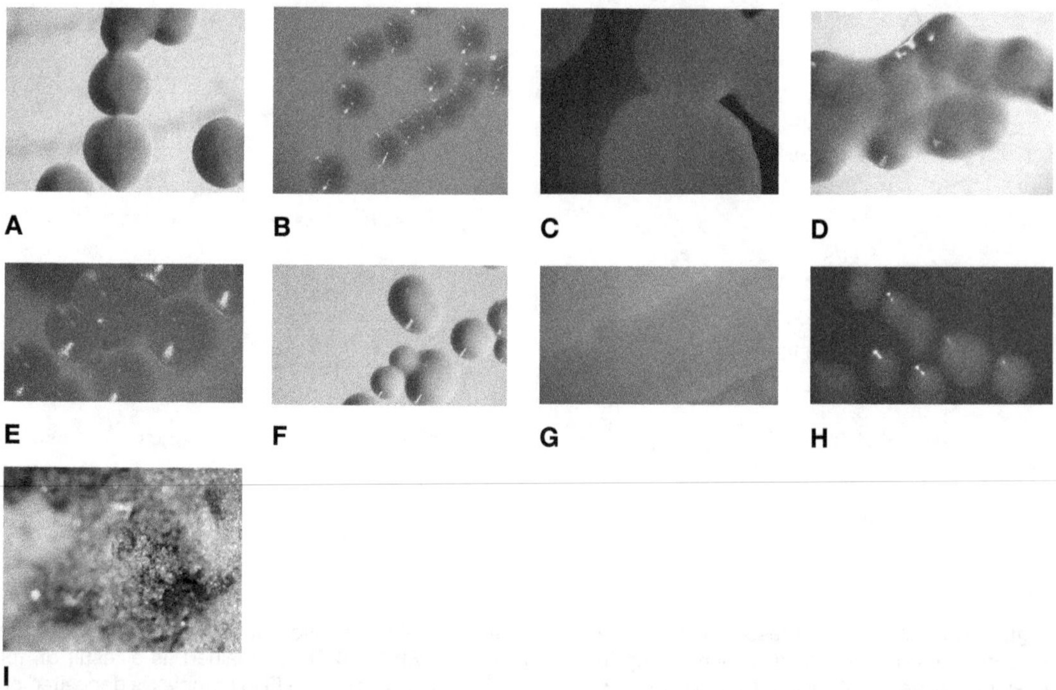

Figure 7.4 Isolates from an MIC tubercle that developed on a copper pipe grown on R2A plates, showing diversity and EPS production (A–H). Panel I shows an epifluorescence image focusing on the tip of a typical tubercle (630× magnification), stained with the Baclight™ viability probe; with the cells hardly distinguishable amongst the EPS and corrosion products. The isolate shown in panel C is a yeast.

Table 7.1 Differences in copper sensitivity of isolates from an MIC tubercle on a copper pipe

	Cu (ppm)						
Isolate	8.5	17	85	170	341	426	597
A	++	+	+	–	–	–	–
B	++	++	++	++	++	–	–
C	++	++	++	++	+	+	+
D	++	++	–	–	–	–	–
E	++	++	–	–	–	–	–
F	++	–	–	–	–	–	–
G	++	++	++	++	++	+	+
H	++	++	+	+	–	–	–

++: no sensitivity; + notable sensitivity; and – sensitive to copper when compared with untreated controls.

alaskensis tested negative for corrosion and was investigated as a suitable source of EPS for future applications in inhibiting corrosion (Stadler *et al.*, 2008).

The fact is that not all biofilms will induce corrosion on metal surfaces and thus colonization of a surface by a non-corrosion inducing biofilm, may protect a surface from colonization and subsequent corrosion by corrosion inducing microorganisms. This strategy is sometimes referred to as biocompetitive exclusion (Videla and Herrera, 2005). Inhibitors of corrosion may obstruct the cathodic or anodic reactions or a combination of the two (Videla and Herrera, 2009). Deterioration of metal through bacterial metabolites is a common scenario in MIC. Acidic end products are corrosive in nature but also have the capacity to increase the rate of anodic and/or cathodic reactions by adding protons to the environment (Videla and Herrera, 2009). Any factor that interferes with or reduces the acidic concentration is considered as inhibiting the corrosion in that system. EPS could have this effect by impeding dissolution, disassociation and adsorption reactions (Videla and Herrera, 2009). In the oil industry this strategy is currently used. This is generally achieved by the addition of nutrients (nitrate) that promote the growth of specific desirable bacterial biofilms (in this case nitrate-reducing bacteria). The presence of these biofilms effectively prevents the growth of corrosion inducing sulfate-reducing bacteria. The use of this strategy in the oil industry is reviewed by Eckford and Fedorak (2004). Outside the oil industry, this process is not widely used.

Another inhibitory mechanism proposed by several researchers is the production of antimicrobials by some species of bacteria that are bacteriocidal to species causing corrosion (Jayaraman *et al.*, 1999; Zuo *et al.*, 2005). The presence of a Gremicidin producing biofilm was recently shown to inhibit the growth of corrosive biofilms on a laboratory scale (Zuo and Wood, 2004).

In another study, the protective effect of EPS produced by *Pseudomonas cichorii* was demonstrated (Chongdar *et al.*, 2005). This EPS is composed of neutral carbonyl groups which form ferrous iron–EPS complexes. This complex strengthens the phosphate inhibiting film on the metal surface. It would have been interesting if Chongdar *et al.* (2005) investigated the electrochemical fluctuations caused by the increase in EPS-ferrous ion concentration as this may have had a detrimental effect on corrosion. In another study on pitting corrosion on stainless steel caused by SRB, Li *et al.* (2009) demonstrated, through X-ray photoelectron spectroscopy that iron is capable of forming sulfide films that may hinder pitting through polarizing the cathodic reactions.

Enzymes from biofilm and their impact on corrosion

Bacteria produce a broad range of enzymes including hydrolytic and proteolytic enzymes, lyases, peptidases, phosphatises, glycosidases, esterases and oxidoreductases (Beech *et al.*, 2005). The biofilm matrix provides ideal conditions for enzymatic reactions in close proximity to the cells that produce them. Enzymes persist in solution, even without cell presence and they may cause corrosion. Enzymes isolated from biofilm and applied in a pure form to metal surfaces, can cause corrosion (Bergel, 2007). Research of this type on purified enzymes from biofilm is not widely reported with most work focussed on enzymatic activity associated with the bacterial cell. This may be due to difficulties in extracting the enzymes from biofilm, sensitivity of the enzymes to harsh corrosive processes and the complex nature of microbial metabolism. Studies of individual enzymes may provide only a limited view of the activity that occurs in the natural environment (Landoulsi *et al.*, 2008).

SRB are regarded as the most important bacteria causing metal corrosion, mainly by increasing the cathodic reaction rate (Xu *et al.*, 2007). Enzymes produced by SRB have been identified as being instrumental in causing metal degradation. The hydrogenases are proposed as being integral in causing cathodic depolarization (Bryant and Laishley, 1990). Hydrogenases are a class of enzymes that catalyse the reversible oxidation of molecular hydrogen. Their activity in

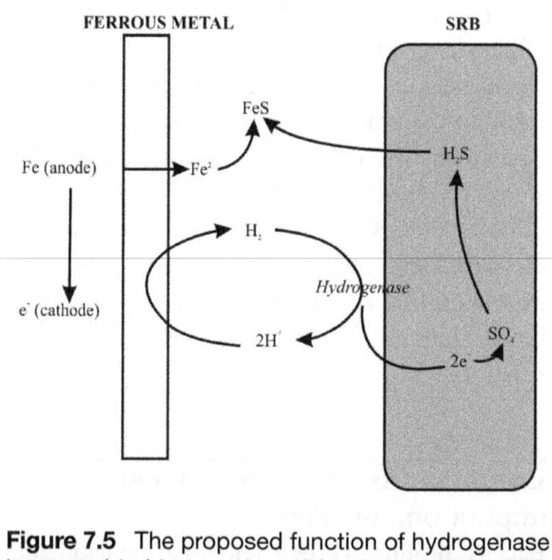

Figure 7.5 The proposed function of hydrogenase in anaerobic biocorrosion of ferrous metal by SRB (Bryant and Laishley, 1990).

anaerobic processes is considered an indicator of biocorrosion.

Fig. 7.5 shows a simplified mechanism of hydrogenase activity using SRB and iron as an example. Early evidence suggested a clear relationship between hydrogenase in SRB and corrosion (Bryant *et al.*, 1991), whereas more recent data has shown hydrogenase negative strains exhibiting more severe corrosion than hydrogenase positive strains (Kumar *et al.*, 1999). These discrepancies are not unexpected when the experimental methods of each study are compared. Bryant *et al.* (1991) used mixed populations from corroded and non-corroded pipelines, whereas Kumar *et al.* (1999) used two hydrogenase positive strains, *Desulfovibrio desulfuricans* and *D. vulgaris*, and two hydrogenase-negative strains, *Desulfotomaculum nigrificans* and *D. orientis*. The use of single strain populations for biofilm experiments is questionable as biofilms usually comprise diverse microbial populations and the interactions between the different genera are crucial to biofilm functionality. In contrast, Bryant *et al.* (1991) used scrapings from an existing biofilm which would harbour all species required for corrosion. Bryant *et al.* (1991) conducted experiments over 64 days in a Robbins device [a flowing system (pipe) containing ports for biofilm sampling] using Butlin's salts medium (a basic salt solution to which they added 0.1% casamino acids and 1% tryptone) while Kumar *et al.* (1999) incubated mild steel coupons in Webster and Newman medium (a solution of salts with 0.05% yeast extract) in still jars for 28 days. The major difference here is that Bryant *et al.* used a flowing system, while Kumar *et al.* used a static system.

Most earlier experiments investigating hydrogenase activity were in the presence of phosphate, which is implicated in augmenting corrosion by deprotonation on the metal surface, as seen in Fig. 7.6A (Mehanna *et al.*, 2008).

This effect is due to the consumption of dihydrogen species at the cathode which further promotes corrosion reactions. Mehanna *et al.* (2008) obtained contrasting results to the scenario presented in Fig. 7.6A as corrosion was observed in the absence of phosphate. Mehanna *et al.* concluded that catalysis by adsorbed hydrogenase by the direct reduction of proton or solvent,

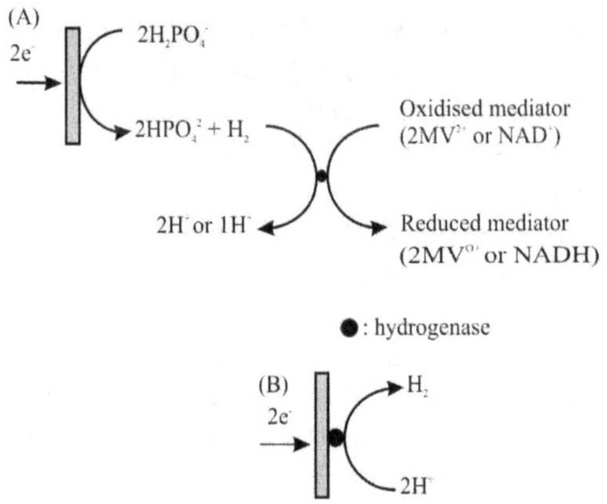

Figure 7.6 Possible cathodic reactions catalysed by hydrogenase that may enhance corrosion (Mehanna *et al.*, 2008). (A) Deprotonation in phosphate buffer. (B) Direct reduction of proton.

is the probable mechanism to explain their results (Fig. 7.6B). This novel hypothesis illustrates the hitherto lack of understanding of this mechanism and begs the question as to what other processes remain yet undiscovered in MIC.

Corrosion in medicine

Biofilm growth on metallic surfaces within the human body can cause accelerated corrosion of metal (Bryers, 2008). This is of concern as the proliferation of microorganisms in certain parts of the body is potentially life threatening and as metal ions and bacterial metabolites directly attributed to growth on metal surfaces can also cause harmful effects. This release of biomaterials from compromised implants can cause bone lysis with aseptic loosening and distribution of harmful particles to distal organs causing local or systemic infection (Al-Saffar and Revell, 1999). The tendency to corrode is an additional factor that must be taken into consideration when selecting prosthetic materials on top of stability, surface texture allowing cellular adhesion, in conjunction with suitably low toxicity, inflammatory activation and immunogenicity (Cadosh *et al.*, 2009). Many metals, such as titanium alloys used in medicine are somewhat resistant to corrosion. However, these alloys cannot always be used in certain medical situations and over time will eventually corrode. The release of metal ions into surrounding tissue and bloodstream has been observed in medical cases with harmful effects (Al-Shaffar and Revell, 1999; Cadosh *et al.*, 2009).

Control of microbially influenced corrosion

The ubiquity of biofilms in natural and engineered systems, combined with the ability to vary their complexity and structure in response to their environment has led to significant efforts aimed at finding ways to control biofilms and protect from MIC and biofouling. Such control strategies may include mechanical, chemical and biological means (Holah and Gibson, 2000). Mechanical methods such as scrubbing of surfaces and the use of high pressure water are effective when the fouled areas are accessible and can tolerate the associated shear. Typically viewed as a last resort, system design often does not make sufficient provision for their use. In cases where biofouling is left untreated for too long, biocides and dispersants usually fail to remove organic and inorganic materials that accumulate in the biofilm matrix with subsequent loss in flow capacity in pipes or heat transfer in cooling systems, and ultimately creating conditions conducive to MIC. An effective method for removing biofouling build-up in pipes involves the use of 'pigs', that are bullet or spherical structures slightly larger than the inside diameter of the pipe to be cleaned. Made of plastic or polyurethane and are pushed through the pipe under pressure, they are flexible and contractible with a large variety of types available to remove biofouling from simple biofilms to encrusted layers, and associated MIC. Huguenin and Colt (2002) and Murthy and Venkatesan (2009) provide relevant information of these and other biofouling control methods in context of system design and operation.

Conventional control of MIC mostly involves application of biocides or sanitizers to prevent the development of biofilms and to inhibit viable microorganisms dispersed by biofilm erosion or mechanical cleaning. These biocides are costly and potentially pose an environmental hazard, and as a result increasingly restrictive environmental regulations dictate their use. In addition, the use of sanitisers to control biofilms is not always effective owing to the naturally protective nature of a biofilm. As a result, biocide application is increasingly combined with dispersants and chelating agents to achieve more effective treatment. Chelating agents are multidentate complexes that bind metals through ionic bonds. Since these ligands bind metals at multiple sites, chelates are more stable than complexes involving unidentate ligands. Chelating agents that are found in natural waters and wastewaters (e.g. humic substances and amino acids) contain a variety of functional groups that are able to donate electrons for bonding to metal ions. Synthetic chelating agents such as sodium tripolyphosphate, sodium ethylenediaminetetraacetate (EDTA), sodium nitrilotriacetate and sodium citrate are commonly used in industrial water treatment to disrupt biofilms (Manahan, 2005). As an example, EDTA binds to metals such as manganese, copper, iron and cobalt via its carboxylate and amino groups, and has been shown to induce biofilm detachment from surfaces through chelation of calcium and magnesium ions, which causes the destabilization of the biofilm matrix (Chen and Stewart, 2000). In addition to combined use with biocides (e.g. Wen *et al.*, 2010), chelators have also been used in combination with other treatment methods in order to ensure maximal removal efficiency. For instance, Oulahal *et al.* (2004) combined ultrasonication with EDTA and demonstrated a synergistic effect on biofilm removal, while Raad *et al.* (2002) demonstrated that a combination of EDTA with the antibiotic minocycline was effective in eradicating catheter surface biofilms.

Monitoring

The methods relevant for the study and measure of biofilm development can provide a reasonable indication of MIC potential because of the strong link between biofilm formation and MIC, although biofilm studies are often performed in simplified systems that lack the highly variable and complex nature of the industrial systems where MIC is of concern. Nevertheless, a better understanding of biofilm form–function relationships, and characterization of their response to interventions such as biocides under controlled conditions when the treatment is the only variable, provides valuable cues for the development of control strategies.

Kroukamp and Wolfaardt (2009) summarized the advances that have been made in the study of biofilms following recognition that surface-associated growth is a dominant form of microbial behaviour (Marshall *et al.*, 1971; Geesey *et al.*, 1978). These include continuous flow systems such as flow cells, the 'roto-torque' reactor (Gjaitema *et al.*, 2004), advanced microscopy, and more recently combining these methods with contemporary molecular techniques. Subsequently, notable progress has been made towards unravelling many of the underlying fundamentals of the biofilm phenomenon, however there are still relatively few methods that are suitable to measure and monitor biofilm function and consequence under operational conditions. In theory, many of the laboratory techniques that are currently available meet the requirements for industrial biofilm monitoring as described by Flemming (2003), namely methods that primarily measure material increases or decreases on surfaces without differentiating microbial cells and products from other components, methods that do distinguish between biotic and abiotic material, and methods that map chemical profiles of biofilms. In addition, Janknecht and Melo (2003) mentioned on-line biofilm monitoring techniques that measure metabolic activity. Widespread use of these techniques (summarized in Table 7.2) to monitor the efficacy of programmes aimed at controlling biofouling and mitigating MIC will probably require various modifications to enhance robustness and reduce dependence on expensive equipment.

Future trends

One interesting alternative idea has been proposed using competitive exclusion through the use of biofilms that protect rather than promote corrosion (Videla and Herrera, 2005). This strategy is used in the oil industry and is achieved by the addition of nutrients (nitrate) to promote the growth of specific desirable bacterial films

Table 7.2 Examples of techniques available to monitor biofilms and biofouling

Measurement	Method	Reference
Material increase/decrease at the interface	Quartz crystal microbalances	Tam *et al.* (2007)
	Large area photometry	Mollica and Cristiani (2003), Bester *et al.* (2005)
	Electrical capacitance	Maurício *et al.* (2006)
	Fibre optical devices	Tamachkiarow and Flemming (2003)
	Differential turbidity measurement devices	Klahre and Flemming (2000)
	Microfluidic biochips	Richter *et al.* (2007)
	Optical coherence tomography	Haisch and Niessner (2007)
	Pressure drop or friction resistance	Lee *et al.* (1998)
	Heat transfer resistance	Ludensky (1998)
Physical and chemical structure of biofilms	Nuclear magnetic resonance	Seymour *et al.* (2003)
	Attenuated total reflectance–Fourier transform infrared spectroscopy	Delille *et al.* (2007)
	Photo acoustic spectroscopy	Schmid *et al.* (2004)
Metabolic activity	DO and redox probes	Lee *et al.* (2007)
	Nuclear magnetic resonance	Wolf *et al.* (2002)
	Substrate consumption	Kappelhof *et al.* (2003), Tanji *et al.* (2007)
	Metabolic products	Vanhooren *et al.* (2000)
	CO_2 production	Kroukamp and Wolfaardt (2009)
	Bioluminescence and fluorometry	Wolf *et al.* (2002)
	MIC	Mollica and Cristiani (2003)

– in this case, nitrate-reducing bacteria. The presence of these biofilms effectively prevents corrosion inducing sulfate-reducing bacteria (Eckford and Fedorak, 2004). This process is not currently used outside the oil industry so there may be potential to look at applying this to other industries.

A novel approach is to prepare a biofilm of protective bacteria on the metal surface (Akid *et al.*, 2008). For this, sol-gel derived coatings were developed to reduce corrosion of metal substrates. Sol-gel materials are prepared by the copolymerisation of functional organosilanes, macromonomers and metal alkoxides. Organic compounds can be encapsulated within the sol-gel coating. Akid *et al.* (2008) applied a hybrid sol–gel coating to aluminium alloy and successfully prevented pitting corrosion of the substrate over a period of 6 months in a tidal estuarine environment.

Material modification that results in a low-fouling or non-sticking surface has recently received notable attention as a promising strategy to mitigate MIC. According to Vladkova (2009), such modification typically alters the chemical composition, morphology, topography and roughness of the surface, or alters the hydrophilic/hydrophobic balance, or surface energy and polarity that leads to stronger repulsion or reduced interaction between surfaces and microorganisms. The author pointed out that in most discussions of such surface modification, microorganisms are considered 'living colloids', which disregard the roles that bacterial appendages such as pili and cell wall components may play in adhesion at the later stages of biofilm formation. However, a number of authors are cited who argue that these structural features are of less significance during the initial stages (1–24 h) of biofilm development and protein adsorption. Majumdar *et al.* (2009) studied the antimicrobial activities of polysiloxane coatings containing chemically bound (tethered) quaternary ammonium salt (QAS) moieties. They were able to identify a specific composition for this coating that provided broad-spectrum antimicrobial activity. Overall, these strategies may provide effective alternatives to conventional chemical treatment: Two challenges common to most biocide/disinfection programs are (1) to ensure sufficient contact time for effective killing of microbial cells and (2) to penetrate the biofilm matrix. As already pointed out, once growing as biofilms, most microbes show significantly higher resistance than the cells in suspension – this increased resistance is often attributed to 'persister' cells at the base regions of biofilms (at the biofilm – surface interface) and/or to the protection offered to the cells by the self-produced EPS matrix in the form of increased mechanical stability or as a diffusion barrier. Conventional chemical and mechanical programs typically fail to reach or inhibit cells in this region; resulting in fast recovery. Further development of 'smart materials' with inherent antimicrobial properties or protective coatings therefore appears to be a realistic goal for improved MIC mitigation. However, as pointed out by Webster *et al.* (2007) a notable challenge to the design of effective non-toxic coatings is the highly diverse range of environments that needs to be rendered antimicrobial. As a result, a large number of synthetic approaches are being explored in a number of laboratories, involving a variety of biofouling resistant coatings with complex composition. The authors further point out that relatively few laboratory assays have been performed to evaluate these coatings and that most of the assays are highly labour intensive. They therefore encourage the development and application of combinatorial and high throughput methods that involve the rapid preparation and screening of a large number of chemically distinct compositions over a wide range of applications.

The role of micro-organisms in corrosion is well established, although we still do not fully understand the mechanisms involved in the diverse microbial communities that constitute biofilms. Irrespective of the mechanism, the involvement of micro-organisms is the key and therefore the development of antimicrobial substrates to prevent microbial colonization will continue to be a focus of future measures to combat microbially influenced corrosion.

Web resources

http://www.biofilm.montana.edu/

References

Akid, R., Wang, H., Smith, T.J., Greenfield, D., and Earthman, J.C. (2008). Biological functionalisation of a sol-gel coating for the mitigation of microbial-induced corrosion. Adv. Funct. Mater. *18*, 203–211.

Al-Saffar, and Revell, P.A. (1999). Pathology of the bone–implant interfaces. J. Long-Term Eff. Med. Implants *9*, 319–347.

Beech, I.B., Sunner, J.A., and Hiraoka, K. (2005). Microbe–surface interactions in biofouling and biocorrosion processes. Int. Microbiol. *8*, 157–168.

Beech, W.B., and Sunner, J. (2004). Biocorrosion: towards understanding interactions between biofilms and metals. Cur. Opin. Biotech. *15*, 181–186.

Bergel, A. (2007). Recent advances in electron transfer between biofilms and metals. In Biohydrometallurgy: From the Single Cell to the Environment, Schippers, A., Sand, W., Glombitza, F., and Willscher, S., eds. (Trans Tech Publications Ltd, Stafa-Surich), pp. 3239–3334.

Bester, E., Wolfaardt, G., Joubert, L., Garny, K., and Saftic, S. (2005). Planktonic-cell yield of a pseudomonad biofilm. Appl. Environ. Microbiol. *71*, 7792–7798.

Bester, E., Edwards, E.A., and Wolfaardt G.M. (2009). Planktonic cell yield is linked to biofilm development. Can. J. Microbiol. *55*, 1195–1206.

Bester, E., Kroukamp, O., Wolfaardt, G.M., Boonzaaier, L., and Liss, S. (2010). Metabolic differentiation in biofilms as indicated by carbon dioxide production rates. Appl. Environ. Microb. *76*, 1189–1197.

Bester, E., Kroukamp, O., Hausner, M., Edwards,E., and Wolfaardt, G.M. (2011). T biofilm form and function: carbon availability affects biofilm architecture, metabolic activity and planktonic cell yield. J. Appl. Microbiol. *110*, 387–398.

Boopathy, R., and Daniels, L. (1991). Effect of pH on anaerobic mild steel corrosion by methanogenic bacteria. Appl. Env. Microbiol. *57*, 2104–2108.

Branda, S.S., Vik, S., Friedman, L., and Kolter, R. (2005). Biofilms: the matrix revisited. Trends Microbiol. *13*, 20–26.

Bryant, R.D., and Laishley, E.J. (1990). The role of hydrogenase in anaerobic biocorrosion. Can. J. Microbiol. *36*, 259–264.

Bryant, R.D., Jansen, W., Boivin, J., Laishley, E.J., and Costerton, J.W. (1991). Effect of hydrogenase and mixed sulfate-reducing bacterial populations on the corrosion of steel. Appl. Env. Microbiol. *57*, 2804–2809.

Bryers, J.D. (2008). Medical biofilms. Biotech. Bioeng. *100*, 1–18.

Cadosch, D., Chan, E., Gautschi, O.P., and Filgueira, L. (2009). Metal is not inert: role of metal ions released by biocorrosion in aseptic loosening – Current concepts. J. Biomedical Materials Res. Part A. *91A*, 1252–1262.

Chan, C.S., De Stasio, G., Welch, S.A., Girasole, M., Frazer, B.H., Nesterova, M.V., Fakra, S., and Banfield, J.F. (2004). Microbial polysaccharides template assembly of nanocrystal fibers. Science *303*, 1656–1658.

Chan, K.Y., Xu, L.C., and Fang, H.H.P. (2002). Anaerobic electrochemical corrosion of mild steel in the presence of extracellular polymeric substances produced by a culture enriched in sulfate-reducing bacteria. Environ. Sci. Tech. *36*, 1720–1727.

Chen, X., and Stewart, P.S. (2000). Biofilm removal caused by chemical treatment. Water Res. *34*, 4229–4233.

Chongdar, S., Cunasekaran, G., and Kumar, P. (2005). Corrosion inhibition of mild steel by aerobic bioflilm. Electrochim. Acta. *50*, 4655–4665.

Coetser, S.E., and Cloete, T.E. (2005). Biofouling and biocorrosion in industrial water systems. Crit. Rev. Microbiol. *31*, 213–232.

Costerton, J.W., Lewandowski, Z., Caldwell, D.E., Korber, D.R., and Lappin-Scott, H. M. (1995). Microbial biofilms. Ann. Rev. Microbiol. *49*, 711–745.

Da Silva, S., Basseguy, R., and Bergel, A. (2004). Electron transfer between hydrogenase and 316L stainless steel: Identification of hydrogenase-catalyzed cathodic reaction in anaerobic microorganisms. J. Electranal. Chem. *561*, 93–102.

Delille, A., Quilès,F., and Humbert, F. (2007). *In situ* monitoring of the nascent *Pseudomonas fluorescens* biofilm response to variations in the dissolved organic carbon level in low-nutrient water by attenuated total reflectance-Fourier transform infrared spectroscopy. Appl. Environ. Microbiol. 73, 5782–5788.

Dowling, N. J.E., Guezennec, J., and White, D.C. (1988). Methods for insight into mechanisms of microbially influenced corrosion. Int. Biodeter. *24*, 404–410.

Dubiel, M., Hsu, C.H., Chien, C.C., Mansfield, F., and Newman, D.K. (2002). Microbial iron respiration can protect steel from corrosion. Appl. Env. Microbiol. *68*, 1440–1445.

Eckford, R.E., and Fedorak, P.M. (2004). Using nitrate to control microbially produced hydrogen sulfide in oil field waters. In Petroleum Biotechnology, Vazquez-Duhalt, R., and Quintero-Ramirez, R., eds. (Amsterdam, The Netherlands), pp. 307–340.

Flemming, H.-C. (2003). Role and levels of real-time monitoring for successful anti-fouling strategies – An overview. Water Sci. Technol. *47*, 1–8.

Ford, T., and Mitchell, R. (1990). The ecology of microbial corrosion. Adv. Microbial Ecol. *11*, 231–262.

Gaines, R.H. (1910). Bacterial activity as a corrosion induced in the soil. J. Engineer Ind. Chem. 2, 128–130.

Garrity, G.M., Bell, J.A., and Lilburn, T. (2005). Acidithobacillales ord. nov. In Bergeys Manual of Systematic Bacteriology, Volume Two, The Proteobacteria Part B, The *Gammaproteobacteria*, Phylum XIV, Class III 60–63, Biomedical Life Sciences, Garrity, G.M., Brenner, D.J., Krieg, N.R., Staley, J.T., eds (Elsevier Science, SpringerLink, USA), pp. 1–2816.

Geesey, G.G., Mutch, R., Costerton, J.W., and Green, R.B. (1978). Sessile bacteria: An important component of the microbial population in small mountain streams. Limnol. Oceanogr. *23*, 1214–1223.

Gehrke, T., Telegdi, J., Thierry, D., and Sand, W. (1998). Importance of extracellular polymeric substances form *Thiobacillus ferrooxidans* for bioleaching. Appl. Env. Microbiol. *64*, 2743–2747.

Gjaitema, A., Arts, P.A.M., van Loosdrecht, M.C.M., Kuenen, J.G., and Heijnen, J.J. (2004). Heterogeneity of biofilms in rotating annular reactors: Occurrence, structure and consequences. Biotech. Bioeng. *44* 194–204.

Gorby, Y.A., and Lovely, D.R. (1991). Electron transport in the dissimilatory iron reducer, GS-15. Appl. Env. Microbiol. *57*, 867–870.

Haisch, C., and Niessner, R. (2007). Visualisation of transient processes in biofilms by optical coherence tomography. Water Res. *41*, 2467–2472.

Hall-Stoodley, L., Costerton, J.W., and Stoodley, P. (2004). Bacterial biofilms: From the natural environment to infectious diseases. Nat. Rev. Microbiol. 2, 95–108.

Hamilton, W.A. (1985). Sulfate-reducing bacteria and anaerobic corrosion. Ann. Rev. Microbiol. *39*, 195–217.

Hamilton, W.A. (1998). Bioenergetics of sulfate-reducing bacteria in relation to their environmental impact. Biodegradation 9, 201–212.

Hamilton, W.A. (2003). Microbially influenced corrosion as a model system for the study of metal microbe interactions: a unifying electron transfer hypothesis. Biofouling *19*, 65–76.

Hamilton, W.A., and Lee, W. (1995). Biocorrosion. In Sulfate Reducing Bacteria, Barton. L.L., ed. (Plenum Press, New York), pp. 243–264.

Heidelberg, J.F., Seshadri, R., Haveman, S.A., Hemme, C.L., Paulsen, I.T., Kolonay, J.F., Eisen, J.A., Ward, N., Methe, B., Brinkac, L.M., *et al.* (2004). The genome sequence of anaerobic sulpfate-reducing bacterium *Desulfovibrio vulgaris* Hildenborough. Nat. Biotechnol. *22*, 554–559.

Holah, J., and Gibson, H. (2000). Food industry biofilms. In Biofilms: Recent advances in their study and control, Evans L.V., ed. (Harwood Academic Publishers, Singapore=), pp 211–236.

Huguenin, J.E., and Colt, J. (2002). Design and Operating Guide for Aquaculture Seawater Systems. (Elsevier, Amsterdam), pp. 328.

Iverson, W. =P. (1987). Microbial corrosion of metals. Adv. Appl. Microbiol. 32, 1–36.

Janknecht, P., and Melo, L.F. (2003). Online biofilm monitoring. Rev. Environ. Sci. Biotechnol. 2, 269–283.

Jayaraman, A., Hallock, P.J., Carson, R.M., Lee, C.C., Mansfield, F.B., and Wood, T.K. (1999). Inhibiting sulfate-reducing bacteria in biofilms on steel with antimicrobial peptides generated *in situ*. Appl. Microbiol. Biotech. 52, 267–275.

Johansson, L.S. and Saastamoinen, T. (1999). Investigating early stages of biocorrosion with XPS:AISI 304 stainless steel exposed to *Burkholderia* species. Appl. Surface Sci. *144–45*, 244–248.

Kappelhof, J. W. N.M., Vrouwenvelder, H.S., Schaap, M., Kruithof, J.C., Van Der Kooij, D., and Schippers, J.C. (2003). An *in situ* biofouling monitor for membrane systems. Water Sci. Technol.: Water Supply 3, 205–210.

Kinzler, K., Gehrke, T., Telegdi, J., and Sand, W. (2003). Bioleaching – a result of interfacial processes caused by extracellular polymeric substances EPSP. Hydrometallurgy *71*, 83–88.

Klahre, J., and Flemming, H.-C. (2000). Monitoring of biofouling in papermill process waters. Water Res. *34*, 3657–3665.

Koch, G.H. (2001). Tests for stress-corrosion. Adv. Mater. Processes, *159*, 36–38.

Kroukamp, O., and Wolfaardt, G.M. (2009). CO_2 production as an indicator of biofilm metabolism. Appl. Environ. Microbiol. 75, 4391–4397.

Kumar, A.V.R., Singh, R., and Nigam, R.K. (1999). Mossbauer spectroscopy of corrosion product of mild steel due to microbiologically influenced corrosion. J. Radioanal. Nucl. Chem. *242*, 131–137.

Landoulsi, J., El Kirat, K., Richard, C., Feron, D., and Pulvin, S. (2008). Enzymatic approach in microbial-influenced corrosion : A review based on stainless steels in natural waters. Env. Sci. Technol. *42*, 2233–2242.

Lee, A.K., and Newman, D.K. (2003). Microbial iron respiration: impacts on corrosion processes. Appl. Microbiol. Biotechnol. *62*, 134–139.

Lee, A.K., Lewandowski, Z., Morrison, M., Characklis, W.G., and Avci, R. (1993a). Corrosion of mild steel underneath aerobic biofilms containing sulfate-reducing bacteria. Part II: at high bulk oxygen concentrations. Biofouling 7, 217–239.

Lee, A.K., Lewandowski, Z., Okabe, S., Characklis, W.G., and Avci, R. (1993b). Corrosion of mild steel underneath aerobic biofilms containing sulfate-reducing bacteria. Part I. at low dissolved oxygen concentration. Biofouling 7, 197–216.

Lee, A.K., Lewandowski, Z., Nielsen, P.H., and Wa, H. (1995). Role of sulfate-reducing bacteria in corrosion of mild steel: A review. Biofouling *8*, 165–194.

Lee, H.J., Han, D.G., Lee, S.H., Yoo, J. W,. Baek, S.H., and Lee, E. K. (1998). On-line monitoring and quantitative analysis of biofouling in low-velocity cooling water system. Korean J. Chem. Eng. *15*, 71–77.

Lee, J.-H., Seo, Y., Lim, T.-S., Bishop, P.L., and Papautsky, I. (2007). MEMS needle-type sensor array for *in situ* measurements of dissolved oxygen and redox potential. Environ. Sci. Technol. *41*, 7857–7863.

Lee, W., and Beer, D. (1995). Oxygen and pH microprofiles above corroded mild steel covered with biofilm. Biofouling, *8*, 273–280.

Lewandowski, Z., Lee, W., Characklis, W.G., and Little, B. (1989). Dissolved oxygen and pH microelectrode measurements at water immersed metal surface. Corrosion 45, 92–98.

Li, F.S., An, M.Z., Liu, G.Z., and Duan, D.X. (2009). Effects of sulfidation of passive film in the presence of SRB on the pitting corrosion behaviors of stainless steels. Materials Chem. Phys. *113*, 971–976.

Lindsay, D., and von Holy, A. (2006). What food safety professionals should know about bacterial biofilms. Brit. Food J. *108*, 27–37.

Lovley, D.R. (1991). Dissimilatory Fe(III) and Mn(IV) reduction. Microbiol. Rev. *55*.<*Please provide page range*>

Lovley, D.R., and Coates, J.D. (2000). Novel forms of anaerobic respiration of environmental relevance. Curr. Opin. Microbiol. 3, 252–256.

Ludensky, M.L. (1998). An automated system for biocide testing on biofilms. J. Ind. Microbiol. Biotechnol. *20*, 109–115.

Majumdar, P., Lee, E., Gubbins, N., Christianson, D.A., Stafslien, S.J., Daniels, J., VanderWal, L., Bahr, J., and Chisholm, B.J. (2009). Combinatorial materials research applied to the development of new surface coatings XIII: An investigation of polysiloxane antimicrobial coatings containing tethered quaternary ammonium salt Groups. J. Comb. Chem. *11*, 1115–1127.

Manahan, S.E. (2005). Environmental Chemistry. 8th edition. (CRC Press, Boca Raton, FL), pp. 67–69.

Marshall, K.C., Stout, R., and Mitchell, R. (1971). Selective sorption of bacteria from seawater. Can. J. Microbiol. *17*, 1413–1416.

Maurício, R., Dias, C.J., and Santana, F. (2006). Monitoring biofilm thickness using a non-destructive, on-line, electrical capacitance technique. Environ. Monit. Assess. *119*, 599–607.

Mehanna, M., Basseguy, R., Delia, M.L., Girbal, L., Demuez, M., and Bergel, A. (2008). New hypotheses for hydrogenase implication in the corrosion of mild steel. Electrochim. Acta *54*, 140–147.

Mollica, A., and Cristiani, P. (2003). On-line biofilm monitoring by 'BIOX' electrochemical probe. Water Sci. Technol. *47*, 45–49.

Murthy, P.S., and Venkatesan, R. (2009). Industrial biofilms and their control. In Marine and Industrial Biofouling, Costerton J.W., ed. (Springer, Berlin), pp. 65–101.

Nielsen, P., Lee, W., Lewandowski, Z., Morrison, M., and Characklis, W.G. (1993). Corrosion of mild steel in an alternating oxic and anoxic biofilm system. Biofouling 7, 267–284.

Nies, D.H. (1999). Microbial heavy-metal resistance. Appl. Microbiol. Biot. *51*, 730–750.

Oulahal, N., Martial-Gros, A., Bonneau, M., and Blum, L.J. (2004). Combined effect of chelating agents and ultrasound on biofilm removal from stainless steel surfaces. Application to *Escherichia coli* milk and *Staphylococcus aureus* milk biofilms. Biofilms *1*, 65–73.

Parsek, M.R. and Fuqua, C. (2004). Biofilms 2003: Emerging themes and challenges in studies of surfaces-associated microbial life. J. Bacteriol. *186*, 4427–4440.

Pitonzo, B.J., Castro, P., Amy, P.S., Southam, G., Jones, D.A., and Ringelberg, D. (2004). Microbiologically influenced corrosion capability of bacteria isolated from Yucca Mountain. Corrosion *60*, 64–74.

Potekhina, J.S., Sherisheva, N.G., Povetkina, L.P., Pospelov, A.P., Rakitina, T.A., Warnecke, F., and Gottschalk, G. (1999). Role of microorganisms in corrosion inhibition of metals in aquatic habitats. Appl. Microbiol. Biotechnol. *52*, 639–646.

Raad, I.I., Chatzinikolaou, I., Chaiban, G., Hanna, H., Hachem, R., Dvorak, T., and Shrerertz, R. (2002). *In vitro* and *ex vivo* activity of minocycline and EDTA (M-EDTA) against microorganisms embedded in biofilm on catheters. Abstr. Intersci. Conf. Antimicrob. Agents Chemother. *42*, abstract no. K-75.

Richter, L., Stepper, C., Mak, A., Reinthaler, A,. Heer, R., Kast, M., Brückl, H., and Ertl, P. (2007). Development of a microfluidic biochip for online monitoring of fungal biofilm dynamics. Lab. Chip *7*, 1723–1731.

Roe, F.L., Lewandowski, Z., and Funk, T. (1996). Simulationg microbiologically influenced corrosion by depositing extracellular biopolymers on steel surfaces. Corrosion *52*, 744–752.

Rohwerder, T., Gehrke, T., Kinzler, K., and Sand, W. (2003). Bioleaching review part A: progress in bioleaching: fundamentals and mechanisms of bacterial metal sulfide oxidation. Appl. Microbiol. Biotechnol. *63*, 239–248.

de Romero, M.F., Duque, Z., de Rincon, O.T., Perez, O., Araujo, I., and Briceno, B. (2002). Microbiological corrosion: Hydrogen permeation and sulfate-reducing bacteria. Corrosion *58*, 429–435.

de Romero, M.F., Duque, Z., de Rincon, O.T., Perez, O., Araujo, I., and Briceno, B. (2003). Study of the cathodic depolarization theory with hydrogen permeation and the bacteria *Desulfonvibrio desulfuricans*. Revista De Metalurgia 182–187.

Sand, W., and Gehrke, T. (2006). Extracellular polymeric substances mediate bioleaching/biocorrosion via interfacial processes involving iron(III) ions and acidophilic bacteria. Res. Microbiol. *157*, 49–56.

Sanders, P.F., and Hamilton, W.A. (1986). Biological and corrosion activities of sulfate-reducing bacteria in industrial process plant. In Biologically Induced Corrosion, Dexter, S.C., ed. (NACE, Houston), pp. 47–68.

Schmid, T., Panne, U., Adams, J., and Niessner, R. (2004). Investigation of biocide efficacy by photoacoustic biofilm monitoring. Water Res. *38*, 1189–1196.

Seymour, J.D., Codd, S.L., Gjersing, E.L., and Stewart, P.S. (2003). Magnetic resonance microscopy of biofilm structure and impact on transport in a capillary bioreactor. J. Magn. Reson. *167*, 322–327.

Sheng, X.X., Ting, U.P., and Pehkonen, S.A. (2007). The influence of sulfate-reducing bacteria biofilm on the corrosion of stainless steel AISI 316. Corrosion Sci. *49*, 2159–2176.

Stadler, R., Fuerbeth, W., Harneit, K., Grooters, M., Woellbrink, M., and Sand, W. (2008). First evaluation of the applicability of microbial extracellular polymeric substances for corrosion protection of metal substrates. Electochem. Acta *54*, 91–99

Stewart, P.S. (2003). Diffusion in biofilms. J. Bacteriol. *185*, 1485–1491.

Tam, K., Kinsinger, N., Ayala, P., Qi, F., Shi, W., and Myung, N.V. (2007). Real-time monitoring of *Streptococcus mutans* biofilm formation using a quartz crystal microbalance. Caries Res. *41*, 474–483.

Tamachkiarow, A., and Flemming, H.-C. (2003). On-line monitoring of biofilm formation in a brewery water pipeline system with a fibre optical device. Water Sci. Technol. *47*, 19–24.

Tanji, Y., Nishihara, T., and Miyanaga, K. (2007). Monitoring of biofilm in cooling water system by measuring lactic acid consumption rate. Biochem. Eng. J. *35*, 81–86.

Valencia-Cantero, E., Pena-Cariales, J.J., and Martinez-Romero, E. (2003). The corrosion effects of sulfate and ferric reducing bacterial consortia on steel. Geomicrobio. J. *20*, 157–169.

Vanhooren, H., Demey, D., Vannijvel, I., and Vanrolleghem, P.A. (2000). Monitoring and modelling an industrial trickling filter using on-line off-gas analysis and respirometry. Water Sci. Technol. *41*, 139–148.

Videla, H.A. (1991). Microbially induced corrosion: an update and overview. In Biodeterioration and Biodegradation, Rossmore, H.W., ed. (Elsevier Applied Science, London), pp. 63–88.

Videla, H.A., and Herrera, L.K. (2009). Understanding microbial inhibition of corrosion. A comprehensive overview. Int. Biodet. Biodegrad. *63*, 896–900.

Vielda, H.A., and Herrera, L.K. (2005). Microbially influenced corrosion: looking to the future. Int. Microbiol. *8*, 169–180.

Vladkova, T. (2009). Surface modification approach to control biofouling. In Marine and Industrial Biofouling Vol. 39, Costerton J.W., ed. (Springer, Berlin), pp. 135–163.

Von Wolzogen Kuhr, C. A.H., and van der Flugt, L.S. (1934). De grafiteering van gietijzer als electrobiochemisch proces in anaerobe grounden. Water (den Haad), *18*, 147–165. [English translation (1961). Corrosion *1917*, 1293–1299].

Webster, D.C., Chisholm, B.J., and Stafslien, S.J. (2007). Combinatorial approaches for the design of novel coating systems. Biofouling *23*, 179–192

Wen, J., Zhao, K., Gu., T., and Raad, I. (2010). Chelators enhanced inhibition of planktonic sulfate-reducing bacterial growth. World J. Microbiol. Biotechnol. *26*, 1053–1057.

Wolf, G., Crespo, J.G., and Reis, M. A. M. (2002). Optical and spectroscopic methods for biofilm examination and monitoring. Rev. Environ. Sci. Biotechnol. *1*,227–251.

Wolfaardt, G.M., and Cloete, T.E. (1992). The effect of some environmental parameters on surface colonization by microorganisms. Wat. Res. *26*, 527–537.

Xu, C.M., Zhang, Y.H., Cheng, G.X., and Zhu, W.S. (2007). Localized corrosion behaviour of 316L stainless steel in the presence of sulpfate-reducing and iron-oxidizing bacteria. Materials Sci. Eng. A. Struct. Mat. Prop. Microst. Proc. *443*, 235–241.

Xu, L.C., Fang, H.H.P., and Chan, K.Y. (1999). Atomic force microscopy study of microbiologically influenced corrosion of milk steel. J. Electrochem. Soc. *146*, 4455–5560.

Zuo, R., and Wood, T.K. (2004). Inhibiting mild steel corrosion from sulfate-reducing and iron-oxidising bacteria using gramicidin-S-producing biofilms. Appl. Microbiol. Biotechnol. *65*, 747–753.

Zuo, R.J., Kus, E., Mansfield, F., and Wood, T.K. (2005). The importance of live biofilms in corrosion protection. Corr. Sci. *47*, 279–287.

Biofilms in Freshwater: Their Importance for the Maintenance and Monitoring of Freshwater Health

8

Gavin Lear, Andrew Dopheide, Pierre-Yves Ancion, Kelly Roberts, Vidya Washington, Jo Smith and Gillian D. Lewis

Abstract

This chapter reviews our current understanding of the roles biofilm-associated microbial communities play in both maintaining and improving the ecological health of freshwater rivers and streams. Biofilms are where most of the bacteria present in freshwater systems are found, and have been identified as major sites for primary production, carbon and nutrient cycling. Advances in various scientific methodologies have recently been used to characterize the enormous diversity of biofilms, in terms of their structural, chemical and biological traits. The microbial life present within most natural biofilms, as well as associated exudates and lysates have been identified as a valuable, nutrient rich food source for a variety of benthic consumers. Furthermore, the diverse metabolic potential of these complex communities, in combination with various protective traits offered by the biofilm 'mode of life', provide biofilms with an excellent ability to degrade, or otherwise transform a vast array of freshwater pollutants. Despite this apparent resilience, we highlight the sensitivity of these poorly studied freshwater biofilm communities to various human activities, and consider their potential as a reliable and sensitive biological indicator of freshwater ecological health.

Introduction

Within freshwater rivers and streams, the surfaces of submerged objects such as rocks, woody debris and macrophytes are cloaked with microbial biofilms. These biofilms typically comprise highly complex communities including bacteria, archaea, fungi, protozoa and other organisms living among a protective matrix of extracellular polymeric substances. These communities play a pivotal role in the functioning of aquatic ecosystems. They are a major source of 'in-stream' primary production (less than 1% of microbial carbon cycling has been observed to occur within the water column of aquatic ecosystems; Buesing and Gessner, 2006), and they play a crucial role in the cycling of key nutrients such as phosphorus and nitrogen within freshwaters. In this chapter, we discuss the importance of biofilm microbial communities for the maintenance of stream ecological health and function, and consider how our knowledge of microbial community structure, composition and diversity can be used to provide a reliable indicator of the condition of freshwater environments.

The importance of biofilms for the maintenance of freshwater ecological health

What are freshwater biofilms?

Freshwater biofilms are architecturally complex microbial communities most often found at solid/water interfaces in aquatic systems. Freshwater biofilms are typically taxonomically diverse, containing a mixture of co-existing species encased in a gelatinous matrix (Costerton *et al.*, 1994). The gel matrix is mostly water (up to 97%; Zhang and Feng, 2001) but also includes other components such as extracellular polysaccharides, extracellular DNA, particulate material and detritus (Costerton *et al.*, 1994; Sutherland, 2001b). The biofilm has interwoven fluid channels that enable the transportation and transformation of vital nutrients, gases, and their associated waste products throughout the structure of the biofilm (de Beer *et al.*, 1994). Microorganisms benefit from adopting the biofilm 'mode-of-life' in several ways including by capitalizing on the efficient use of available resources by the varied metabolic capabilities of different members of the biofilm community (Jefferson, 2004), the recycling of secreted end products (e.g. the usage of organic carbon produced by autotrophs, by heterotrophic bacteria; Romani and Sabater, 1999; Roeselers *et al.*, 2007), buffering against nutrient limitations (Freeman and Lock, 1995), protection from physical factors such as scouring by the rivers flow (Rickard *et al.*, 2004) and protection from predation (Matz and Kjelleberg, 2005).

The composition of freshwater biofilm communities have been investigated using a number of surrogate measures including measurements of total biomass (Bouletreau *et al.*, 2006; Elsdon and Limburg, 2008), chlorophyll *a* concentration (Bouletreau *et al.*, 2006), nitrification rates (Teissier *et al.*, 2007), oxygen consumption (Ardon and Pringle, 2007) and community respiration (Hoellein *et al.*, 2007). Molecular methods are increasingly applied to investigate responses of whole communities in biofilms. Methods commonly used include fluorescence *in situ* hybridization (FISH), terminal restriction fragment length polymorphism (t-RFLP), temperature and denaturing gradient gel electrophoresis (TGGE and DGGE), automated ribosomal intergenic spacer analysis (ARISA) and 16S rRNA gene clone libraries (reviewed in Dorigo *et al.*, 2005). However, to date, only a few studies have yet evaluated the biofilm community structure of epilithic biofilms in fluvial systems (Lyautey *et al.*, 2003; Lear *et al.*, 2008, 2009a).

Despite there being limited research in this area, there currently seems to be a consensus that there are cosmopolitan groups of freshwater bacteria, and that certain bacterial groups appear to be dominant freshwater biofilm communities (Brummer *et al.*, 2000; Glockner *et al.*, 2000; Araya *et al.*, 2003; Brummer *et al.*, 2003; Lyautey *et al.*, 2003; Fazi *et al.*, 2005; Besemer *et al.*, 2007, 2009; Lear *et al.*, 2009b). Bacteria commonly identified as being abundant in freshwater biofilms include the α-, β- and γ-Proteobacteria, Bacteriodetes (formerly the *Cytophaga–Flavobacterium–Bacteriodes* group), the Cyanobacteria, Actinobacteria, Verrucomicrobia, the green non-sulfur bacteria and candidate division OP10 (Glockner *et al.*, 2000; Zwart *et al.*, 2002). However, comparison of diversity among different studies is difficult as they are reliant on the sampling effort, methodological approach and phylogenetic resolution applied to the research in question (Clarke and Warwick, 2001; Magurran *et al.*, 2010). Furthermore, because of the great diversity of organisms present in most natural freshwater biofilms, it is apparent that the majority of studies to date (e.g. using t-RFLP, ARISA, DGGE and 16S rRNA gene clone libraries) have not been sensitive enough to characterize the diversity of the whole community, and thus are skewed to reveal the identity of only the most dominant organisms present. The question whether the 'rare' microorganisms [e.g. those that represent <0.01% of the community (Galand *et al.*, 2009)] are important with regards to stream ecological function is yet to be answered. Such research questions will undoubtedly benefit greatly from the recent advent of pyrosequencing technologies which allow rapid and comparatively cheap assessments of microbial diversity via the parallel analysis of millions of DNA sequence tags.

As advances in scientific practices improve our ability to describe the composition, diversity and function of biologically diverse biofilm communities, it is apparent that human activities are impacting the balance of ecosystem processes in freshwater systems, including those related to biofilm-associated microbial communities. Restoration efforts are increasingly being applied, but to better understand the effectiveness of restoration efforts, it is crucial to first have a base knowledge of the temporal and spatial heterogeneity in bacterial community composition (Magurran *et al.*, 2010). Researchers have tackled this question using biofilms in a number of ways, both by assessing community changes based on anthropogenic impacts in different fluvial systems (Brummer *et al.*, 2000; Hullar *et al.*, 2006; Dorigo *et al.*, 2009), or examining sections of the same river or stream across the length of various pollution gradients (Lyautey *et al.*, 2005; Anderson-Glenna *et al.*,

2008). Studies have also characterized changes in biofilm bacterial community structure related to anthropogenic impacts across a range of time scales, ranging from a few months (Lear *et al.*, 2008) to a number of years (Anderson-Glenna *et al.*, 2008). In both the studies referenced, temporal factors explained the highest degree of variability within the epilithic biofilm. In the following sections, we describe how studies of biofilm bacterial community diversity, structure and function have improved our understanding of the importance of biofilms in freshwater systems, and how human disturbance may impair their vital contribution for the maintenance of freshwater ecological health.

Biofilms as a basal food resource in aquatic ecosystems

The ecological importance of biofilms has only recently been recognized. These microbial communities are a key component of freshwater ecosystems, acting as sites of critical ecological processes such as primary production, carbon and nitrogen fixation and nutrient cycling. Biofilms are where most bacteria occurring in freshwater ecosystems are found (Geesey *et al.*, 1978; Edwards *et al.*, 1990), and the majority of trophic interactions involving bacteria must therefore take place in biofilms. The feeding interactions of diverse heterotrophic organisms provide an important pathway for the transfer of nutrients and energy to higher trophic levels. For example, the importance of algal biofilm material is now well recognized as being a valuable food resource (McNeely and Power, 2007) for a variety of macroinvertebrates and fish (Lamberti, 1996). Bacteria, and to a lesser extent fungi, tend to be enriched in both nitrogen and phosphorus compared with the organic material that they degrade, and bulk periphyton material (Cross *et al.*, 2005). This suggests that biofilm-associated bacterial biomass, in particular, may constitute a valuable and nutritious food resource for benthic consumers. Indeed, detritus-associated bacteria are a major carbon source for mayfly (Edwards and Meyer, 1990) and chironimid larvae have been observed to increase in accordance with bacterial density, highlighting that bacterial sources of nutrition may be more important to some stream invertebrates than previously envisaged (Fuller *et al.*, 2004). Experiments with radiolabelled epilithic biofilms showed that the grazing snail *Elmia clara* assimilated bacterial material in near equivalent amounts to algae (Morales and Ward, 2000), while a recent study concluded that detritivorous stream invertebrates assimilated microbial biofilms present on wood with greater efficiency than leaf detritus (Eggert and Wallace, 2007). Other studies highlight the importance of extracellular polysaccharides produced by bacterial biofilms for invertebrate nutrition (Couch *et al.*, 1996; Lawrence *et al.*, 2002). Stream biofilm extracellular polysaccharide material contains a variety of carbohydrates and adsorbed molecules (Neu *et al.*, 2001), and may represent a major component of stream biofilm biomass (Geesey *et al.*, 1978; Couch *et al.*, 1996). Lawrence *et al.* (2002) showed that grazing by snails, mayfly and ostracods all significantly reduced the amount of extracellular polymer but not necessarily the bacteria present in riverine biofilms, suggesting that polymer components are an important food source. The same study showed that invertebrates with different morphologies had contrasting effects on biofilm composition, indicating that different components of the biofilm are likely to have differential nutritional importance to various invertebrate taxa. Nevertheless, most invertebrate feeding groups consume biofilm in some form by either grazing the biofilm directly, or filtering and collecting particles that have been dislodged from the biofilm. As biofilm composition is high in lipids and proteins compared with the detritus (Lamberti, 1996), even invertebrates feeding on decaying leaves may obtain a significant proportion of their required nutritional needs from the biofilm growing on leaves. This is supported by the work of Hall Jr. and Meyer (1998), which showed that biofilm bacteria are an important food source for a variety of stream invertebrates and that >20% of the carbon in most macroinvertebrate taxa is derived from bacteria, irrespective of functional feeding group.

The trophic interactions of heterotrophic biofilm-associated microeukaryotes with either bacteria or macroinvertebrates in lotic systems are not well characterized. Biofilm communities in rivers and streams typically host abundant and diverse heterotrophic protozoa (Arndt *et al.*, 2003). Protozoa have a significant role as predators of bacteria in both marine and freshwater systems, influencing bacterial abundance, community composition, morphology and physiology, and contributing to carbon transfer and nutrient cycling processes throughout the water column (Berninger *et al.*, 1991; Sanders *et al.*, 1992; Sherr and Sherr, 1994; Finlay and Esteban, 1998; Sherr and Sherr, 2002). Heterotrophic flagellates are considered the most important consumers of bacteria in aquatic systems, although mixotrophic flagellates and ciliates also contribute to the consumption (Caron, 2000). According to Wetzel (2001), protists are largely responsible for 90% of organic carbon mineralization and nutrient recycling in ecosystems, and are the most important consumers of microbes in aquatic and terrestrial ecosystems. Most of the data supporting these conclusions derive from studies of protistan bacterivory in pelagic habitats however, and many fewer data are available from benthic habitats such as freshwater biofilms. A number of recent studies, described below, have examined protozoan–bacterial biofilm feeding interactions in model biofilm systems however, clarifying the potential for protozoans to prey on biofilm bacteria and acting as a link between biofilms and organisms at higher trophic levels.

Protozoa comprise considerable diversity in terms of morphology, motility and feeding behaviour, leading to a diverse range of potential trophic interactions between them and their prey. A growing number of studies indicate that amoebae, ciliates and flagellates may all feed on biofilm bacteria within the constraints imposed by their motility, morphology and feeding modes, as reviewed by Parry (2004). While heterotrophic flagellates are considered major predators of bacteria in planktonic systems, it is less clear how efficiently flagellates can feed upon surface-attached biofilm bacteria. Morphological diversity among the flagellates is relatively low (compared with ciliates), but flagellates nonetheless exhibit a variety of different feeding modes (Boenigk and Arndt, 2002). For example, the benthic raptorial-feeding flagellate *Rhynchomonas nasuta* feeds on bacteria present on a substrate but is unable to ingest firmly attached cells, suggesting a limited ability to feed upon biofilm cells (Boenigk and Arndt, 2000). Similarly, the suspension feeding flagellate *Bodo* sp. had little impact on microcolony biofilms of *Serratia marcescens* (Queck *et al.*, 2006). In contrast, Böhme *et al.* (2009) recently investigated the impacts of a variety of stream-biofilm-associated protozoa on multispecies biofilms, finding that the interception-feeding *Spumella* spp. increased biofilm porosity and reduced biofilm thickness, possibly using its flagellum to extract and consume cells from within biofilm structures.

Amoebae are morphologically adapted to a benthic mode of existence and restricted to feeding on surface-associated prey. They have been found to consume fluorescently labelled bacteria in sediments at rates ranging from fewer than one cell per amoeba per hour to over 1000 cells per amoeba per hour (Rogerson *et al.*, 1996; Butler and Rogerson, 1997; Mayes *et al.*, 1997). A recent study estimated that *Acanthamoeba castellanii* consumed 90 bacterial cells per amoeba per hour from a mixed bacterial biofilm, reducing overall bacterial abundance and becoming integrated into the biofilm (Huws *et al.*, 2005). Similarly, Bohme *et al.* (2009) investigated the impacts of stream-biofilm-associated protozoa on multispecies bacterial biofilms, finding that the amoeba *Vanella* spp. reduced the maximal biofilm thickness and the abundance of microcolonies. However, the ability of amoebae to consume bacterial biofilms appears to vary depending on the prey organisms involved. For example, *Acanthamoeba polyphaga* was found to have a significant impact on a microcolony-forming *Pseudomonas aeruginosa* biofilm but no effect on a toxin-producing *P. aeruginosa* strain (Weitere *et al.*, 2005), while Queck *et al.* (2006) demonstrated that *Serratia marcescens* biofilm microcolony structures are susceptible to grazing by *A. polyphaga* but filamentous structures formed by *S. marcescens* biofilms under flow conditions are resistant to grazing by this amoeba.

The ciliates include organisms with a wide variety of morphologies and feeding modes. Browsing ciliates may ingest tens to hundreds of fluorescently labelled bacteria per ciliate per hour in sediments (Kemp, 1988; Epstein and Shiaris, 1992), but attachment of prey to surfaces may significantly reduce ingestion rates (Albright *et al.*, 1987). For example, *Tetrahymena* spp. ingested

1382 ± 1029 *Pseudomonas putida* cells per ciliate per hour , but this rate halved when *P. putida* cells were attached to glass beads and further decreased when a current was initiated (Eisenmann *et al.*, 1998). Lawrence and Snyder (1998) meanwhile determined that the crawling ciliate *Euplotes* spp. removed surface-attached *Vibrio natriegens* and *Pseudomonas fluorescens* cells at rates of 120 and over 800 cells per ciliate per hour respectively, clearing the substrate of bacteria at rates of 0.02 and 0.03 mm^2 per ciliate per hour. Huws *et al* (2005) clearly demonstrated that free-swimming filter-feeding ciliates can consume biofilm bacteria, with *Colopoda maupasi* reducing the thickness of a multispecies bacterial biofilm by 60%. Böhme *et al.* (2009) showed that ciliates with different feeding modes and morphology have contrasting effects on biofilms, and presumably differ in their ability to feed upon biofilm cells. *Dexiostoma campylum*, another free-swimming filter-feeder, caused a reduction of maximal biofilm thickness but did not affect biofilm volume, porosity or thickness of the basal biofilm layer, while the crawling raptorial ciliate *Chilodonella uncinata* (in combination with a flagellate) reduced biofilm volume, increased porosity and reduced the thickness of the basal biofilm layer by 30.5% (Böhme *et al.*, 2009). Similarly, the authors investigations of ciliate grazing upon stream biofilm-forming bacteria showed that filter-feeding *Tetrahymena* spp. drastically reduced biofilms composed of the stream-biofilm-forming bacterium *Serratia plymuthica*, but stimulated formation of grazing-resistant microcolonies in *Pseudomonas costantinii* biofilms. In contrast, *Chilodonella* spp. had less clear impact on *S. plymuthica* biofilm and did not stimulate microcolony formation in *P. costantinii* biofilms (Dopheide, unpublished data). These studies have all been carried out in simplified model systems, but suggest that protozoans must consume significant quantities of bacteria from natural stream biofilms.

To date, only a limited number of investigations have illuminated the trophic interactions involving protozoa in natural stream and riverine biofilms. In a recent study in the Rhine River in Germany, biofilms were established in flow cells supplied by river water, and the effect on resulting biofilm communities of excluding certain groups of organisms (flagellates, ciliates, and metazoans) by filtration of the inflowing water was monitored (Wey *et al.*, 2008). The presence of flagellates caused the reduced abundance of individual bacterial cells and enhanced the development of microcolonies. The addition of ciliates had a negative effect on flagellate abundance but enhanced the abundance of single bacterial cells, probably due to a flagellate-mediated trophic cascade.

The abundance and diversity of protozoa in natural biofilms suggests trophic interactions involving these organisms are quantitatively important. For example, a recent long-term study of biofilm composition in the Rhine River found that the average abundance of flagellates in riverine biofilms was 21×10^3 cells/cm^2, while the average ciliate abundance was 3.4×10^3 ciliates/cm^2. Mean bacterial abundance was considerably higher at 9.1×10^7 cells/cm^2. However, ciliates accounted for half of the total biofilm biovolume, while bacteria accounted for only 2.8% of the biovolume, and algae and flagellates less than 1%. Macroinvertebrates occurred in only low numbers, but accounted for 43% of biovolume (Ackermann *et al.*, 2011). While this represents an unusually high abundance of ciliates, it suggests that metazoans grazing upon biofilms are likely to consume protozoan biomass in significant quantities. Investigations of microbial trophic interactions in planktonic systems have shown that carnivorous protozoa, rotifers and macroinvertebrates may negatively affect protozoan assemblages through predation (Pace and Funke, 1991; Muylaert *et al.*, 2000; Kneitel and Chase, 2004), while Stoecker and McDowell-Capuzzo (1990) review evidence indicating that planktonic protozoa are an important source of nutrition for many invertebrate zooplankton and fish larvae. While protozoan predation is considered to contribute significantly to the transfer and cycling of energy and nutrients in planktonic microbial food webs (Sherr and Sherr, 2002), there is very little information available on the nature of trophic interactions between biofilm-associated protozoa and macroinvertebrates, although one study found evidence of invertebrates negatively affecting ciliates in a stream biofilm (McCormick, 1991). The quantitative importance of transfer of nutrients and energy through bacteria-protozoan-metazoan trophic links in stream food webs is also little studied, although an extensive study of a European allochthonous debris-based stream food web suggests 10% of bacterial production is consumed directly by invertebrates while 36% is consumed by protozoa, making otherwise inaccessible nutrients available to metazoan consumers (Marxsen, 2006).

Other biofilm-associated organisms represent probable sources of nutrition, but have similarly been subjected to very little study. Caddisflies have been found to preferentially feed on fungal-colonized patches on leaves, suggesting that fungal hyphae, in combination with fungus-modified leaf tissue, may constitute a favourable source of nutrition for these organisms (Arsuffi and Suberkropp, 1985). Schmid-Araya and Schmid (2000) review the role of meiofauna, such as rotifers, nematodes, gastrotrichs and oligochaetes, in lotic food webs. Meiofauna feed upon particles of various sizes including various microorganisms and biofilm material and in turn represent a food source for larger organisms such as macroinvertebrates. For example, Bott and Borchardt (1999) found that meiofauna abundance in streambed sediments ranged from 4 to 1094 animals per 10 cm, and that diatoms were consumed at rates of several hundred cells per meiofauna per hour and protozoa were ingested at rates of 46–153 cells per meiofauna per hour, equivalent to between 2.1% and 3.0% of protozoan abundance per hour. Similarly, rotifers, cladocerans and copepods reduced the abundance of ciliates and large flagellates in a pond littoral system through competition and predation (McCormick and Cairns Jr, 1991). Meiofauna are likely to be important components of lotic food webs, but understanding of their trophic relationships is currently very limited owing to a scarcity of data.

It is clear that stream biofilms play a major role in the bottom-up supply of nutrients to organisms at a higher trophic level, potentially affecting food webs throughout the stream ecosystem (Sheldon and Walker, 1997). However, the quantitative importance of trophic interactions involving protozoa and meiofauna in stream food webs requires further research.

Role in the cycling of carbon and key nutrients

Biofilms are major sites of carbon cycling in streams (Romani *et al.*, 2004). Carbon derived from terrestrial sources (allochthonous carbon) represents a major source of energy for stream biota, particularly in smaller, shaded streams (typically of many natural headwaters) where in-stream productivity is low (Vannote *et al.*, 1980). In the shaded headwater, the organic material in the stream is dominated by woody debris and particulate organic matter originating from outside of the stream (allochthonous carbon). As further described in Chapter 9, biofilms are 'hot-spots' for the decomposition of complex and polymeric organic matter, owing to the diverse range of extracellular enzyme activities present in the matrix of most biofilms. These extracellular enzymes, which may be found at high concentrations within freshwater biofilms and associated extracellular polymeric substances (EPS), are the primary mechanism for the degradation of complex organic matter. As a result, nutrient uptake in the biofilm is expected to be correlated with extracellular enzyme activity as only lower-weight molecules can cross the membranes of bacterial cells and be used for their nutrition (Hill *et al.*, 2010). The enzymatic degradation of organic material means that progressing towards the river mouth, quantities of fine particulate organic matter typically increase as a consequence of the upstream processing of dead leaves and woody debris. The large biomass and activity of microbial communities associated with stream biofilms therefore play a vital role in the 'river continuum' (first described by Vannote *et al.*, 1980) in which the sources, quantity and quality of organic material change, in a predictable manner, from the headwaters to the mouths of streams and rivers. In addition, as the width of the channel characteristically increases downstream, the relative importance of autochthonous-derived carbon produced by algae and cyanobacteria (as well as macrophytes) becomes greater (Fig. 8.1).

The diversity of organisms present within most aquatic biofilms and the spatial organization between them offer the potential for a great number of important biochemical interactions to occur. The interface between the surface of

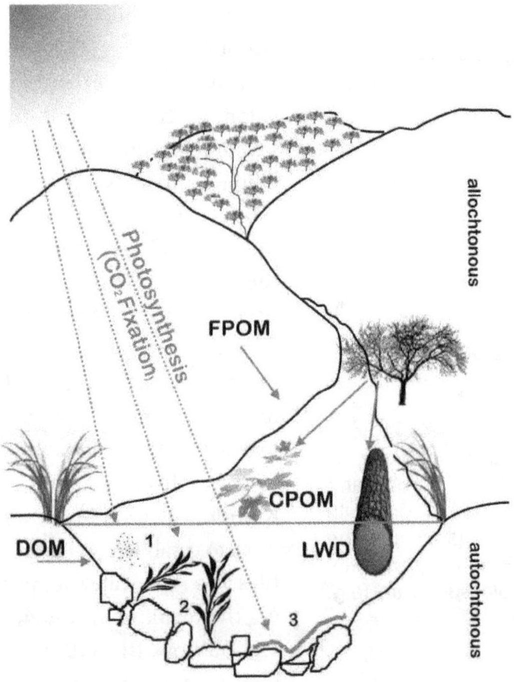

Figure 8.1 Major sources of carbon within freshwater ecosystems. Allochtonous carbon (LWD, large woody debris; CPOM, coarse particulate organic Matter; FPOM, fine particulate organic matter; DOM, dissolved organic matter). Autochthonous carbon [(1) Phytoplankton in water column; (2) Macrophytes, (3) Phytoplankton in biofilm community].

a biofilm and the wider aquatic environment is crucial, providing sharp gradients in a variety of key physicochemical parameters (e.g. pH, O_2). Of particular importance, cellular oxidation reactions (e.g. phosphorylation and biosynthetic processes) are strongly linked to the availability of molecular oxygen, which typically declines with depth in the biofilm. Below, we discuss the role of biofilms for the fixation and transformation of C, N and S in aquatic systems, processes which are highly influenced by the thickness and structure of the biofilm, because of the gradients in available oxygen that they provide [see Fig. 8.2 and refer to Paerl and Pinckney (1996) for an excellent review].

Biofilms play a key role in aquatic carbon cycling, actively excreting large quantities of extracellular polysaccharides, resulting in the increased availability of 'non-living' organic carbon which is a reliable source of energy for organisms residing in both the biofilm and surrounding environment (i.e. the water column). However, as already mentioned, the biofilm-mediated cycling of carbon is strongly regulated by the availability of oxygen as (i) the autotrophic production of macromolecules by photosynthetic and chemisynthetic processes is mediated by functional groups of microorganisms (e.g. oxygenic phototrophs, anoxygenic photosynthesizers) with varying tolerance to oxygen, and (ii) the availability of oxygen tightly regulates oxidative (i.e. aerobic mineralization, methane oxidation) and reductive (i.e. fermentation, methanogenesis) processes. The oxygen gradients provided by biofilm communities allow a diversity of microorganisms to undertake a wide-range of biochemical processes in 'microzones' within the biofilm from aerobic respiration at the surface to methane oxidation, anaerobic respiration and fermentation occurring in zones of increased oxygen depletion.

Biofilm-associated microorganisms also play an important role in the cycling of nitrogen within aquatic communities. The assimilation of nitrate, ammonia and ammonium can occur in oxic zones

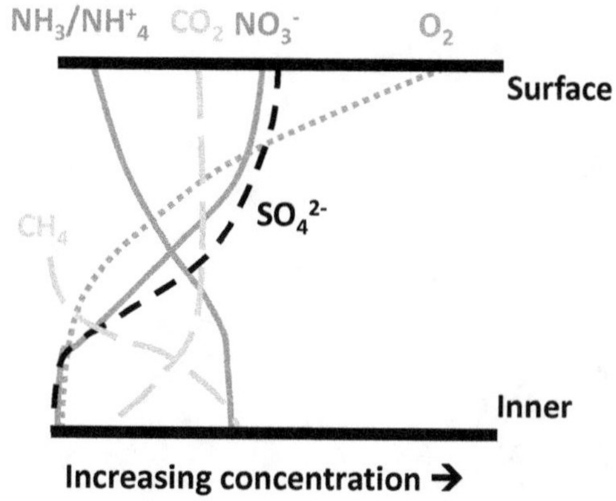

Figure 8.2 Schematic representation of microscale biogeochemical gradients in nitrogen, carbon and sulfur in a surficial biofilm. The surface represents the interface between the biofilm matrix and the bulk aqueous environment. The inner represents the interface with an impermeable substrate (e.g. a rock).

of the biofilm, whereas nitrification and denitrification occur at the aerobic/anaerobic interface. For example, the decomposition of organic matter in O_2-depleted biofilm zones leads to the production of ammonia/ammonium, which can be converted to nitrate only under oxic conditions. For this reason, rates of nitrification are maximized at oxic/anoxic interfaces within the biofilm community. Similarly, microzones with steep oxygen gradients provide ideal habitats for nitrogen fixation. The close association of aerobic and anaerobic microzones provided by the complex three dimensional structure of surficial biofilms are therefore of great importance for maintaining the balance of the nitrogen cycle within aquatic systems. A range of other biochemical gradients can be detected within the microscale-structure of freshwater biofilms. For example, biogeochemically important sulfate reduction occurs only within anoxic zones of the biofilm, whereas the exterior, aerobic zones provide ideal conditions for bacteria to reoxidize S^{2-} to elemental sulfur, sulfate or thiosulfate.

In comparison to the bacteria, the role of micro-eukaryotes in carbon and nutrient cycling in biofilms is far less well understood. The presence of bacterivorous protozoans has been shown to reduce bacterial abundance and yet promote bacterially associated decomposition processes in various habitats (Jiang, 2007). For example, Ribblett *et al.* (2005) demonstrated that the presence of bacterivorous flagellates in streams significantly enhanced the decomposition and presumed mineralization of leaf detritus. A modelling approach suggests that grazing by protozoa results in nutrient release into otherwise nutrient-limited habitats, thereby enhancing growth of the remaining bacteria and facilitating increased decomposition (Wang *et al.*, 2009). Additionally, recent research has shown that sessile filter-feeding ciliates and rotifers, which typically occur in high abundance in riverine and stream biofilms, may act as an important pathway for importation of carbon from the water column into the biofilm. Kathol *et al.* (2011) allowed natural riverine biofilms to establish on basalt tiles, before placing the tiles into river water-supplied flow cells. By analysing the planktonic content in the flow cell inflow and outflow, the authors determined that sessile filter-feeding ciliates (mainly peritrichs plus the heterotrich *Stentor*) and to lesser extent rotifers, imported between 15 and 137 mg carbon/m^2/day from the water column into the biofilm, demonstrating the existence of an important and dynamic trophic link between the benthos and the plankton. Thus, microeukaryote organisms in biofilm communities appear to contribute significantly to carbon and nutrient availability in aquatic ecosystems.

The effect of human activities on stream biofilm communities

Human activities cause multiple and complex impacts on rivers and streams world-wide (Ward and Trimble, 2003), with many freshwater systems being impacted by multiple physical (e.g. channel straightening), chemical (contamination by organic and inorganic pollutants) and biological (e.g. introduction of invasive species) stresses. The nature of anthropogenic disturbances varies within catchments affected by different land uses. For example, agricultural land uses may increase inputs of non-point source sediments and pollutants (e.g. fertilizers and pesticides), while in urban areas, the geomorphology of streams is generally simplified and the riparian vegetation removed (Paul and Meyer, 2001). The concentrations of numerous contaminants entering streams such as suspended sediments, heavy metals, hydrocarbons and oil, organic chemicals, nutrients and litter typically increase with the level of urban and rural development in the upstream catchment (Burton and Pitt, 2002), and may modify the health and distribution of the aquatic organisms within them (Allan, 2004).

It is evident that microorganisms associated with freshwater biofilms can respond rapidly to their surrounding environmental conditions. Changes caused by human activities can therefore alter the structure and the composition of biofilm communities (Findlay and Sinsabaugh, 2006; Lear and Lewis, 2009), which in turn may contribute to the degradation of whole stream ecosystems since, as already described, biofilms constitute a major food resource for organisms at a higher trophic level, and play key roles in nutrient cycling.

Of particular concern for urban streams is the effect of metal contaminants associated with runoff (or stormwater), particularly zinc, copper and lead, owing to their common abundance in urban runoff (Kennedy, 1999; Kayhanian *et al.*, 2007). Ancion *et al.* (2010) recently investigated the absorption rates of Zn, Cu and Pb in freshwater biofilm and assessed whether biofilm bacterial populations were affected by exposure to environmentally relevant concentrations of these metals. Zn, Cu and Pb were circulated in highly controlled flow chamber microcosms containing biofilms, at concentrations simulating either moderately, or highly contaminated urban runoff. Significant differences in bacterial community structure (assessed using automated ribosomal intergenic spacer analysis – ARISA) were detected within only 3 days of exposure to metals and increased over time until the longest period of exposure of 21 days. In a second experiment investigating the recovery of bacterial communities after exposure to pollutants, the biofilm was exposed for 7 days to metals then transferred to uncontaminated water. After 14 days of 'recovery' the concentrations of biofilm-associated metals remained elevated and differences in bacterial community structure could still be detected. Such studies suggest that biofilms can fix contaminants and play an important role in the transfer of contaminants throughout the stream food web. The sensitivity of microbial communities and their relatively slow recovery also suggest that exposure to moderate levels of contaminant could have longer lasting impacts on stream ecosystems than previously considered.

Excessive loads of inorganic nutrients and dissolved organic matter (DOM) also pose significant problems in both urban and rural catchments where they may be present in sufficient concentration to encourage eutrophic conditions. Olapede and Leff (2005) examined the response of biofilm bacterial communities to DOM (i.e. glucose, leaf leachates and algal exudates) and inorganic nutrients (nitrate and phosphate) using nutrient diffusing clay substrates to deliver a 'steady stream' of nutrients. Substrates were incubated in a stream for 3 weeks before biofilm communities were sampled from the substrates, and bacterial responses determined by fluorescence *in situ* hybridization. Shifts in bacterial community structure and composition confirmed the sensitivity of biofilms communities to changes in concentration of DOM and inorganic nutrients. The authors also suggested that the variations among bacterial taxa observed in the response to the DOM and the inorganic nutrient sources may be associated with differences in the species' metabolic capabilities.

A variety of other human activities have been observed to affect the structure, diversity and activity of biofilm-associated microbial communities. Deforestation, particularly of riparian margins, typically reduces allochthonous (terrestrial) inputs of organic matter (including leaf litter and woody debris), while increasing the importance of autochthonous (photosynthetic) carbon as a nutrient source as more light enters the stream (Fig. 8.1). Recently, Lear *et al.* (2009c) investigated the effect of varying light regimes (i.e. elevated light, ambient light, or darkness) on the taxa composition of stream biofilm bacterial communities. The authors used a combination of nucleic acid analysis and stable isotope probing (first undertaken by Radajewski *et al.*, 2000) to distinguish the metabolically active fraction (i.e. able to assimilate [^{13}C]-acetate) of the microbial community. In this study, the organisms actively utilizing acetate, added as a proxy for dissolved organic matter, differed to that of the general bacterial community, revealing that very specific populations of the biofilm community were stimulated by the addition of this labile source of carbon. Contrary to expectation, a complete loss of chlorophyll-containing organisms occurred in treatments exposed to elevated light. The authors suggest that increased exposure to light, in combination with the provision of an abundant simple carbon source (acetate) led to a decline in photosynthetic function, perhaps caused by photosystem II quinine receptors becoming highly reduced. These findings challenge current thinking that elevated light promotes communities dominated by photoautotrophs in nutrient-enriched environments.

A limited number of examples also provide direct evidence of the impact of human activities (mostly related to anthropogenic pollution) on other biofilm dwelling organisms. For example, Shukla and Gupta (2001) compared lake habitats that were moderately and heavily polluted (with high inputs of domestic waste and organic material), finding more protozoan species at less polluted sites, which also supported larger ciliate species (Shukla and Gupta, 2001). More recently, Dopheide *et al.* (2009) characterized ciliate diversity in four streams subjected to different impacts by human activity using both microscopy and terminal restriction fragment length polymorphism (T-RFLP) of 18S rRNA sequences, amplified using a novel primer set (Dopheide *et al.*, 2008). The data suggest the existence of widely distributed, resilient biofilm-associated ciliates as well as ciliate taxa restricted to sites with particular environmental conditions. For example, predatory ciliates related to *Litonotus* sp., were found in greatest abundance in the most impacted stream. While relationships between protozoan populations and various environmental factors have been investigated to some extent, the biochemical and physiological mechanisms which underlie such changes in protozoan abundance and community structure appear to be unexplored and unknown. Alterations in protozoan abundance and diversity may have consequences for bacteria and algae, and also for organisms at higher trophic levels, but these potential effects remain largely unstudied.

Biofilm-mediated degradation and transformation of freshwater pollutants

As further described in Chapter 5, the communal lifestyle adopted by biofilm bacteria expands the array of biochemical capabilities available to the population, thus improving the ability of these organisms to tolerate, transform and degrade a wide-range of pollutants. Advantages of the biofilm lifestyle for participating microorganisms include (i) sharing of the metabolic burden, in which substrate-driven spatial arrangements of microbial cells allow the efficient breakdown of complex substances by processes such as cometabolism and commensalism; (ii) enhanced levels of bacterial gene transfer via transformation, transduction and conjugation enabling the rapid spread of positive traits (e.g. encoding for greater resistance to heavy metals); (iii) physical protection, whereby soluble EPS and other microbial cells play a role as a protective barrier. The sorption of various metal and organic pollutants can provide a protective strategy against toxic chemical effects (Decho, 2000) whilst offering increased resistance to physical stresses (e.g. shear forces) and biological pressures (e.g. protozoan grazing) (Jefferson, 2004).

The diversity of microorganisms typically present within freshwater biofilms, along with their associated enzymes, mean that these communities are of great importance in the natural attenuation of freshwater organic pollutants by biodegradation and biotransformation. Whilst certain microbial species may show a high resistance to, or ability to degrade particular substances, typically mixed species consortia are best suited for this purpose (Pelz *et al.*, 1999; Cowan *et al.*, 2000). Studies have found that biofilms, through both spatial organization and community structure, are able to enhance degradation, and withstand higher concentrations of a range of organic pollutants including phenol (Bajaj *et al.*, 2008), 2-chloroethanol, *p*-cresol (Cowan *et al.*, 2000), and linear alkylbenzenesulfonates (Takada *et al.*, 1994). For example, Takada *et al* (1994) examined the natural removal of linear alkylbenzenesulfonates (LAS), which are toxic to aquatic biota at more than 0.1 mg/l (Lewis, 1991), in the Nagawa River, Tokyo. LAS concentrations decreased from ~1 mg/l in the upper stream to less than 0.05 mg/l only 6 km downstream. Batch culture experiments incubated either with, or without biofilm previously grown on concrete slabs within the stream bed, indicated significant acceleration of LAS degradation in cultures where biofilm was present. Since concentrations of LAS in the biofilm were also monitored, this study revealed that biofilm-mediated degradation, and not sorption, was responsible for the loss of LAS in this experiment. Microbially mediated degradation of organic compounds is clearly of key importance to removal of pollutants within the freshwater environment; however, this study is one of relatively few to examine the remediative capability of natural biofilms to 'cleanse' waters contaminated by anthropogenic pollutants. The findings of such studies have significant implications for improving our understanding of the fate of contaminants within freshwater rivers and streams, and of how we should aim to remediate specific contamination events in the future (i.e. if additional human intervention is even necessary). Furthermore, if the specific bacteria, or bacterial consortia, capable of the efficient degradation of environmental pollutants can be identified, they may be capitalised upon, for example, by use in *ex situ* bioreactors designed for the treatment of contaminated wastes, or as efficient inocula for *in situ* bioremediation strategies to treat sites in which the natural environmental microflora are not capable of efficiently degrading the pollutant.

Freshwater biofilms are ideal agents for the binding and sequestration of minerals, through both active and passive processes. The composition of EPS is complex and may vary greatly depending on the bacterial strain and environmental conditions (Sutherland, 2001a; van Hullebusch *et al.*, 2003); but typically contains a wide-range of ionizable functional groups such as carboxyl, phosphoryl, hydroxyl, and amino acid groups that enable EPS to fix metals (Jang *et al.*, 2001; van Hullebusch *et al.*, 2003). The surface properties of microbial cells also aid metals binding (Douglas and Beveridge, 1998; Wuertz *et al.*, 2000), and metals may be taken into cells and form complexes or chelates in the bacterial cytoplasm (Sterritt and Lester, 1980; van Hullebusch *et al.*, 2003). As a result, concentrations of metals including cadmium, lead, arsenic, zinc, and copper are typically significantly higher within biofilm than other stream compartments, including the water column, colloidal fraction, and at times sediments (Holding *et al.*, 2003; Toner *et al.*, 2005; Farag *et al.*, 2007). Active deposition of minerals within freshwater biofilms may occur through the action of bacterial exoenzymes, which can result in the concentration of specific minerals to high levels. The enzymatic oxidation of manganese and subsequent incorporation of the metal oxides into biofilms and cell structures is one such example (Ghiorse and Ehrlich, 1992; Lundsdorf *et al.*, 1997; Hill and Larsen, 2005), highlighted by a recent study undertaken by the chapter authors which identified unusual biofilm-associated structures, including a high inorganic component, within a number of anthropogenically impacted urban streams. Microscopic analysis revealed these structures to be microcolonies of a Gram-negative bacterium enclosed by an outer 'ring' enriched in iron and manganese (Fig. 8.3). Ring formation appears to involve oxidation of manganese and iron from the water column and their consequent deposition within a bacterially produced exopolymer matrix. The purpose of this manganese deposition is currently unknown, but may provide an example of a biofilm community employed in the removal of potential toxins from aqueous systems, by concentrating and binding ions from the passing water. Additionally to metal species, silica minerals can bind or associate preferentially with biofilm (Lünsdorf *et al.*, 2000; Alimova *et al.*, 2009; Brydie *et al.*, 2009; Konhauser *et al.*, 2011), and form cell–clay complexes (Chaerun *et al.*, 2005).

The incorporation of certain minerals within freshwater biofilms provides a mechanism for the sequestration and transformation of additional elements. The layered structure of many silica minerals provides a large capacity for adsorption of various compounds, as well as enzymes (Tietjen and Wetzel, 2003; Tietjen *et al.*, 2005). This property makes clay colloids an important vehicle for the transportation of organic pollutants, such as polychlorinated biphenyls, to, and immobilisation within, freshwater biofilms (Lünsdorf *et al.*, 2000; Abraham *et al.*, 2003). Additionally studies have shown that clays and

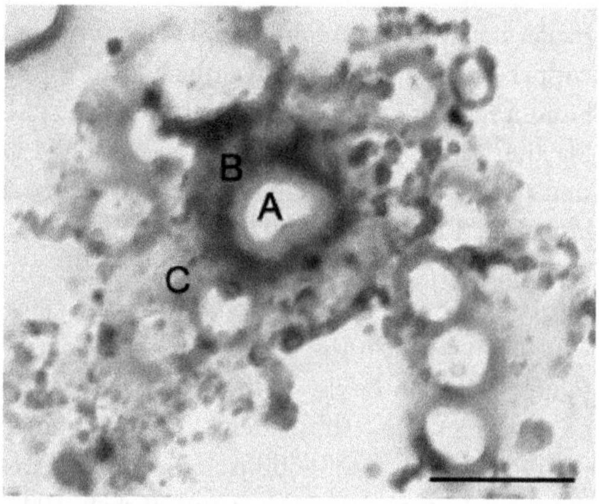

Figure 8.3 Typical 'ring' formation of Fe/Mn enriched biofilm structures showing (A) pore, (B) ring, and (C) edge. Biofilms were grown on glass microscope slides submerged in natural stream water for 14 days, and viewed under bright field illumination. Bar is 10 µm.

humic acids can both enhance or inhibit the degradation of a variety of organic pollutants, including phenanthrene (Ortega-Calvo and Saiz-Jimenez, 1998) LAS, linear alcohol ethoxylate, stearate and stearyl trimethylammonium chloride (Knaebel *et al.*, 1994), in part due to chemical–mineral interactions, enabled through biofilm association. Manganese oxides also have considerable sorptive capacities, and due to their high redox potentials are involved in transformation reactions of a wide range of elements (Tebo *et al.*, 2004), with biogenic manganese minerals showing higher sorptive capacities than their abiotic counterparts (Toner *et al.*, 2006). Biofilm-associated manganese oxides have roles in the cycling, sorption, speciation and mobility of numerous metals and metalloids, including nickel, cobalt, chromium, lead and zinc, within diverse freshwater environments (Saunders *et al.*, 1997; Fuller and Harvey, 2000; Olivie-Lauquet *et al.*, 2001; Wilson *et al.*, 2001; Haack and Warren, 2003; Manceau *et al.*, 2003). Thus, the specific accumulation of compounds such as silica and manganese in freshwater biofilms can affect cycling and transformation of a diverse range of compounds, including many environmental pollutants.

The ability of freshwater biofilms to bind, transform and degrade freshwater pollutants implicates these communities for use in bioremediation and wastewater treatment, as is discussed in Chapters 5 and 6. However, not all bacterially mediated transformations of pollutants result in a reduction in toxicity. Examples such as the mobilization of more toxic chromium species by manganese oxides (Kim *et al.*, 2002), or bacterially mediated reduction of As(V) resulting in release of mobile As(III) species (Héry *et al.*, 2008) exemplify the importance of reaching a clear understanding of the complex interactions between biofilms, minerals and other compounds within freshwater systems.

The use of biofilms to monitor freshwater ecological health

Current status of global freshwater environment monitoring

In the remainder of this chapter, we discuss the utility of biofilms as a reliable and sensitive indicator for monitoring freshwater ecological health. Environmental monitoring involves the systematic collection of data primarily for the purpose of recording the 'state of the environment' (Bruns and Wiersma, 2004). Such monitoring processes often incorporate the use of integrative data collection systems, which provide long-term measurements of air, water, soil and biota present within the specific ecosystem. The resulting data are scientifically analysed, used for ecosystem characterization and ultimately in regulatory processes and policy-making. Such programs provide guidelines to establish baseline reference conditions of water quality, which can then be used to identify emerging issues and threats, especially those in immediate response to anthropogenic impact (Bruns and Wiersma, 2004; Neale, 2010). Consistent and integrative monitoring can provide information on spatial and temporal trends in water quality, enable the measurement of ecosystem responses to remedial measures, and assist in regulatory decisions. The necessity for rigorous environmental monitoring systems is well established, especially because of the worldwide concern arising from increases in toxicant and pollutant concentrations, and from the effects of global warming. Increasing awareness of environmental monitoring has alerted ecologists to encourage the use of sampling programs which are minimally invasive to the environment under study, and provide an opportunity for the re-establishment of mixed-species biological communities, which might have been lost/dislodged because of destructive sampling.

Traditional Indicators of ecological health

Successful monitoring of the health of aquatic environments requires the ability to describe ecological change accurately using appropriately sensitive, quantitative indicators (Ryder and Miller, 2005). Most commonly, a suite of physicochemical measurements are taken, which may include concentrations of dissolved oxygen, organic carbon and nitrogen, suspended solids and pH (Tsegaye *et al.*, 2006). However, the large spatial and temporal variability in the data may preclude the assessment of sufficient sample numbers to adequately monitor stream ecosystem health. In addition, the data collected is frequently highly specific and therefore the outcomes of monitoring are limited to conclusions based on only the subset of water quality parameters chosen for use in that particular study. For this reason, biological indicators may provide a useful surrogate, as inferences can be made from the presence/absence, condition, relative abundance and community structure of groups of organisms, to provide a 'score' of overall ecosystem health (Marchant, 2007; Lear *et al.*, 2009a). The presence and abundance of different organisms in a community is the result of a cumulative response to past conditions. The use of biological indicators therefore provides a broader temporal aspect than traditional chemical and hydrological monitoring techniques, which require more frequent sampling to characterize trends in highly variable water quality parameters.

Many groups of macroorganisms have been proposed, and used, as indicators of ecological change, including fishes (Seilheimer and Chow-Fraser, 2006), macroinvertebrates (Feld and Hering, 2007), birds (Morrison *et al.*, 2007) and plants (Denniston *et al.*, 1993). Macroinvertebrates are most commonly exploited and a well-researched body of evidence now support their use (Feld and Hering, 2007; Trigal *et al.*, 2007) since they are comparatively easy to sample, and various taxa are now associated with different levels of water quality (Boothroyd and Stark, 2000). However, there remain numerous limitations for the use of fishes and macroinvertebrates as biological indicators of stream health (summarized in Table 8.1), mainly relating to the ease of sample collection and analysis (specifically taxon identification which largely relies on the visualization identification of individual organisms and requires significant expertise). In the following section, we discuss the potential for biofilms to provide a useful biological indicator of the ecological health of freshwater rivers and streams, and also outline the potential advantages and disadvantages of integrating biofilm-based approaches into current freshwater monitoring strategies.

Biofilm-based bacterial community approaches to monitor aquatic ecological health

In contrast to fishes and macroinvertebrates, biofilm-associated bacterial communities are easy to sample and to characterize in terms of taxa richness, using modern molecular methodologies. However, very few studies have yet investigated the use of whole bacterial communities as a tool to monitor variability in freshwater systems. Most of the studies undertaken to date (some examples include Lyautey *et al.* 2003, 2005; Langenheder *et al.*, 2006b; and Lear *et al.*, 2009a) have used DNA fingerprinting techniques such as denaturing gradient gel electrophoresis (Lyautey *et al.*, 2003), terminal restriction fragment length polymorphism (T-RFLP; Fierer *et al.*, 2007) and automated ribosomal intergenic spacer analysis (ARISA; Lear *et al.*, 2009a) to characterize the variability in bacterial community structure. In general, the major focus of these studies has been to determine the biogeography and biodiversity of biofilm-associated microbial species across a wide-range of both spatial and temporal scales. Only a few studies have yet sought to identify links between the heterogeneity in biofilm community structure, composition, or diversity and diverse environmental factors. In one such study, Lear and Lewis (2009) used ARISA (a community DNA 'fingerprinting' approach) as a high

Table 8.1 Comparison of the use of microorganisms and macroorganisms for assessment of the ecological health of freshwaters

Parameter	Microorganisms	Macroorganisms
Sampling effort	Easy. Requires very little sampling effort to collect a swab of biofilm, or sediment. Sufficient number of individuals (millions of cells) are typically present in only one gram of biofilm, or sediment	Can be challenging. Many species adapt their behaviour to avoid capture and require significant sampling effort. May require sorting from debris (e.g. leaf litter) to permit identification, or estimates of abundance. Small sample sizes are common and further reduce statistical accuracy
Likelihood of indicators response to be linked to local conditions	Microorganisms associated with the sediment, and particularly biofilms, are relatively sessile	May exhibit significant season or diurnal migration, such that the abundance/activity/health of these organisms may not closely reflect the environment in which they are captured
Site disturbance	Minimal disturbance associated with collecting small volumes of sediment or biofilm swabs	Sampling can be highly destructive, and prevent regular sampling of a single site (e.g. kick sampling for invertebrates)
Ease of taxa identification	Advances in molecular techniques allow rapid phylogenetic identification/characterization of communities	Identification to species level may be difficult
Speed of response to environmental change	Rapid life cycle means microorganisms may be rapid indicators of environmental change	Macroorganisms typically have slower life cycles meaning that effects of fecundity may not be measurable for weeks, months or even years
Known links between freshwater ecological health and taxon presence, abundance, activity and health	Largely unknown since our knowledge is currently limited to only a few small-scale studies	Relationships between the presence and abundance of various fish and macroinvertebrate taxa well studied, and in many cases, highly correlated

throughput, low cost molecular method to assess differences in stream biofilm bacterial community structure and diversity among 18 streams surrounding the city of Auckland, New Zealand. The streams under analysis were selected to provide a broad geographic coverage, representing the major land use classes in the Auckland Region. In this study, significant differences in biofilm bacterial community structure were identified between streams impacted by different land uses (rural and urban), and correlated to a number of informative physicochemical characteristics including concentrations of phosphorus, nitrogen, chloride, and turbidity. The study concluded that bacterial community DNA fingerprinting methodologies such as ARISA may provide a cost-effective and rapid approach to monitor land use impacts on freshwater biofilms, such as the influence of encroaching urban development on the ecological health of rural streams. In a separate study, the authors used ARISA to assess the relationship between biofilm bacterial community structure and the level of impact associated with acid mine drainage (AMD), sampling sites ranging from relatively pristine streams to those highly impacted by AMD (Lear *et al.*, 2009b). Despite the presence of high concentrations of dissolved metals, such as aluminium and zinc, and deposits of iron hydroxide in some streams studied, pH (which ranged from pH 2.8 to 8.3) was determined to be the most important determinant of biofilm bacterial community variability. Moreover, this study identified that there was partial 'recovery' of biofilm communities in an AMD contaminated stream that had been amended with lime to alleviate some of the effects of the acidic drainage. The findings of this biofilm-based study indicate that restoration activities in such systems (affected by acid mine drainage) should focus on dealing with pH as a first priority. Notably, Fierer *et al.* (2007) used a different DNA-fingerprinting approach (T-RFLP) to characterize the bacterial community structure within the fine benthic organic matter of 23 stream locations in New Hampshire, USA. Although the streams in this study exhibited only moderate variation in pH (from between from pH 4 to 6.3), pH was again determined to be the single variable most correlated to bacterial community structure (although other factors also contributed such as the amount of dissolved organic carbon and nitrogen and the quality of fine benthic organic matter). The authors suggest that pH may be a universal and primary factor influencing the biogeography of aquatic bacterial communities, and in doing so further highlight that bacterial communities may provide a cost-effective and rapid approach to monitor land use impacts on the ecological health of freshwaters. If the biodiversity and distribution of biofilm bacterial communities are indeed *predictably* regulated by particular environmental variables (e.g. temperature, pH, concentrations of dissolved organic nitrogen) then this provides exciting opportunities for the use of biofilm-associated microbial communities as reliable indicators of aquatic ecological health.

In one of the largest studies of bacterial biogeography undertaken to date, we recently used ARISA to analyse the biofilm bacterial communities structure of 1270 stream samples (five biofilm samples collected from each of 254 streams). These samples were obtained from New Zealand streams under the care of seven different regional authorities from north of Auckland (36°20′33 S; North Island) to south of Timaru (44°55′57 S; South Island), and pass through a wide variety of catchment types, from 100% native forest to 100% urban development in the upstream catchment. To understand the regional impact on stream biofilm bacterial communities, stream sites were classified into the seven geographical regions (regional council districts) that the stream samples were obtained from. A permutational analysis of variance (PERMANOVA; Anderson, 2001, McArdle and Anderson, 2001) of the ARISA data revealed significant differences in bacterial community structure among regions. However, by restricting our analysis to samples taken from a single region (e.g. Fig. 8.4, showing data obtained from a separate study of the Greater Auckland Region), it is evident that on a local scale,

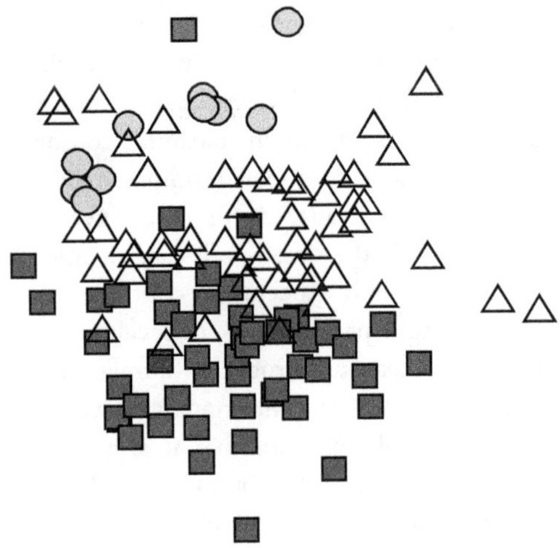

Figure 8.4 Comparison of bacterial community profiles from biofilm samples within catchments comprising predominantly (○) native forest; (△) rural development; (■) urban development. Plot is derived from non-metric multidimensional scaling of ARISA traces using a Manhattan distance matrix, 2D stress, 0.29. Five biofilm samples were removed from each of 24 different streams in the Greater Auckland Region of New Zealand.

bacterial communities are affected by land use and human interference in a somewhat predictable manner. Using data derived from this study, we are developing a predictive model to 'score' stream health from bacterial community (ARISA) data, somewhat analogous to the Macrobenthic Community Index (Stark, 1993, 1998; Stark and Maxted, 2007), where the relative abundance of ARISA peaks are used to generate a quantitative descriptor of stream ecological health. It is hoped that such biofilm-derived indices might allow the drivers of any observed community change/ecological degradation to be predicted from simple assessments of bacterial community structure, as is already routinely done using freshwater macroinvertebrate communities (e.g. RIVPACS; Wright *et al.*, 1993).

Future trends

To date, even where detailed taxonomic information has been gathered, the different functional roles of bacteria within complex environmental communities remain poorly understood (Langenheder *et al.*, 2006a), as does the relative sensitivity of different bacterial taxa to various anthropogenic disturbances. Nevertheless, in addition to the advantages of biofilm bacterial community analysis already described (compared with the more traditional analysis of macroinvertebrate communities), the speed and scale of microbial sample analysis is expected to continue improving at a rapid pace, owing to developments in both molecular and robotic sample handling technologies. Most importantly, advances in molecular biology mean that it is possible to rapidly characterize the abundance and diversity of key 'functional groups' of bacteria, such as those conferring resistance to heavy metals, or the ability to degrade various organic pollutants. The development molecular tools targeting specific functional groups of microorganisms promises to revolutionize the application of biological indicators, such that the analysis of biofilm bacterial communities may not only be used to provide an indication of general ecological stress, but to identify the specific drivers of environmental degradation within that community.

Originally, studies sought to link the presence of specific microbial taxa to a small number of functional processes. For example, increased abundances of members of the *Cytophaga–Flavobacteria* group have been linked to low nutrient and substrate concentrations (Battin *et al.*, 2001) owing to their ability to gain energy from complex macromolecules, and bacteria affiliated with the family *Geobacteriaceae* have been associated with Fe-reducing conditions (Lear *et al.*, 2010). However, describing the phylogenetic composition and diversity of biofilm communities is only the first step in understanding the complex functional roles undertaken by the wider community. A key advance in this field of research has been the recent explosion of 'omics' related studies, in which researchers have seemingly begun to catalogue the entirety of life around us using the information contained within the DNA barcode (metagenomics), RNA molecules (metatranscriptomics), proteins (metaproteomics) and small molecule metabolites (metametabolomics). Originally, metagenomic studies largely sought to link microbial functions to phylogenetic sources. However, more recently, a number of studies have used these approaches to determine the proteins encoded by the community DNA, and to reveal insights into the functional capability of these complex communities. For example, Tringe *et al* (2005) compared the functional characteristics of diverse samples including agricultural soils, whale carcasses, Sargasso Sea surface waters and an acid mine drainage biofilm community. The authors identified informative functional differences among these communities. For example, DNA fragments encoding for cellobiose and phosphorylase, involved in the degradation of plant material were found in the soil, but not the surface waters. In contrast, homologues of the light-driven proton pump bacteriorhodopsin were found in the surface waters, but not on the (deep sea) whale carcases, or in soil. While such variability in the abundance of putative functional genes may appear self-explanatory, the expected future increases in the volume of metagenomic data, and associated environmental 'meta-data', will inevitably revolutionize our understanding microbial community functions. Such approaches offer great promise for us to better understand the complex interactions occurring between biofilm microbial communities and the environments in which they live and to develop biological indicators of environmental health based on environmentally relevant functional data, rather than phylogenetic data.

Other significant recent advances include the development of functional gene microarrays such as the GeoChip (He *et al.*, 2007, 2010), which have been designed to provide information on the presence and relative abundance of a wide range of environmentally relevant functional genes. The most recent version of the GeoChip array (version 3.0) contains approximately 57,000 probes, targeting variants of 292 genes involved in carbon, nitrogen, phosphorus and sulfur cycles, energy metabolism, antibiotic resistance, metal resistance and organic contaminant degradation (He *et al.*, 2010). Whilst some initial studies have used this approach to report on the functional diversity of microbial communities within groundwater systems (Van Nostrand *et al.*, 2009; Waldron *et al.*, 2009) very few studies have yet examined the microbial communities within surface waters [e.g.

in the hypersaline Great Salt Lake, USA (Parnell *et al.*, 2010) and river basins contaminated with hexachlorobenzene (Tas *et al.*, 2009)]. To our knowledge, no publications have yet described the use of this methodology to characterize the functional diversity of freshwater biofilm-based communities.

The development and application of new, more sensitive biological indicators of environmental health will be a key new area for biofilm-based research. Novel methods including 'meta-omics' and microarray-based techniques offer significant promise for the development of multipurpose tools that can be used, in a single analysis, to monitor the effect of the whole gamut of stressors on biofilm communities. It is hoped that such information could be used not only to monitor the health of aquatic ecosystems around the world, but to identify the key drivers of change within degraded ecosystems.

Conclusions

It is evident that biofilms are in integral component of stream ecological health (i) providing a basal food resource for organisms at a higher trophic level, (ii) contributing to changes in detritus/nutrient quality observed in the 'river continuum', (iii) playing a role in the cycling and transformation of key nutrients including nitrogen and sulfur, (iv) altering the toxicity of both natural and anthropogenic contaminants by microbially mediated pollutant sorption, degradation and transformation. Using novel molecular methods, we are only just beginning to understand the enormous diversity of microbial species and functions associated with freshwater biofilms. Already it is evident that human disturbance is altering the structure of these sensitive, and often overlooked communities, with unknown consequences for stream ecological health and function. Whilst such impacts are of concern, the sensitivity of biofilm bacterial communities and their speed of response to environmental change make them excellent candidates as biological indicators of the changing health of our rivers and streams. Continued rapid developments in DNA-based methodologies are providing new ways for scientists to identify not only the level of general ecological stress, but to identify the specific drivers of environmental degradation within any site. As more and more pressures are placed on our global freshwaters, it seems that we may begin to seek the 'advice' of these unassuming biofilm microbial communities as to the nature and extent of anthropogenic impacts on this vital resource.

References

Abraham, W.-R., Lünsdorf, H., Strömpl, C., Nogales, B., Moore, E.R.B., and Timmis, K.N. (2003). Microbial communities in composite biofilms participating in the degradation of PCB. Water Air Soil Poll. *3*, 57–64.

Ackermann, B., Esser, M., Scherwaß, A., and Arndt, H. (2011). Long-term dynamics of microbial biofilm communities of the River Rhine with special references to ciliates. Int. Rev. Hydrobiol. *96*, 1–19.

Albright, L.J., Sherr, E.B., Sherr, B.F., and Fallon, R.D. (1987). Grazing of ciliated protozoa on free and particle-attached bacteria. Mar. Ecol. Prog. Ser. *38*, 125–129.

Alimova, A., Katz, A., Steiner, N., Rudolph, E., Wei, H., Steiner, J.C., and Gottlieb, P. (2009). Bacteria–clay interaction: structural changes in smectite induced during biofilm formation. Clays Clay Min. *57*, 205–212.

Allan, J.D. (2004). Landscape and riverscapes: The influence of land use on stream ecosystems. Ann. Rev. Ecol. Evol. Syst. *35*, 257–284.

Ancion, P.-Y., Lear, G., and Lewis, G.D. (2010). Three common metal contaminants of urban runoff (Zn, Cu & Pb) accumulate in freshwater biofilm and modify embedded bacterial communities. Environ. Poll. *158*, 2738–2745.

Anderson, M.J. (2001). A new method for non-parametric multivariate analysis of variance. Austral. Ecol.*26*, 32–46.

Anderson-Glenna, M.J., Bakkestuen, V., and Clipson, N.J.W. (2008). Spatial and temporal variability in epilithic biofilm bacterial communities along an upland river gradient. FEMS Microbiol. Ecol. *64*, 407–418.

Araya, R., Tani, K., Takagi, T., Yamaguchi, Y., and Nasu, M. (2003). Bacterial activity and community composition in stream water and biofilm from an urban river determined by fluorescent in site hybridization and DGGE analysis. FEMS Microbiol. Ecol. *43*, 111–119.

Ardon, M., and Pringle, C.M. (2007). The quality of organic matter mediates the response of heterotrophic biofilms to phosphorus enrichment of the water column and substratum. Freshwat. Biol. *52*, 1762–1772.

Arndt, H., Schmidt-Denter, K., Auer, B., and Weitere, M. (2003). Protozoans and biofilms. In Fossil and Recent Biofilms, Krumbein, W.E., Paterson, D.M., and Zavarzin, G.A., eds. (Dordrecht, Kluwer Academic Publishers), pp. 161–180.

Arsuffi, T.L., and Suberkropp, K. (1985). Selective feeding by stream caddisfly (Trichoptera) detritivores on leaves with fungal-colonized patches. Oikos *45*, 50–58.

Bajaj, M., Gallert, C., and Winter, J. (2008). Biodegradation of high phenol containing synthetic wastewater by an aerobic fixed bed reactor. Biores. Technol. *99*, 8376–8381.

Battin, T.J., Wille, A., Sattler, B., and Psenner, R. (2001). Phylogenetic and functional heterogeneity of sediment biofilms along environmental gradients in a glacial stream. Appl. Environ. Microbiol. *67*, 799–807.

Berninger, U.G., Finlay, B.J., and Kuuppo-Leinikki, P. (1991). Protozoan control of bacterial abundances in freshwater. Limnol. Oceanogr. *36*, 139–147.

Besemer, K., Singer, G., Limberger, R., Chlup, A.K., Hochedlinger, G., Hodl, I., Baranyi, C., and Battin, T.J. (2007). Biophysical controls on community succession in stream biofilms. Appl. Environ. Microbiol. 73, 4966–4974.

Besemer, K., Singer, G., Hodl, I., and Battin, T.J. (2009). Bacterial community composition of stream biofilms in spatially variable-flow environments. Appl. Environ. Microbiol. 75, 7189–7195.

Boenigk, J., and Arndt, H. (2000). Comparative studies on the feeding behavious of two heterotrophic nanoflagellates: the filter-feeding choanoflagellate *Monosiga ovata* and the raptorial-feeding kinetoplastid *Rhynchomonas nasuta*. Aquat. Microb. Ecol. *22*, 243–249.

Boenigk, J., and Arndt, H. (2002). Bacterivory by heterotrophic flagellates: Community structure and feeding strategies. Anton Leeuwen Int. J. Gen. Mol. Microbiol. *81*, 465–480.

Böhme, A., Risse-Buhl, U., and Küsel, K. (2009). Protists with different feeding modes change biofilm morphology. FEMS Microbiol. Ecol. *69*, 158–169.

Boothroyd, I.K.G., and Stark, J.D. (2000). Use of invertebrates in monitoring. In New Zealand Stream Invertebrates: Ecology and Implications for Management, Collier, K.J., and Winterbourn, M.J., eds. (Christchurch, New Zealand Limnological Society), pp. 344–373.

Bott, T.L., and Borchardt, M.A. (1999). Grazing of protozoa, bacteria and diatoms by meiofauna in lotic epibenthic communities. J. N. Am. Benthol. Soc. *18*, 499–513.

Bouletreau, S., Garabetian, F., Sauvage, S., and Sanchez-Perez, J.-M. (2006). Assessing the importance of a self-generated detachment process in river biofilm models. Freshwat. Biol. *51*, 901–912.

Brummer, I.H.M., Fehr, W., and Wagner-Dobler, I. (2000). Biofilm community structure in polluted rivers: Abundance of dominant phylogenetic groups over a complete annual cycle. Appl. Environ. Microbiol. *66*, 3078–3082.

Brummer, I.H.M., Felske, A., and Wagner-Dobler, I. (2003). Diversity and seasonal variability of beta-proteobacteria in biofilms of polluted rivers: Analysis by temperature gradient gel electrophoresis and cloning. Appl. Environ. Microbiol. *69*, 4463–4473.

Bruns, D.A., and Wiersma, G.B. (2004). Conceptual basis of environmental monitoring systems: A geospatial perspective. In Environmental Monitoring, Wiersma, G.B., ed. (CRC Press, Boca Raton, Florida).

Brydie, J.R., Wogelius, R.A., Boult, S., Merrifield, C.M., and Vaughan, D.J. (2009). Model system studies of the influence of bacterial biofilm formation on mineral surface reactivity. Biofouling *25*, 463–472.

Buesing, N., and Gessner, M.O. (2006). Benthic and fungal productivity and carbon turnover in a freshwater marsh. Appl. Environ. Microbiol. 72, 596–605.

Burton, G.A., and Pitt, R.E. (2002). Stormwater Effects Handbook: A Toolbox For Watershed Managers, Scientists And Engineers (Lewis Publishers, Boca Raton).

Butler, H., and Rogerson, A. (1997). Consumption rates of six species of marine benthic naked amoebae (Gymnamoebia) from sediments in the Clyde Sea area. J. Mar. Biol. Assoc. U. K. 77, 989–997.

Caron, D.A. (2000). Symbiosis and mixotrophy among pelagic microorganisms. In Microbial Ecology of the Oceans, Kirchman, D.L., ed. (Wiley-Liss, New York), pp. 495–523.

Chaerun, S.K., Tazaki, K., Asada, R., and Kogure, K. (2005). Interaction between clay minerals and hydrocarbon-utilizing indigenous microorganisms in high concentrations of heavy oil: implications for bioremediation. Clay Min. *40*, 105–114.

Clarke, K.R., and Warwick, R.M. (2001). Changes in marine communities: An approach to statistical analysis and interpretation. Plymouth, Primer-E Ltd.

Costerton, J.W., Lewandowski, Z., Debeer, D., Caldwell, D., Korber, D., and James, G. (1994). Biofilms, The customised microniche. J. Bacteriol. *176*, 2137–2142.

Couch, C.A., Meyer, J.L., and Hall, R.O. (1996). Incorporation of bacterial extracellular polysaccharide by blackfly larvae (*Simullidae*). J. N. Am. Benthol. Soc. *15*, 289–299.

Cowan, S.E., Gilbert, E., Liepmann, D., and Keasling, J.D. (2000). Commensal interactions in a dual-species biofilm exposed to mixed organic compounds. Appl. and Environ. Microbiol. *66*, 4481–4485.

Cross, W.F., Benstead, J.P., Frost, P.C., and Thomas, S.A. (2005). Ecological stoichiometry in freshwater benthic systems: Recent progress and perspectives. Freshwat. Biol. *50*, 1895–1912.

Debeer, D., Stoodley, P., and Lewandowski, Z. (1994). Liquid flow in heterogeneous biofilms. Biotechnol. Bioeng. *44*, 636–641.

Decho, A.W. (2000). Microbial biofilms in intertidal systems: an overview. Cont. Shelf Res. *20*, 1257–1273.

Denniston, W.C., Orth, R.J., Moore, K.A., Stevenson, J.C., Carter, V., Kollar, S., Bergstron, P.W., and Batiuk, R.A. (1993). Assessing water quality with submerged aquatic vegetation. Bioscience *43*, 86–94.

Dopheide, A., Lear, G., Stott, R., and Lewis, G.D. (2008). Molecular characterization of ciliate diversity in stream biofilms. Appl. Environ. Microbiol. *74*, 1740–1747.

Dopheide, A., Lear, G., Stott, R., and Lewis, G.D. (2009). Relative diversity and community structure of ciliates in stream biofilms according to molecular and microscopy methods. Appl. Environ. Microbiol. *75*, 5261–5272.

Dorigo, U., Volatier, L., and Humbert, J.F. (2005). Molecular approaches to the assessment of biodiversity in aquatic microbial communities. Water Res. *39*, 2207–2218.

Dorigo, U., Lefranc, M., Leboulanger, C., Montuelle, B., and Humbert, J.F. (2009). Spatial heterogeneity of periphytic microbial communities in a small pesticide-polluted river. FEMS Microbiol. Ecol. *67*, 491–501.

Douglas, S., and Beveridge, T.J. (1998). Mineral formation by bacteria in natural microbial communities FEMS Microbiol. Ecol. *26*, 79–88.

Edwards, R.T., and Meyer, J.L. (1990). Bacterivory by deposit-feeding mayfly larvae (*Stenonema* spp.). Freshwat. Biol. *24*, 453–462.

Edwards, R.T., Mayer, J.L., and Findlay, S.E.G. (1990). The relative contribution of benthic and suspended bacteria to system biomass, production, and metabolism in a low-gradient blackwater river. J. N Am. Benthol. Soc. *9*, 216–228.

Eggert, S.L., and Wallace, J.B. (2007). Wood biofilm as a food resource for stream detritivores. Limnol. Oceanogr. *52*, 1239–1245.

Ehrlich, H.L. (1999). Microbes as geological agents: their role in mineral formation. Geomicrobiol. J. *16*, 135–153.

Eisenmann, H., Harms, H., Meckenstock, R., Meyer, E.I., and Zehnder, A.J.B. (1998). Grazing of a *Tetrahymena* sp. on adhered bacteria in percolated columns monitored by *in situ* hybridization with fluorescent oligonucleotide probes. Appl. Environ. Microbiol. *64*, 1264–1269.

Elsdon, T.S., and Limburg, K.E. (2008). Nutrients and their duration of enrichment influence periphyton cover and biomass in rural and urban streams. Mar. Freshwat. Res. *59*, 467–476.

Epstein, S.S., and Shiaris, M.P. (1992). Size-selective grazing of coastal bacterioplankton by natural assemblages of pigmented flagellates, colorless flagellates, and ciliates. Microb. Ecol. 23, 211–225.

Farag, A.M., Nimick, D.A., Kimball, B.A., Church, S.E., Harper, D.D., and Brumbaugh, W.G. (2007). Concentrations of metals in water, sediment, biofilm, benthic macroinvertebrates, and fish in the Boulder River watershed, Montana, and the role of colloids in metal uptake. Arch. Environ. Contam. Toxicol. *52*, 397–409.

Fazi, S., Amalfitano, S., Pernthaler, J., and Puddu, A. (2005). Bacterial communities associated with benthic organic matter in headwater stream microhabitats. Environ. Microbiol. 7, 1633–1640.

Feld, C.K., and Hering, D. (2007). Community structure or function: effects of environmental stress on benthic macroinvertebrates at different spatial scales. Freshwat. Biol. 52, 1380–1399.

Fernandes, S.O., Krishnan, K.P., Khedekar, V.D., and Loka Bharathi, P.A. (2005). Manganese oxidation by bacterial isolates from the Indian Ridge System. Biometals *18*, 483–492.

Fierer, N., Morse, J.L., Berthrong, S.T., Bernhardt, E.S., and Jackson, R.B. (2007). Environmental controls on the landscape-scale biogeography of stream bacterial communities. Ecology *88*, 2162–2173.

Findlay, S., and Sinsabaugh, R. (2006). Large-scale variation in subsurface stream biofilms: a cross-regional comparison of metabolic function and community similarity. Microb. Ecol. *52*, 491–500.

Finlay, B.J., and Esteban, G.F. (1998). Freshwater protozoa: Biodiversity and ecological function. Biodivers. Conserv. *7*, 1163–1186.

Freeman, C., and Lock, M.A. (1995). The biofilm polysaccharide matrix – a buffer against changing organic substrate supply. Limnol. Oceanogr. *40*, 273–278.

Fuller, C.C., and Harvey, J.W. (2000). Reactive uptake of trace metals in the hyporheic zone of a mining-contaminated stream, Pinal Creek, Arizona. Environ. Sci. Technol. *34*, 1150–1155.

Fuller, R.L., Kennedy, B.P., and Nielson, C. (2004). Macroinvertebrate responses to algal and bacterial manipulations in streams. Hydrobiologia *523*, 113–126.

Galand, P.E., Casamayor, E., Kirchman, D.L., and Lovejoy, C. (2009). Ecology of the rare microbial biosphere of the Arctic Ocean. P. Natl. Acad. Sci. U.S.A. *106*, 22427–22432.

Geesey, G.G., Mutch, R., and Costerton, J.W. (1978). Sessile bacteria: an important component of the microbial population in small mountain streams. Limnol. Oceanogr. 23, 1214–1223.

Ghiorse, W.C. (1984). Biology of iron- and manganese-depositing bacteria. Ann. Rev. Microbiol. *38*, 515–550.

Ghiorse, W.C., and Ehrlich, H.L. (1992). Microbial biomineralisation of iron and manganese. Catena Suppl. *21*, 75–99.

Glockner, F.O., Zaichikov, E., Belkova, N., Denissova, L., Pernthaler, J., Pernthaler, A., and Amann, R. (2000). Comparative 16S rRNA analysis of lake bacterioplankton reveals globally distributed phylogenetic clusters including an abundant group of actinobacteria. Appl. Environ. Microbiol. *66*, 5053–5065.

Haack E.A. and Warren, L.A. (2003). Biofilm hydrous manganese oxyhydroxides and metal dynamics in acid rock drainage. Environ. Sci. Technol. 37, 4138–4147.

Hall, R.O., Jr., and Meyer, J.L. (1998). The trophic significance of bacteria in a detritus-based stream food web. Ecol. Lett. 79, 1995–2012.

He, Z., Deng, Y., Van Nostrand, J.D., Tu, Q., Xu, M., Hemme, C.L., Li, X., Wu, L., Gentry, T.J., Yin, Y., *et al.* (2010). GeoChip 3.0 as a high-throughput tool for analyzing microbial community composition, structure and functional activity. ISME J. *4*, 1167–1179.

He, Z.L., Gentry, T.J., Schadt, C.W., Wu, L.Y., Liebach, J., Chong, S.C., Huang, Z.J., Wu, W.M., Gu, B.H., Jardine, P., *et al.* (2007). GeoChip: a comprehensive microarray for investigating biogeochemical, ecological and environmental processes. ISME J. *1*, 67–77.

Héry, M., Gault, A.G., Rowland, H.A.L., Lear, G., Polya, D.A., and Lloyd, J.R. (2008). Molecular and cultivation-dependent analysis of metal-reducing bacteria implicated in arsenic mobilisation in south-east asian aquifers. Appl. Geochem. 23, 3215–3223.

Hill, B.H., McCormick, F.H., Harvey, B.C., Johnson, S.H., Warren, M.L., and Elonen, C.M. (2010). Microbial enzyme activity, nutrient uptake and nutrient limitation in forested streams. Freshwat. Biol. *55*, 1005–1019.

Hill, W.R., and Larsen, I.L. (2005). Growth dilution of metals in microalgal biofilms. Environ. Sci. Technol. *39*, 1513–1518.

Hoellein, T.J., Tank, J.L., Rosi-Marshall, E.J., Entrekin, S.A., and Lamberti, G.A. (2007). Controls on spatial and temporal variation of nutrient uptake in three Michigan headwater streams. Limnol. Oceanogr. *52*, 1964–1977.

Holding, K.L., Gill, R.A., and Carter, J. (2003). The relationship between epilithic periphyton (biofilm) bound metals and metals bound to sediments in freshwater systems. Environ. Geochem. Health *25*, 87–93.

Hullar, M.A.J., Kaplan, L.A., and Stahl, D.A. (2006). Recurring seasonal dynamics of microbial communities in stream habitats. Appl. Environ. Microbiol. *72*, 713–722.

van Hullebusch, E.D., Zandvoort, M.H., and Lens, P.N.L. (2003). Metal immobilisation by biofilms: Mechanisms and analytical tools. Rev. Environ. Sci. Biotechnol. *1*, 9–33.

Huws, S.A., McBain, A.J., and Gilbert, P. (2005). Protozoan grazing and its impact upon population dynamics in biofilm communities. J. Appl. Microbiol. *98*, 238–244.

Jang, A., Kim, S.M., Kim, S.Y., Lee, S.G., and Kim, I.S. (2001). Effect of heavy metals (Cu, Pb and Ni) on the compositions of EPS in biofilms. Water Sci. Technol. *43*, 41–48.

Jefferson, K.K. (2004). What drives bacteria to produce a biofilm? FEMS Microbiol. Lett. *236*, 163–173.

Jiang, L. (2007). Negative selection effects suppress relationships between bacterial diversity and ecosystem functioning. Ecology *88*, 1075–1085.

Kathol, M., Fischer, H., and Weitere, M. (2011). Contribution of biofilm-dwelling consumers to pelagic-benthic coupling in a large river. Freshwat. Biol. *56*, 1160–1172.

Kayhanian, M., Suverkropp, C., Ruby, A., and Tsay, K. (2007). Characterization and prediction of highway runoff constituent event mean concentration. J. Environ. Manage. *85*, 279–295.

Kemp, P.F. (1988). Bacterivory by benthic ciliates: significance as a carbon source and impact on sediment bacteria. Mar. Ecol Prog. Ser. *49*, 163–169.

Kennedy, P. (1999). The effects of road transport on freshwater and marine ecosystems. Technical Report (Ministry of Transport New Zealand, Wellington).

Kim, J.G., Dixon, J.B., Chusuei, C.C., and Deng, Y. (2002). Oxidation of chromium(III) to (VI) by manganese oxides. Soil Sci. Soc. Am. *66*, 306.

Knaebel, D.B., Federle, T.W., McAvoy, D.C., and Vestal, J.R. (1994). Effect of mineral and organic soil constituents on microbial mineralization of organic compounds in a natural soil. Appl. Environ. Microbiol. *60*, 4500–4508.

Kneitel, J.M., and Chase, J.M. (2004). Disturbance, predator, and resource interactions alter container community composition. Ecology *85*, 2088–2093.

Konhauser, K.O., Fisher, Q.J., Fyfe, W.S., Longstaffe, F.J., and Powell, M.A. (2011). Authigenic mineralization and detrital clay binding by freshwater biofilms: The Brahmani river, India. Geomicrobiol. J. *15*, 209–222.

Lamberti, G.A. (1996). The role of periphyton in benthic food webs. In Algal Ecology: Freshwater Benthic Ecosystems, Stevenson, R., Bothwell, M.L., and Lowe, R.L., eds. (Academic Press, San Diego), pp. 533–572.

Langenheder, S., Lindstrom, E.S., and Tranvik, L.J. (2006a). Structure and function of bacterial communities emerging from different sources under identical conditions. Appl. Environ. Microbiol. *72*, 212–220.

Langenheder, S., Sobek, S., and Tranvik, L.J. (2006b). Changes in bacterial community composition along a solar radiation gradient in humic waters. Aquat. Sci. *68*, 415–424.

Lawrence, J.R., and Snyder, B.D. (1998). Feeding behaviour and grazing impacts of a *Euplotes* sp. on attached bacteria. Can. J.Microbiol. *44*, 623–629.

Lawrence, J.R., Scharf, B., Packroff, G., and Neu, T.R. (2002). Microscale evaluation of the effects of grazing by invertebrates with contrasting feeding modes on river biofilm architecture and composition. Microb. Ecol. *44*, 199–207.

Lear, G., Anderson, M.J., Smith, J.P., Boxen, K., and Lewis, G.D. (2008). Spatial and temporal heterogeneity of the bacterial communities in stream epilithic biofilms. FEMS Microbiol. Ecol. *65*, 463–473.

Lear, G., and Lewis, G.D. (2009). Impact of catchment land use on bacterial communities within stream biofilms. Ecol. Indic. *9*, 848–855.

Lear, G., Boothroyd, I.K.G., Turner, S.J., Roberts, K., and Lewis, G.D. (2009a). A comparison of bacteria and benthic invertebrates as indicators of ecological health within streams. Freshwat. Biol. *54*, 1532–1543.

Lear, G., Niyogi, D., Harding, J., Dong, Y., and Lewis, G.D. (2009b). Biofilm bacterial community structure in streams affected by acid mine drainage. Appl. Environ. Microbiol. *75*, 3455–3460.

Lear, G., Turner, S.J., and Lewis, G.D. (2009c). Effect of light regimes on the utilisation of an exogenous carbon source by freshwater biofilm bacterial communities. Aquat. Ecol. *43*, 207–220.

Lear, G., McBeth, J.M., Boothman, C., Gunning, D.J., Ellis, B.L., Lawson, R.S., Morris, K., Burke, I.T., Bryan, N.D., Brown, N.D., *et al.* (2010). Probing the biogeochemical behaviour of technetium using a novel nuclear imaging approach. Environ. Sci.Technol. *44*, 156–162.

Lewis, M.A. (1991). Chronic and sublethal toxicities of surfactants to aquatic animals: a risk and review assessment. Water Res. *25*, 101–113.

Lundsdorf, H., Brummer, I., Timmis, K.N., and Wagner-Dobler, I. (1997). Metal selectivity of *in situ* microcolonies in biofilms of the Elbe River. J. Bacteriol. *179*, 31–40.

Lundsdorf, H., Erb, R.W., Abraham, W.R., and Timmis, K.N. (2000). 'Clay hutches': a novel interaction between bacteria and clay minerals. Environ. Microbiol. 2, 161–168.

Lyautey, E., Teissier, S., Charcoset, Y.-V., Rols, J.-L., and Garabetian, R. (2003). Bacterial diversity of epilithic biofilm assemblages of an anthropised river section, assessed by DGGE analysis of a 16S rDNA fragment. Aquat. Microb. Ecol. *33*, 217–224.

Lyautey, E., Jackson, C.R., Cayrou, J., Rols, J.-L., and Garabetian, F. (2005). Bacterial community succession in natural river biofilm assemblages. Microbiol. Ecol. *50*, 589–601.

Magurran, A.E., Baillie, S.R., Buckland, S.T., Dick, J.M., Elston, D.A., Scott, E.M., Smith, R.I., Somerfield, P.J., and Watt, A.D. (2010). Long-term datasets in biodiversity research and monitoring: assessing change in ecological communities through time. Trends Ecol. Evol. *25*, 574–582.

Manceau, A., Tamura, N., Celestre, R.S., MacDowell, A.A., Geoffroy, N., Sposito, G., and Padmore, H.A. (2003). Molecular-scale speciation of Zn and Ni in soil ferromanganese nodules from loess soils of the Mississippi Basin. Environ. Sci. Technol 37, 75–80.

Marchant, R. (2007). The use of taxonomic distinctness to assess environmental disturbance of insect communities from running water. Freshwat. Biol. *52*, 1634–1645.

Marxsen, J. (2006). Bacterial production in the carbon flow of a central European stream, the Breitenbach. Freshwat. Biol. *51*, 1838–1861.

Matz, C., and Kjelleberg, S. (2005). Off the hook – how bacteria survive protozoan grazing. Trends Microbiol. *13*, 302–307.

Mayes, D.F., Rogerson, A., Marchant, H., and Laybourn-Parry, J. (1997). Growth and consumption rates of bacterivorous Antarctic naked marine amoebae. Mar. Ecol. Prog. Ser. *160*, 101–108.

McArdle, B.H., and Anderson, M.J. (2001). Fitting multivariate models to community data: a comment on distance-based redundancy analysis. Ecology *82*, 290–297.

McCormick, P.V. (1991). Lotic protistan herbivore selectivity and its potential impact on benthic algal assemblages. J. N. Am. Benthol. Soc. *10*, 238–250.

McCormick, P.V., and Cairns Jr, J. (1991). Effects of micrometazoa on the protistan assemblage of a littoral food web. Freshwat. Biol. *26*, 111–119.

McNeely, C., and Power, M.E. (2007). Spatial variation in caddisfly grazing regimes within a northern Carolina watershed. Ecology *88*, 2609–2619.

Morales, J.B.T., and Ward, A.K. (2000). Differential incorporation of algae and bacteria by Elimia clara (Prosobranchia:Pleuroceridae): A study using dual-labeled epilithon. J. N. Am. Benthol. Soc. *19*, 289–297.

Morrison, N., Bearden, D., Bundy, J.G., Collette, T., Currie, F., Davey, M.P., Haigh, N.S., Hancock, D., Jones, O.A.H., Rochfort, S., *et al.* (2007). Standard reporting requirements for biological samples in metabolomics experiments: environmental context. Metabolomics 3, 203–210.

Muylaert, K., Van Mieghem, R., Sabbe, K., Tackx, M., and Vyverman, W. (2000). Dynamics and trophic roles of heterotrophic protists in the plankton of a freshwater tidal estuary. Hydrobiologia *432*, 25–36.

Neale, M. (2010). State of the Environment Monitoring: River Water Quality Annual Report 2008. Technical Report 2010/016. (Auckland Regional Council, Auckland, New Zealand).

Nealson, K.H. (2006). The manganese-oxidising bacteria. In The Prokaryotes Volume 5: Proteobacteria: Alpha and Beta Subclases (Springer, New York).

Neu, T.R., Swerhone, G.D.W., and Lawrence, J.R. (2001). Assessment of lectin-binding analysis for *in situ* detection of glycoconjugates in biofilm systems. Microbiology *147*, 299–313.

Van Nostrand, J.D., Wu, W.M., Wu, L., Deng, Y., Carley, J., Carroll, S., He, Z., Gu, B., Luo, J., Criddle, C.S., *et al.* (2009). GeoChip-based analysis of functional microbial communities during reoxidation of a bioreduced uranium-contaminated aquifer. Environ. Microbiol. *11*, 2611–2626.

Olapede, O., and Leff, L.G. (2005). Seasonal response of stream biofilm communities to dissolved organic matter and nutrient enrichments. Appl. Environ. Microbiol. *71*, 2278–2287.

Olivie-Laquet, G., Grau, G., Riou, C., Jaffrezic, A., and Henin, O. (2001). Release of trace elements in wetlands: role of seasonal variability. Water Res. *35*, 943–952.

Ortega-Calvo, J.-J., and Saiz-Jimenez, C. (1998). Effect of humic fractions and clay on biodegradation of phenanthrene by a *Pseudomonas fluorescens* strain isolated from soil. Appl. Environ. Microbiol. *64*, 3123–3126.

Pace, M.L., and Funke, E. (1991). Regulation of planktonic microbial communities by nutrients and herbivores. Ecology 72, 904–914.

Paerl, H.W., and Pinckney, J.L. (1996). A mini-review of microbial consortia: their roles in aquatic production and biogeochemical cycling. Microb. Ecol. *31*, 225–247.

Parnell, J.J., Rompato, G., Latta, L.C., Pfrender, M.E., Van Nostrand, J.D., He, Z., Zhou, J., Andersen, G., Champine, P., Ganesan, B., *et al.* (2010). Functional biogeography as evidence of gene transfer in hypersaline microbial communities. PLoS One 5, e12919.

Parry, J.D. (2004). Protozoan grazing of freshwater biofilms. Adv. Appl. Microbiol. *54*, 167–196.

Paul, M.J., and Meyer, J.L. (2001). Streams in the urban landscape. Ann. Rev. Ecol. System. *32*, 333–365.

Pelz, O., Tesar, M., Wittich, R.-M., Moore, E.R.B., Timmis, K.N., and Abraham, W.-R. (1999). Towards elucidation of microbial community metabolic pathways: unravelling the network of carbon sharing in a pollutant-degrading bacterial consortium by immunocapture and isotopic ratio mass spectrometry. Environ. Microbiol. *1*, 167–174.

Queck, S.Y., Weitere, M., Moreno, A.M., Rice, S.A., and Kjelleberg, S. (2006). The role of quorum sensing mediated developmental traits in the resistance of *Serratia marcescens* biofilms against protozoan grazing. Environ. Microbiol. *8*, 1017–1025.

Radajewski, S., Ineson, P., Parekh, N.R., and Murrell, J.C. (2000). Stable isotope probing as a tool in microbial ecology. Nature *403*, 646–649.

Ribblett, S.G., Palmer, M.A., and Coats, D.W. (2005). The importance of bacterivorous protists in the decomposition of stream leaf litter. Freshwat. Biol. *50*, 516–526.

Rickard, A.H., McBain, A.J., Stead, A.T., and Gilbert, P. (2004). Shear rate moderates community diversity in freshwater biofilms. Appl. Environ. Microbiol. *70*, 7426–7435.

Roeselers, G., Van Loosdrecht, M.C.M., and Muyzer, G. (2007). Heterotrophic pioneers facilitate phototrophic biofilm development. Microbiol. Ecol. *54*, 578–585.

Rogerson, A., Hannah, F., and Gothe, G. (1996). The grazing potential of some unusual marine benthic amoebae feeding on bacteria. Eur. J. Protistol. *32*, 271–279.

Romani, A.M., and Sabater, S. (1999). Effect of primary producers on the heterotrophic metabolism of a stream biofilm. Freshwat. Biol. *41*, 729–736.

Romani, A.M., Guasch, H., Munoz, I., Ruana, E., Vilalta, T., Schwartz, F., Emtiazi, F., and Sabater, S. (2004). Biofilm structure and function and possible implications for riverine DOC dynamics. Microb. Ecol. *47*, 316–328.

Ryder, D.S., and Miller, W. (2005). Setting goals and measuring success: linking patterns and processes in stream restoration. Hydrobiol *552*, 147–158.

Sanders, R.W., Caron, D.A., and Berninger, U.G. (1992). Relationships between bacteria and heterotrophic nanoplankton in marine and fresh waters: an inter-ecosystem comparison. Mar. Ecol. Prog. Ser. *86*, 1–14.

Saunders, J.A., Pritchett, M.A., and Cook, R.B. (1997). Geochemistry of biogenic pyrite and ferromanganese coatings from a small watershed: a bacterial connection? Geomicrobiol. J. *14*, 203–217.

Schmid-Araya, J.M., and Schmid, P.E. (2000). Trophic relationships: Integrating meiofauna into a realistic benthic food web. Freshwat. Biol. *44*, 149–163.

Seilheimer, T.S., and Chow-Fraser, P. (2006). Development and use of the Wetland Fish Index to assess the quality of coastal wetlands in the Laurentian Great Lakes. Can. J. Fish. Aquat. Sci. *63*, 354–366.

Sheldon, F., and Walker, K.F. (1997). Changes in biofilms induced by flow regulation could explain extinctions of aquatic snails in the lower River Murray, Australia. Hydrobiologia *347*, 97–108.

Sherr, E.B., and Sherr, B.F. (1994). Bacterivory and herbivory: key roles of phagotrophic protists in pelagic food webs. Microb. Ecol. *28*, 223–235.

Sherr, E.B., and Sherr, B.F. (2002). Significance of predation by protists in aquatic microbial food webs. Anton Leeuwen Int. J. Gen. Mol. Microbiol. *81*, 293–308.

Shukla, U., and Gupta, P.K. (2001). Assemblage of ciliated protozoan community in a polluted and non-polluted environment in a tropical lake of central Himalaya: Lake Naini Tal, India. J. Plankton Res. *23*, 571–584.

Skinner, H.C.W., and Fitzpatrick, R.W. (1992). Iron and manganese biomineralisation. Catena Suppl. *21*, 1–6.

Stark, J.D. (1993). Performance of the Macroinvertebrate Community Index: Effects of sampling method, sample replication, water depth, current velocity, and substratum on index values. NZ. J. Mar. Freshwat. Res. *27*, 463–478.

Stark, J.D. (1998). SQMCI: A biotic index for freshwater macroinvertebrate coded-abundance data. NZ. J. Mar. Freshwat. Res.*32*, 55–66.

Stark, J.D., and Maxted, J.R. (2007). A biotic index for New Zealand's soft-bottomed streams. NZ. J. Mar. Freshwat. Res.*41*, 43–61.

Sterritt, R.M., and Lester, J.N. (1980). Interactions of heavy metals with bacteria. Sci. Total Environ. *14*, 5–17.

Stoecker, D.K., and McDowell Capuzzo, J. (1990). Predation on protozoa: its importance to zooplankton. J. Plankton Res. *12*, 891–908.

Stuetz, R.M., Greene, A.C., and Madgwick, J.C. (1996). The potential use of manganese oxidation in treating metal effluents. Min. Eng. *9*, 1253–1261.

Sutherland, I.W. (2001a). Biofilm exopolysaccharides: a strong and sticky framework. Microbiology *147*, 3–9.

Sutherland, I.W. (2001b). The biofilm matrix – an immobilized but dynamic microbial environment. Trends Microbiol. *9*, 222–227.

Takada, H., Mutoh, K., Tomita, N., Miyadzu, T., and Ogura, N. (1994). Rapid removal of linear alklbenzenesulfonates (LAS) by attached biofilm in an urban shallow stream. Water Res. *28*, 1953–1960.

Tas, N., van Eekert, M.H.A., Schraa, G., Zhou, J., de Vos, W.M., and Smidt, H. (2009). Tracking functional guilds:*'Dehalococcoides'* spp. in European river basins contaminated with hexachlorobenzene. Appl. Environ Microbiol. *75*, 4696–4704.

Tebo, B.M., Bargar,J.R., Clement, B.G., Dick, G.J., Murray, K.J., Parker, D., Verity, R., and Webb, S.M (2004). Biogenic manganese oxides: Properties and mechanisms of formation. Annu. Rev. Earth Planet. Sci. *32*, 287–328.

Tebo, B.M., Johnson, H.A., McCarthy, J.K., and Templeton, A.S. (2005). Geomicrobiology of manganese(II) oxidation. Trends Microbiol. *13*, 421–428.

Teissier, S., Torre, M., Delmas, F., and Garabetian, F. (2007). Detailing biogeochemical N budgets in riverine epilithic biofilms. J. N. Am. Benthol. Soc. *26*, 178–190.

Tietjen, T., and Wetzel, R.G. (2003). Extracellular enzyme-clay mineral complexes: enzyme adsorption, alteration of enzyme activity, and protection from photodegredation. Aquat. Ecol. *37*, 331–339.

Tietjen, T., Vähätalo, A.V., and Wetzel, R.G. (2005). Effects of clay mineral turbidity on dissolved organic carbon and bacterial production. Aquat. Sci. *67*, 51–60.

Toner, B., Fakra, S., Villalobos, M., Warwick, T., and Sposito, G. (2005). Spatially resolved characterization of biogenic manganese oxide production within a bacterial biofilm. Appl. Environ. Microbiol. *71*, 1300–1310.

Trigal, C., Garcia-Criado, F., and Alaez, C.F. (2007). Macroinvertebrate communities of mediterranean ponds (North Iberian Plateau): importance of natural and human-induced variability. Freshwat. Biol. *52*, 2042–2055.

Tringe, S.G., von Mering, C., Kobayashi, A., Salamov, A.A., Chen, K., Chang, H.W., Podar, M., Short, J.M., Mathur, E.J., Detter, J.C., *et al.* (2005). Comparative metagenomics of microbial communities. Science *308*, 554–557.

Tsegaye, T., Shappard, D., Islam, K.R., Johnson, A., Tadesse, W., Atalay, A., and Marzen, L. (2006). Development of a chemical index of a measure of in-stream water quality in response to land-use cover changes. Water Air Soil Poll. *174*, 161–179.

Vannote, R.L., Minshall, G.W., Cummins, K.W., Sedell, J.R., and Cushing, C.E. (1980). The river continuum concept. Can. J. Fish. Aquat. Sci. *37*, 130–137.

Waldron, P.J., Wu, L., Van Nostrand, J.D., Schadt, C.W., He, Z., Watson, D.B., Jardine, J.M., Palumbo, A.V., Hazen, T.C., and Zhou, J. (2009). Functional gene array-based analysis of microbial community structure in groundwaters with a gradient of contaminant levels. Environ. Sci. Technol. *43*, 3529–3534.

Wang, H., Jiang, L., and Weitz, J.S. (2009). Bacterivorous grazers facilitate organic matter decomposition: A stoichiometric modeling approach. FEMS Microbiol. Ecol. *69*, 170–179.

Ward, A.D., and Trimble, S.W. (2003). Environmental Hydrology, 2nd edn (Lewis Publishers, Boca Raton).

Weitere, M., Bergfeld, T., Rice, S.A., Matz, G., and Kjelleberg, S. (2005). Grazing resistance of *Pseudomonas aeruginosa* biofilms depends on type of protective mechanism, developmental stage and protozoan feeding mode. Environ. Microbiol. *7*, 1593–1601.

Wey, J.K., Scherwass, A., Norf, H., Arndt, H., and Weitere, M. (2008). Effects of protozoan grazing within river biofilms under semi-natural conditions. Aquat. Microb. Ecol. *52*, 283–296.

Wilson, A.R., Lion, L.W., Nelson, Y.M., Shuler, M.L., and Ghiorse, W.C. (2001). The effects of pH and surface composition on Pb adsorption to natural freshwater biofilms. Environ. Sci. Technol. *35*, 3182–3189

Wright, J.F., Furse, M.T., and Armitage, P.D. (1993). RIVPACS – A technique for evaluating the biological control of rivers in the U.K. Eur. Water Poll. Contr. 3, 15–25.

Wuertz, S., Muller, E., Spaeth, R., Pfleiderer, P., and Flemming, H.C. (2000). Detection of heavy metals in bacterial biofilms and microbial flocs with the fluorescent complexing agent Newport Green. J. Ind. Microbiol. Biotechnol. *24*, 116–123.

Zhang, T., and Fang, H.H.P. (2001). Quantification of extracellular polymeric substances in biofilms by confocal laser scanning microscopy. Biotechnol. Lett. *23*, 405–409.

Zwart, G., Crump, B.C., Agterveld, M., Hagen, F., and Han, S.K. (2002). Typical freshwater bacteria: an analysis of available 16S rRNA gene sequences from plankton of lakes and rivers. Aquat. Microb. Ecol. *28*, 141–155.

9 Extracellular Enzymes in Aquatic Biofilms: Microbial Interactions versus Water Quality Effects in the Use of Organic Matter

Anna M. Romaní, Joan Artigas and Irene Ylla

Abstract

Biofilms in aquatic ecosystems colonize various compartments (sand, rocks, leaves) and play a key role in the uptake of inorganic and organic nutrients. Owing to their extracellular enzyme capabilities, biofilm microorganisms are able to use organic matter from the surrounding water and increasing activities are related to the availability of biodegradable organic carbon. The most common extracellular enzymes analysed are those involved in the decomposition of polysaccharides, peptides and organic phosphorus compounds, and changes in enzyme expression have been related to the use of different sources of organic matter available in the ecosystem (i.e. during drought-storm and/or pollution episodes). Enzymes important for microbial acquisition of nitrogen and phosphorus also respond to nutrient content and/or imbalances in the flowing water. Additionally, biofilm extracellular enzyme activities are modified by the internal recycling of organic matter and microbial interactions (competition/synergism) within the biofilm, such as algal-bacterial and fungal–bacterial interactions. Although an extensive knowledge of the biofilm structure is required for the interpretation of extracellular enzyme activities in aquatic biofilms, they give a very useful, integrative measure of the biofilm community function in relation to organic matter use and cycling.

Introduction

Freshwater biofilms play a key role in the use and decomposition of available organic matter, being an important place for energy transduction (Lock *et al.*, 1984) and ecosystem carbon cycling (Battin *et al.*, 2003). Because of their autotrophic and heterotrophic metabolism, biofilm communities are major sites for the uptake, storage and transformation of fluvial dissolved organic matter and nutrients (Bott *et al.*, 1984; Romaní *et al.*, 2004a,b). Significant extracellular enzyme activities have been measured in freshwater biofilms characterizing them as hot spots of organic matter cycling.

Bioavailable, low-molecular-weight organic matter compounds can be transported directly from the surrounding medium into microbial cells, but these compounds are present in low concentration in most freshwater environments, and competition for such resources may exist (Arnosti, 2003). The bulk of organic compounds in aquatic environments is complex and polymeric and cannot be directly incorporated into bacterial cells (Hoppe *et al.*, 1993). Extracellular enzymes are the primary mechanism for the degradation of polymeric and macromolecular organic matter into low-molecular-weight molecules which can then cross bacterial cell membranes (Rogers, 1961), becoming available for bacterial growth and nutrient cycles. This microbial activity protects the environment against the excessive accumulation of detrital organic matter, supplies photosynthetic organisms with nutrients, and provides clues to the growth and developmental needs of aquatic bacterial communities (Gajewski and Chróst, 1995).

The activity of extracellular enzymes is, in many cases, a limiting factor for substrate decomposition and bacterial growth. These enzymes represent the 'toolboxes' which enable heterotrophic bacteria to obtain suitable carbon and nutrients from a diverse array of compounds (Arnosti, 2003). The rate of organic matter decomposition is dependent on this enzymatic activity (Lock, 1990) which thus plays a key role in carbon cycling in aquatic environments (Turley, 1994). Among the metabolic metrics used for microorganisms, extracellular enzyme activities reflect the microbial nutrient demands and decomposition activity of dissolved organic matter in the environment (Allison and Vitousek, 2005).

Extracellular enzymes are inducible catabolic enzymes, their rates of synthesis and activity being strongly dependent on the presence of suitable polymeric substrates not continuously available in the environment (Chróst, 1990). Inducible enzymes are synthesized at a low basal rate in the absence of the appropriate substrates, whereas when the substrate becomes available in the environment, there is a huge increase in the production rate of the particular enzyme (Chróst, 1990). Since a specific enzyme will be synthesized only when the available substrate is present, the enzymatic activities could, in part, inform about the quality of the organic matter in the habitat under study. At the same time, extracellular enzyme synthesis is also regulated by catabolic repression. The production of enzymes is repressed by the end product that is derived from the substrate and accumulates in the cell or in the surrounding environment (Chróst, 1990). A well-known example is the repression of alkaline phosphatase synthesis when inorganic phosphate is available (Allison and Vitousek, 2005). In general, the synthesis of extracellular enzymes is repressed when there are sources of readily utilizable organic matter available. These two regulation mechanisms of enzyme synthesis (inducible by the substrate and repression by end products) avoid any unnecessary waste of energy expended in enzyme synthesis when utilizable dissolved organic compounds are available and/or when there are no substrates available for enzyme reaction (Chróst, 1991). Enzymatic activity and bacterial uptake are closely coupled processes (Hoppe *et al.*, 1988; Chróst, 1989, 1990; Chapell and Goulder, 1995), thus the microorganisms with the capacity to synthesize extracellular enzymes have a competitive advantage over other organisms (Wetzel, 1991; Gajewski and Chróst, 1995) owing to their ability to use molecules which are not directly available. Extracellular enzymes are particularly important for the survival of bacteria in oligotrophic conditions (Chróst, 1990).

The expression of a certain set of enzymes provides a 'fingerprint' of the composition of the organic matter pool in use (Hopkinson *et al.*, 1998). The utilization of extracellular enzyme activity in this manner relies on the assumption that as the composition of the bulk dissolved organic matter pool fluctuates, bacteria have the capacity to express a different set of enzymes to gain carbon from the new mixture (Chróst, 1991). Thus, extracellular enzyme activities are good proxies to characterize the nature and quantity of available dissolved organic carbon (DOC) and nutrients (N and P).

The biofilm complexity, mainly owing to its three-dimensional structure [defined by its specific thickness, density, content of extracellular polymeric substances (EPS), and microbial community composition; Romaní, 2010)], determines an extra complexity for the correct measurement of extracellular enzyme activities. Also, the different biofilm microorganisms present (bacteria, fungi, protozoa, algae) might modulate biofilm enzyme expression by microbial interactions and internal recycling processes within the biofilm, hampering the correct interpretation of results.

In this chapter, first a revision of the actual methods for measuring extracellular enzyme activity is provided, including the most relevant methodological tools to monitor enzyme activity in biofilms. Second, the sensitivity of extracellular enzyme activities to changing conditions both inside and outside biofilms is discussed.

Measuring extracellular enzymes in biofilms: from colorimetric approaches to functional diversity analysis

The importance of the bacterial community for the degradation of organic matter by means of extracellular enzymes was first stressed in the soil habitat (Burns, 1983). The first studies in aquatic environments showed the potential of the methodology and revealed links between enzymatic and bacterial activity in the marine planktonic microbial loop (Hoppe, 1983; Somville and Billen, 1983) and in lake waters (Reichardt *et al.*, 1967; Jones, 1971). Wider studies of these activities have since been performed in the freshwater planktonic environment (i.e. Chróst, 1991).

Hydrolases form the majority of extracellular enzymes in aquatic environments; among them the main enzymes are those which degrade polysaccharides, proteins and organophosphoric esters, linked to C, N, and P recycling, respectively. In Table 9.1, the most relevant enzymes measured in freshwater biofilms and their role in organic matter degradation is indicated. Most enzymes measured in ecological studies catalyse the last step of degradation of polymeric compounds, obtaining assimilable products although previous degradation by other enzymes is needed. Degradation of cellulose involves cellulase enzymes (acting on native cellulose), endo-β-1,4-glucanases and exo-β-1,4-glucanases (such as the cellobiohydrolase, converting the partially degraded cellulose molecules to cellobiose), and β-glucosidase (Atlas and Bartha, 1987). This last enzyme, widely measured in ecological studies, degrades cellobiose and other relatively small oligomers containing β-d-glucose linkages by splitting off the terminal β-d-glucose residue (Desphande and Eriksson, 1988). The β-xylosidase, involved in hemicellulose decomposition, hydrolyses xylobiose or xylooligosaccharides to form d-xylose as the end product (Lacke, 1988). The leucine aminopeptidase is an exopeptidase which breaks down peptide molecules from its endings, releasing amino acids (Jones and Lock, 1991). Phosphatase activity

Table 9.1 Summary of extracellular enzymes most commonly analysed in aquatic biofilms. The enzyme names (including the Enzyme Commission codes), their function and the artificial substrates used for their measurement are included

Enzyme	Involved in the metabolism of	Artificial substrate commercially available
β-D-Glucosidase (EC 3.2.1.21)	Carbon (last step of cellulose decomposition, decomposition of cellobiose or small oligomers containing β-D-glucose linkages)	MUF-β-D-glucoside
β-D-Xylosidase (EC 3.2.1.37)	Carbon (last step of hemicellulose decomposition, decomposition of xylobiose or xylooligosaccharides)	MUF-β-D-xyloside
Cellobiohydrolase (EC 3.2.1.91)	Carbon (exocellulase, decomposition of partially degraded cellulose into cellobiose)	MUF-cellobioside
Leucine-aminopeptidase (EC 3.4.11.1)	Peptide decomposition (exopeptidase, removing an amino acid from the end of a peptide chain)	Leu-AMC (l-leucine-4-methyl-7-coumarinylamide)
Lipase (EC 3.1.1.3)	Lipid decomposition (hydrolysis of ester bonds in water-insoluble lipid substrates)	MUF-palmitate
β-*N*-acetylglucosaminidase (EC 3.2.1.52)	Chitin decomposition (exoglycosidase, removing glucosamine from the end of a chitin chain, chitobiose hydrolysis, i.e. degradation and synthesis of fungal cell wall and arthropod exoskeletons)	MUF-*N*-acetyl-β-D-glucosaminide
Chitinase (EC 3.2.1.14)	Chitin decomposition (endochitinase)	MUF-*N*-acetyl-β-D-triacetylchitotrioside
Phosphatase (EC 3.1.3.1–2)	Phosphorus (decomposition of phosphomonoesters obtaining inorganic phosphorus)	MUF-phosphate
Phenol oxidase (EC 1.10.3) and peroxidase (EC 1.11.1)	Degradation of lignin, oxidation by using oxygen (phenol oxidases) or peroxide oxygen (peroxidases) as an electron acceptor	L-DOPA (L-3,4-dihydroxyphenylalanine)

indicates the degradation of phosphomonoesters to obtain inorganic phosphorus (Jansson *et al.*, 1988) and has been especially useful for assessing impacts of nutrients on biofilm communities (Scott *et al.*, 2009). The enzymes enabling the degradation and utilization of chitin (β-glucosaminidases and chitinases) may also cause lysis of fungal cell walls (degradation of fungal cell wall for fungal growth and/or fungal lysis by bacterial action; Wohl and McArthur, 2001) or arthropod exoskeletons. Oxidative enzymes, such as phenol oxidase and peroxidase are relevant for the degradation of lignin and are involved in the decomposition of woody plant material (Sinsabaugh, 2010).

Measurements of extracellular enzyme activities in freshwater biofilms began in the early nineties. The first studies mostly focused on the description of enzyme activities in different benthic habitats (sediment, rocks and cobbles, leaves, wood, peat) and the evaluation of the respective role of each habitat in relation to the organic compounds mostly used by the biofilm (Sinsabaugh *et al.*, 1991a; Romaní and Sabater, 2001). In sediments, the phosphatase activity was relevant for phosphorus recycling and the use of organic phosphorus compounds (Marxsen and Schmidt, 1993) while the relevance of sediment habitat for C turnover has also been stressed (Sinsabaugh and Findlay, 1995; Romaní *et al.*, 1998). In epilithic biofilms, seasonality of polysaccharidic and peptidic and phosphatase enzyme activities indicated changes in the organic matter availability and use for the microbial biofilm and were related either to water quality parameters (conductivity, pH, temperature, DOC) and/or microbial biofilm biomass (Jones and Lock, 1989; Sinsabaugh *et al.*, 1991b; Chapell and Goulder, 1992, 1995; Romaní and Sabater, 1999a; Romaní and Sabater, 2000a,b). Analyses in wood and leaf detritus showed the decomposition of mostly recalcitrant and plant material being linked to particulate organic matter decomposition in the stream habitat (Golladay and Sinsabaugh, 1991; Sinsabaugh *et al.*, 1993).

Although comparisons between enzyme activities may be treated with caution and a high variability exist (as will be evident in the following sections), it seems clear that biofilm heterotrophs have a greater potential to actively use peptides followed by simple polysaccharides, and a lower potential capacity to decompose more complex polysaccharides such as cellulose and hemicellulose (i.e. Romaní and Marxsen, 2002). This tendency would be related to both the concentration of substrate available for the enzymatic degradation (greater peptides and polysaccharides within the biofilm and/or in the flowing water) and greater energy obtained from these compounds after enzyme activity.

The use of fluorescent substrates for enzyme activity measurements

The measurement of extracellular enzyme activities by utilizing fluorescent -substrate analogues was first developed by Hoppe (1983) within the marine environment. The technique for measuring such enzymatic activities is based on an incubation of the natural sample with an artificial substrate that is linked to the substrate that the enzyme is able to hydrolyse. This link connects the organic molecule of interest with an artificial molecule (e.g. methylumbelliferone, MUF, or aminomethyl coumarin, AMC) which emits fluorescent light when it is free of any linkage (Fig. 9.1). The measurement of this fluorescence after the incubation time is directly related to the proportion of artificial substrate molecules cleaved by the natural enzyme activity. In the first studies of extracellular enzyme activities using this approach, chromogenic artificial substrates were used (e.g. Sinsabaugh and Linkins, 1988) measuring the absorbance of the coloured compound at the end of the incubation. Nowadays, in aquatic environments fluorogenic substrates are preferred because of their greater sensitivity (Chróst and Krambeck, 1986; Münster *et al.*, 1989). This feature permits a short incubation time, avoiding the problems of possible nutrient depletion and/or changes in the bacterial community, and is capable of detecting even very low enzyme activities. Special attention should be paid to pH, incubation time and presence of humic substances (Chróst and Krambeck, 1986; Münster *et al.*, 1989).

The validity of using artificial MUF-substrates in natural waters was demonstrated by

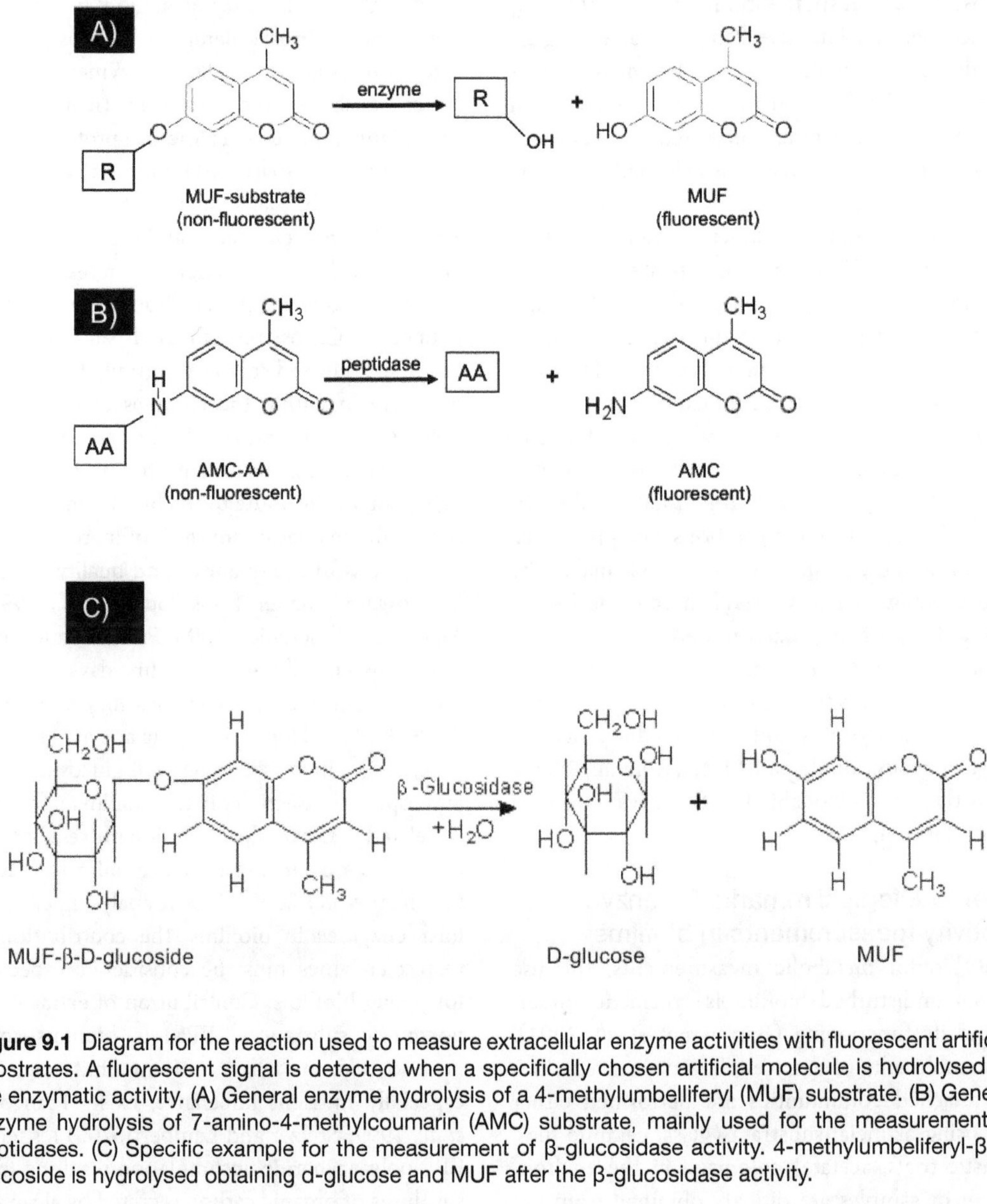

Figure 9.1 Diagram for the reaction used to measure extracellular enzyme activities with fluorescent artificial substrates. A fluorescent signal is detected when a specifically chosen artificial molecule is hydrolysed by the enzymatic activity. (A) General enzyme hydrolysis of a 4-methylumbelliferyl (MUF) substrate. (B) General enzyme hydrolysis of 7-amino-4-methylcoumarin (AMC) substrate, mainly used for the measurement of peptidases. (C) Specific example for the measurement of β-glucosidase activity. 4-methylumbeliferyl-β-d-glucoside is hydrolysed obtaining d-glucose and MUF after the β-glucosidase activity.

Hoppe (1983), Somville (1984) and Chróst (1990) who found a competitive inhibition of enzymatic activity (increase in hydrolysation time and apparent Km) when adding natural substrate analogues (cellobiose for MUF-β-d-glucoside and phosphogluconic acid or glucose-6-phosphate for MUF-phosphate). Bögershausen and Marxsen (1995) demonstrated a high specificity of polysaccharidic enzymes and aminopeptidase and phosphatase activities for the MUF-substrates.

Several constraints on the use of MUF-substrates are worth noting. First, the utilization of artificial substrates for measuring enzyme activity may overestimate the real activity because of the simplicity of their molecules (dimeric substrate), in contrast to the natural substrates (oligomeric or polymeric) found in the natural environment (King, 1986; Chróst and Rai, 1993). Second, most studies measuring extracellular enzyme activities in biofilms are carried out by monitoring potential enzyme activities (adding the artificial substrate at substrate-saturation concentration) which permits intensive sampling in ecological studies. However, this is a potential measurement which can be far from the activity observed in the ecosystem. A much more realistic approach to the enzymatic activities is achieved when the kinetic behaviour of each enzyme is measured, therefore providing the kinetic parameters (V_{max} and K_m). The V_{max} and K_m are related to natural substrate availability (Chróst and Rai, 1993) and allow the calculation of the turnover time, related to the efficiency of substrate utilization (Hoppe *et al.*, 1988). Moreover, the difficulty of measuring enzymatic velocities of natural microbial assemblages at low substrate concentration (Lewis *et al.*, 1988) together with the usually unknown natural substrate concentrations (Hoppe *et al.*, 1988; Marxsen and Fiebig, 1993) make the kinetic approach very useful in ecological studies, although being much more time consuming than the potential measurements. By the kinetic approach, turnover time values can be compared between ecosystems and indicate differences in organic matter biodegradability (Romaní, 2000), sensitive to microhabitat differences (Romaní and Marxsen, 2002).

Methodological remarks for enzyme activity measurements in biofilms

For biofilm metabolic measurements, the use of an undisturbed biofilm is crucial to understand its functioning (Barranguet *et al.*, 2004). Measurements of extracellular enzyme activities in attached communities are performed either by using artificial substrata (glass, ceramic tiles, plastic trials, acetate) colonized in the environment or samples are directly obtained from the field. In this latter case, most often biofilms are scrapped and a biofilm suspension is obtained. The disruption of the biofilm might overestimate the enzyme activity since the three-dimensional biofilm structure and its related gradients are lost. Specifically, Smucker *et al.* (2009) demonstrate significantly higher extracellular enzyme activities measured in a disrupted than in an intact biofilm, also underlying the importance of performing the measurements on the same sampling day.

The complex characteristics of the biofilm structure also have implications for the definition of the extracellular enzyme activity being measured. From Chróst's definition (Chróst, 1990), ectoenzymes are extracellular enzymes bound to the cell surface of microorganisms (mainly bacteria and fungi but also in algae and protozoa) or in the periplasmic space in Gram-negative bacteria (Chróst, 1991b) that act outside the cell (Chróst, 1990). In this chapter and in most scientific literature we use the general term 'extracellular enzyme' to refer to the cell-bound ectoenzymes defined by Chróst but which also includes 'free enzymes' released in the environment, but that may remain within the confines of the biofilm polymeric matrix and/or be bound to detritus. Extracellular enzyme activity free in the biofilm matrix may contribute to the biofilm metabolism and might be relevant for the biofilm resistance to fluctuations in the quantity, and quality of available organic matter (Sinsabaugh *et al.*, 1991b; Marxsen and Schmidt, 1993; Freeman and Lock, 1995). Specifically, during the first days of biofilm formation, matrix free enzymes may account to 65–81% of total biofilm enzyme activity (Romaní *et al.*, 2008). Increased values of β-glucosidase and aminopeptidase activity in the biofilm matrix may be related to the biofilm adhesion process but also to cell lysis occurring during biofilm formation (Romaní *et al.*, 2008). When analysing extracellular enzymes in biofilms, the contribution of matrix enzymes must be considered, especially for young biofilms. Contribution of extracellular polymeric substances (EPS) in biofilms could be also relevant when nutrient limitation, and especially nutrient imbalance, exist (Francoeur *et al.*, 1999; Stelzer and Lamberti, 2001), such as EPS-polysaccharide accumulation resulting from surpluses of organic carbon secreted by algae and bacteria (Mohamed *et al.*, 1998).

Aquatic biofilms are highly dynamic structures. In rivers and streams, biofilm attachment, detachment and recolonization processes occur periodically, mainly owing to discharge, grazing and release of dead cells when reaching the maximum carrying capacity (Stoodley *et al.*, 2001). The biofilm structure changes with respect to its metabolism, thickness, density and composition depending on its own evolution (Jenkinson and Lappin-Scott, 2001). Extracellular enzyme activity changes also take place throughout biofilm formation, associated with various structural changes (Jones and Lock, 1989; Sabater and Romaní, 1996; Romaní *et al.*, 2004b). Specifically, proteolytic and polysaccharidic enzymes are positively associated with the accrual of algal biomass and with the photosynthetic activity in the biofilm (Romaní and Sabater, 1999b; Francoeur and Wetzel, 2003).

The relevance of the three-dimensional biofilm structure might also determine the spatial localization of the enzyme activity which is most probably not equally distributed throughout the biofilm. The specific technique of using ELF (enzyme labelled fluorescence) substrata together with the confocal laser scanning microscopy (CLSM) could provide a powerful tool to analyse and localize the enzyme activity within the intact biofilm. The measurement of cell enzyme activities has been developed for planktonic environments by the use of ELF substrata that crystallize at the site of enzyme activity. This methodology has been mainly developed for the measurement of phosphatase activity of attached bacteria (Van Ommen Kloeke and Geesey, 1999) and planktonic algae (Nedoma *et al.*, 2003). A few studies of cell-enzyme activity in attached communities (cyanobacterial mat, activated sludge) have been performed by disrupting the biofilm structure to obtain a biofilm suspension (Sirová *et al.*, 2006; Schade and Lemmer, 2006). The application of this technique in biofilm studies has huge potential but it has not yet used extensively, probably owing to (1) the cost of CLSM, (2) the few ELF substrata designed and commercially available, and (3) the difficulty of relating the measurements of such a fine scale to whole biofilm metabolism (characterized by its high heterogeneity). The technique could be relevant for young biofilm studies.

Enzyme multifunctionality

The analysis of extracellular enzyme activities in biofilms inform about the use of specific organic compounds that might be available for the microbial community. At the same time, the presence of a wide range of enzyme capabilities might determine a greater enzyme multifuntionality in a biofilm and thus a greater competitive advantage over other biofilm communities. Most organic compounds need the action of diverse enzymes. A higher level of enzyme multifunctionality could be related to a higher microbial diversity (Cardinale *et al.*, 2006) and environmental heterogeneity, at the same time, affecting the whole ecosystem functioning (Langenheder *et al.*, 2010).

The analysis of a large range of enzyme capabilities in order to measure enzyme multifunctionality in a biofilm can be performed by using the single enzyme biofilm assays (as the fluorescence method discussed) for all commercially available artificial substrates. However, the integrated physiological and metabolic potential of the microbial community can be also analysed by the so called community-level physiological profile, when several potential enzymes are analysed at once for the same sample. This profile can be obtained by incubating samples in Biolog Ecoplates™ (Hayward, CA, USA) (Salomo *et al.*, 2009). This Biolog redox technology is a rapid, economic and effective community-level approach for assessing patterns of sole carbon source utilization by mixed microbial samples. The method involves direct inoculation of environmental samples into Biolog microtitre plates, and uses colour formation from reduction of a tetrazolium dye to assess utilization of separate sole carbon sources during a 2- to 7-day incubation period. The Biolog-Ecoplate contains 31 carbon sources in triplicate (Insam, 1997), including polymers, carbohydrates, carboxylic acids, phenolics and amino acids. Analyses of colour responses can be quantified from digitized images or from spectrophotometry readings and reveal distinctive degradation patterns among microbial communities (Garland and Mills, 1991; Garland, 1997). This approach has been effective at providing information on the functional diversity and estimation of bacterial use of dissolved organic compounds in marine plankton (Hollibaugh, 1994; Jellet *et al.*, 1996; Sala *et al.*, 2008), in freshwater (Grover and Chrzanowski, 2000; Sinsabaugh and Foreman, 2001), and in sediments (Douterelo *et al.*, 2010). This technique nevertheless has some disadvantages. It is a culture-based method, and as such, the biases of enrichment may render the results unrepresentative of the native microbiota (Konopka *et al.*, 1998; Campbell *et al.*, 2003). However, the potential of analysing community level physiological profile of biofilms remains highly relevant owing to the key role of biofilms in the cycling of organic matter.

Sensitivity of biofilm extracellular enzymes to water nutrient content and pollutants

Biofilms colonizing inorganic and organic surfaces are partially reliant on organic material and inorganic nutrients supplied by the surrounding water. Consequently, biofilms show shifts in metabolic activity with variation dependent on the quantity and quality of solutes in transport (Bott *et al.*, 1984). Extracellular enzyme activities are good proxies to determine the nutrient demands of microorganisms, as well as to characterize the nature and quantity of available dissolved organic carbon (DOC) and nutrients (N and P). However, relationships between nutrient resources and extracellular enzyme activity expression are not always direct because of the imbalance between nutrient availability and the quality of dissolved organic carbon (chemical composition and biodegradability, Fig. 9.2).

Biofilm extracellular enzyme responses to DOC availability (quantity and quality)

The amount (quantity) and origin (quality) of organic matter is one of the most important factors in determining the microbial metabolism and community structure of a riverine ecosystem (Bott *et al.*, 1984; Judd *et al.*, 2006). As a general response, biofilm extracellular enzyme activity increases with dissolved organic carbon (DOC) availability, although this will be highly modulated by the nature of the available organic molecules (quality) enhancing specific enzyme activities. A higher content of biodegradable DOC enhanced biofilm extracellular enzyme activities related to lipid, polysaccharide and peptide decomposition (Romaní *et al.*, 2004a). However, in freshwater and specifically in river ecosystems, DOC quantity and quality vary largely in their content and composition. DOC quantity in streams is mainly affected by land use and riparian flow path, whereas DOC quality depends on riparian vegetation and exposure to ultraviolet light (Findlay *et al.*, 2001).

The diversity of DOC within a specific reach of a stream is dependent on the potential number of sources (Koetsier *et al.*, 1997). Sources of carbon include autochthonous (carbon derived from in-stream processes) and allochthonous sources (carbon that enters the stream from outside the channel, mostly from hill slopes). Primary producers can be important sources of DOC in streams and rivers, releasing mainly low molecular weight and labile matter as exudates and from cell lysis (Bott *et al.*, 1984; Bertilsson and Jones, 2003). This organic C source may be available mostly during episodes of high primary production when exudates are produced, and within biofilms where exudates and products of cell lysis become concentrated. During springtime biofilm blooms, stream DOC concentrations increase as much as 37%, apparently due to extracellular release by algae (Kaplan and Bott, 1989). The autochthonous organic matter provides the biofilm with high-quality compounds excreted from the algae (Kaplan and Bott, 1989) although decaying algae and/or decaying heterotrophic organisms as well as possible heterotrophic exudates may also be a source of autochthonous material. Allochthonous sources are mainly composed of terrestrial plant materials (leachates from surroundings soils, grasses and inputs from riparian trees; Graça, 1993; Boyer and Groffman, 1996) as well as from DOC inputs from precipitation (Willey *et al.*, 2000). This allochthonous material is considered to be more refractory and of lower quality (Kaplan and Newbold, 1995; Joffre *et al.*, 2001). The major DOC compounds identified in river waters are carbohydrates (100–2000 μg/l), amino acids (50–1000 μg/l) and fatty acids (50–500 μg/l), the rest (ca. 75%) being humic, fulvic acids and hydrophilic acids (Thurmann, 1985).

The input of dissolved organic matter quantity and composition changes between seasons (Lock and Ford, 1986), but is generally associated with autumn leaf fall (Cummins *et al.*, 1983) and discharge variations (Fisher and Likens, 1973). Daily shifts in the organic matter pool have also been observed (Kuserk *et al.*, 1984) and related to algal activity (Kaplan and Bott, 1982). In temperate rivers, seasonal changes in extracellular enzyme activity reflect the major organic matter source available being used. Greater use of peptides was

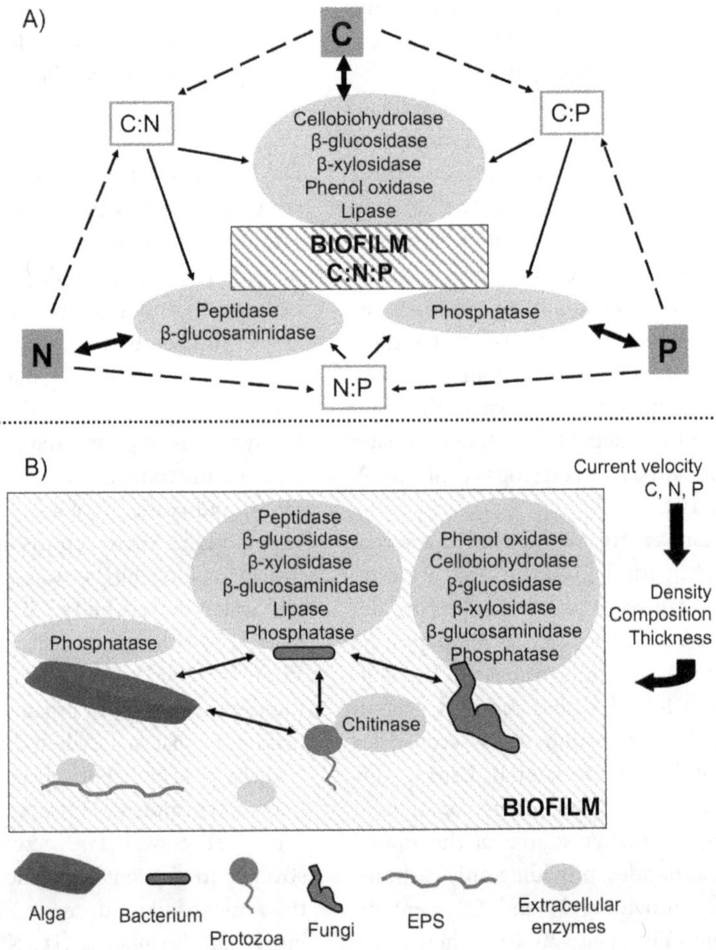

Figure 9.2 Combined response of the biofilm extracellular enzyme activities to water nutrient content (carbon, nitrogen and phosphorus compounds, panel A), and to the respective capabilities of biofilm microorganisms and microbial interactions, favouring internal nutrient cycling (panel B). In the upper panel (A) the biofilm is considered as a black box, the enzymes involved in the decomposition of C, N, and P compounds (large ovals) being regulated by the respective quantity and quality of available C, N, and P compounds (thick double-edge arrows). All enzymes are further regulated by nutrient molar ratios (C:N, N:P, C:P) both from the flowing water and from the main biofilm (thin arrows). In the lower panel (B), the main contribution of the extracellular enzyme activities from the biofilm microorganisms (large ovals) and their interactions (arrows) are shown. Bacteria mainly produce extracellular enzymes involved in the degradation of simple polysaccharides, peptides and lipids, their activity being highly regulated by the availability of algal released compounds and/or compounds previously degraded by fungi and/or interactions with protozoa. Fungi are the major organisms responsible for the decomposition of plant material such as cellulose and hemicellulose and lignin compounds, and their activity can be further regulated by interactions with bacteria. Algae usually show the capacity to decompose organic phosphorus compounds to obtain inorganic phosphorus. Protozoa show a relevant capacity to decompose chitin and they may acquire the availability to express enzyme activities from the digested bacteria. The enzymes linked to each microbial group in this figure are the major ones but specific conditions may lead the microorganisms to express further activities, such as the described capacity of several bacteria to decompose lignin, the availability of heterotrophic algae to use simple polysaccharides, and the capacity of fungi to use peptides. Thus, within the biofilm it is not often possible to clearly allocate each enzyme capability to a specific microbial group. Also, free extracellular enzymes or those linked to the extracellular polymeric substances (EPS) may also exist providing the biofilm microorganisms with a pool of shared enzymes.

measured in spring and linked to higher primary production. Decomposition of cellulose and hemicellulose was higher in autumn coinciding with the largest input of plant material (Artigas *et al.*, 2009; Sinsabaugh and Shah, 2010).

Decomposition of lignin (phenol oxidase and peroxidase activities) becomes dominant in sandy biofilms in periods when fine detritus accumulates (i.e. in late spring and late autumn, Artigas *et al.*, 2009). These activities, however, are very low in epilithic biofilms, probably because of the low availability of substrata for these activities in cobble habitats. In contrast, phenol oxidase, peroxidase and cellobiohydrolase activities are dominant in leaves, where the main organic substratum is degraded by fungi (aquatic hyphomycetes) and bacteria (Artigas *et al.*, 2009; Sinsabaugh, 2010).

In Mediterranean streams exposed to a severe summer drought period, DOC changes are even more drastic (Vázquez *et al.*, 2007). During the drought, a decrease in leucine aminopeptidase activity and increase in β-glucosidase in the benthic biofilm was linked with a reduced protein content, but higher proportion of polysaccharides in the water and benthic material. During this process, microbial heterotrophs were constrained to use the organic matter source of the lowest quality (polysaccharides, providing only carbon), since peptides (providing N and C) were no longer available (Ylla *et al.*, 2010). When water flow resumed in autumn, rainfall events leached leaf litter accumulated in the streambed and riparian forest, and the highest DOC values were measured (Vázquez *et al.*, 2007). Stimulation of cellulolytic and hemicellulolytic activities as well as peptidase activity coincide during these periods because of the greater availability of large molecules (>10 kDa) of highly biodegradable nature (Romaní *et al.*, 2006a; Marxsen *et al.*, 2010).

A large proportion of DOC reaching aquatic environments is made up of recalcitrant high-molecular-weight material and humic substances that requires of complex hydrolysis and oxidation and may inhibit extracellular enzyme activities in biofilms (Freeman and Lock, 1992). Humic-enzyme complexation can result in stabilization by restricting enzyme degradation while retaining high activities (Espeland and Wetzel, 2001). Ultraviolet light may have a photolytic effect, enhancing the degradation of such complex compounds, but also may damage heterotrophic microorganisms and slow degradation by inactivating microbial extracellular enzymes (Espeland and Wetzel, 2001).

Biofilm extracellular enzyme responses to N and P availability

Freshwater biofilm metabolism might respond to phosphorus (P) availability as P controls microbial growth and activity (Mohamed *et al.*, 1998; Stelzer and Lamberti, 2001; Dodds, 2003), as well as being the most limiting nutrient in aquatic ecosystems (Hutchinson, 1957). P is necessary for the synthesis of phospholipids, nucleic acids and ATP in microorganisms (Healey and Hendzel, 1979) and is therefore a key element for microbial growth. Many reports have described the inverse relationship between phosphatase activity and water P availability (Sinsabaugh and Moorhead, 1994; Romaní *et al.*, 2004b; Allison and Vitousek, 2005). Phosphatase activity mediates the decomposition of organic P compounds into orthophosphates, so when dissolved inorganic P availability is high, phosphatase activity decreases in the environment. However, many other studies state that phosphatase activity responds more strongly to nutrient imbalances (high N:P ratio) than to absolute decreases in the content of inorganic phosphorus (i.e. Klotz, 1992). Sala *et al.* (2001) proposed that the nutrient limitation of microbial communities could be examined *in situ* by assessing the ratio of alkaline phosphatase activity: aminopeptidase activity.

In the case of nitrogen (N), leucine-aminopeptidase activity, involved in the acquisition of N by decomposing simple peptides, would be related to the availability of water dissolved organic nitrogen and balanced by dissolved inorganic nitrogen concentrations. Also, ammonia and amino acids, which can be assimilated directly by the microorganisms, reduce peptidase activity (Kirchman, 2003). Enzymes targeting N have a high diversity of fates for the end products in microbial cells; (i) they may act as direct N sources, (ii) be sources of energy via catabolism of amino acids, or (iii) incorporate intact amino acids for further protein synthesis in the cell (Findlay and

Sinsabaugh, 2003). According to this, the regulation of peptidase activity will ultimately depend on the physiological status of microbial cells. Studies from Cottrell and Kirchman (2000) in coastal waters suggested that specific bacterial groups can show preferences for dissolved organic nitrogen decomposition; the group *Cytophaga–Flavobacteria* is important for the degradation of proteins, while the α-proteobacteria group become relevant for amino acid uptake. However, the capacity to acquire N and amino acids is sufficiently distributed across bacterial taxa and does not necessarily imply changes in bacterial community structure, as observed by Findlay and Sinsabaugh (2003) in N enrichment experiments in rivers.

N availability regulates lignin degrading activities, such as peroxidases and phenol oxidases (Keyser *et al.*, 1978). Sinsabaugh *et al.* (2005) found that N enrichment depletes peroxidase, phenol oxidase and cellobiohydrolase activities in leaf litter. Then, high N availability could reduce litter decomposition rates in streams, especially the most recalcitrant parts of decomposing materials.

Although theoretically an individual reduction in easily uptaken C, N and P sources would lead to an increase in the enzymatic use of respective available polymeric compounds, in natural ecosystems, organic matter use is highly regulated by the stoichiometry theory (Sterner and Elser, 2002; Allen *et al.*, 2009). Thus, nutrient imbalance in water or within the biofilm might regulate the enzyme expression (Fig. 9.2). Sinsabaugh *et al.* (2009) developed the concept of 'ecoenzymatic stoichiometry' which describes that enzyme activities would reflect the equilibrium between the elemental composition of microbial biomass and organic matter resources and the efficiencies of microbial nutrient content and growth. In streams, nutrient molar ratios of biofilms were close to the type of organic matter resources available at each type of benthic community (Artigas *et al.*, 2008). For instance, microbial communities associated with leaves and branches have higher C:N and lower N:P molar ratios and higher polysaccharide degrading activities. Instead, biofilms on sand and gravel, where algae accumulate and fine particulate material were more available, have lower C:N and higher N:P molar ratios and greater ligninolytic and peptidase activities.

The response of the biofilm to nutrient availability needs the parallel understanding of the whole biofilm structure and composition since not only the described relationships between the biofilm –as a black box- and the flowing water, explain the variability of the enzymes related to nutrient acquisition (Fig. 9.2). C, N and P compounds recycling within the biofilm exist and this can reduce the dependence of biofilms on resources from the water flow (Freeman and Lock, 1995; Romani *et al.*, 2004a). This recycling is highly dependent on biofilm structure (thickness, EPS content) and microbial interactions between biofilm compartments (algal–bacterial–fungal–protozoal), as discussed in the following section.

Biofilm extracellular enzyme responses to water pollutants

The degradation efficiency of pollutants in aquatic environments is mainly controlled by the action of microorganisms (Meyer, 1994) and can be detected with enzyme activity tests (Montuelle and Volat, 1998). Changes in extracellular enzyme activity patterns across catchment urbanization gradients in streams have been observed (Harbott and Grace, 2005). Organic matter carried by urban wastewater is a complex mixture of single carbohydrates, amino acids, alcohols and volatile fatty acids mixed with polymers and heteropolymers including proteins, polysaccharides and lipids (Raunkjaer *et al.*, 1994) and may determine increases in the polysaccharidic, peptidic and lipidic enzyme activities. Harbott and Grace (2005) found that less urbanized sites were characterized by high contribution of β-*N*-acetyl glucosaminidase and β-xylosidase activities highlighting the significance of microbial detrital material as a carbon source and the processing of plant-derived xylooligosaccharide substrates. In comparison, highly urbanized catchments recorded higher peptidase and esterase activities which were related to the higher levels of nutrient pollution and algal production. Increases in extracellular enzyme activities, i.e. peptidase, in waste water treatment plant (WWTP) effluents have also been described (Montuelle and Volat, 1998;

Ainsworth and Goulder, 2000). Discharges from sewage treatment works and fish farms have been observed to contain extracellular enzymes which might cause transient increases in enzyme activity in recipient rivers (Nybroe *et al.*, 1992; Davis and Goulder, 1993; Chappell and Goulder, 1994).

Extracellular enzymes are also involved in the biodegradation of high molecular weight compounds by sewage biofilms (Confer and Logan, 1997) and activated sludge flocs (Conrad *et al.*, 2000) in WWTP. Bacterial laccase (a class of phenol oxidase with copper atoms at its reaction centre) activity efficiently removes recalcitrant and emerging pollutants from waste water, becoming an important decontamination agent (Majeau *et al.*, 2009). Fungal laccase activity is also employed for bioremediation roles in industrial waste waters (Ryan *et al.*, 2005). Laccase substrates are diverse, and include phenols, dyes, pesticides, endocrine disrupters and polycyclic aromatic hydrocarbons, some of which can be oxidized by extracellular fungal or bacterial laccase. However, despite the enormous decontaminant potential of these enzymes, their production costs remain expensive.

Effects of specific pollutants, such as heavy metals and pharmaceuticals, on biofilm enzyme activities have also been reported. Phosphatase activity has been largely used in toxicological studies of heavy metals, because of the interaction between phosphates and metals. For instance, exposure of biofilms to zinc and copper concentrations induces phosphorus depletion due to the precipitation of zinc and copper phosphate, and, thus, leads to increased phosphatase activity (Chapell and Goulder, 1994; Paulsson *et al.*, 2002). Higher algal metal tolerance in communities grown under more eutrophic conditions may be attributed to polyphosphate bodies (PPB), which act as a metal detoxification strategy (Serra *et al.*, 2009).

The herbicide Diuron inhibits photosynthesis and enhances bacterial phosphatase activity (Espeland and Wetzel, 2001). Indirect effects of Diuron, after inhibiting algal photosynthesis, include a significant increase of bacterial mortality and peptidase activity in an epilithic biofilm (Ricart *et al.*, 2009). Emergent pollutants such as propanolol, and especially atenolol, have negative impacts on peptidase activity (Bonnineau *et al.*, 2010). Atenolol cause mortality in bacteria and results in the depletion of peptidase activity of biofilms.

Extracellular enzymes as indicators of microbial interactions within the biofilm

Within the biofilm, bacteria, fungi, algae and protozoa are involved in the cycling of organic matter either by their release of organic compounds, direct uptake or enzyme decomposition (Fig. 9.2B). However, the close contact between biofilm microorganisms within the biofilm microecosystem may cause interactions in organic matter use such as the bacterial utilization of algal release compounds (Romaní 2010). The relative contribution of the different microbial groups as well as the contribution of the EPS conferring a specific biofilm architecture might determine changes in their interaction and therefore in the use of biofilm organic matter (Fig. 9.2B).

Algal–bacterial interactions

An early study on epilithic biofilms demonstrated that algae may supply bacteria with DOC excreted during photosynthesis (Haack and McFeters 1982a, 1982b). Other studies of epilithic biofilms also concluded that algal exudates are a major carbon source for bacteria (Geesey *et al.*, 1978; Kaplan and Bott, 1989; Hepinstall and Fuller, 1994). Healthy algae release a proportion of the carbon they fix during photosynthesis (Bjoernsen, 1988; Sell, 1994), which is related to availability of inorganic nutrients. Algae release extracellular organic carbon during active metabolism (enhanced during photosynthetic production) or following cell lysis and natural senescence. These algal exudates contribute positively to the development of a nutrient-rich zone around themselves which could enhance the growth of bacteria (Murray *et al.*, 1986). Algal extracellular products include low molecular weight compounds (Bjoernsen, 1988) including carbohydrates, lipids, peptides, organic phosphates, volatile substances, vitamins, toxins and antibiotics (Fogg, 1966). Algal accumulation in the epilithic biofilm influences the use of organic matter by the heterotrophic community by increasing the amount of organic substrate available for bacteria (Romaní and Sabater, 1999b, 2000b).

The activity of leucine-aminopeptidase has been positively linked to the activity of photosynthetic primary producers (Francoeur and Wetzel, 2003), and to the use of algal-released proteinaceous compounds by microbial heterotrophs (Romaní *et al.*, 2004b). On the other hand, β-glucosidase activity has been correlated with the degradation of either organic compounds of algal origin (algal exudates and EPS) or polysaccharidic organic compounds dissolved in flowing water (Jones and Lock, 1993). In biofilm communities grown on artificial substrata, algal photosynthesis simultaneously increased the activities of extracellular α-glucosidase, β-glucosidase and β-xylosidase (Espeland *et al.*, 2001). Furthermore, bacterial phosphatase activity is also stimulated by the amount of extracellular organic carbon accumulated within the biofilm matrix during photosynthesis (Espeland and Wetzel, 2001). Thus, the photosynthetic activity of benthic algae has been recognized as a potentially important stimulator of extracellular enzyme activities within biofilms (Francoeur *et al.*, 2006; Rier *et al.*, 2007).

A direct consequence of the internal carbon recycling in light grown biofilms is the greater independency of the biofilm from the DOC flowing in the water column. Differences between light grown and dark grown biofilms (Sekar, 2002) stress the relative importance of algal-derived compounds for heterotrophic organisms in aquatic biofilms (Sinsabaugh and Linkins, 1988; Romaní and Sabater, 1999b; Romaní and Sabater, 2000b). For example, the greater use of β-glucosidase compared with β-xylosidase indicates the larger utilization of authochtonous than allochthonous compounds in light grown biofilms (Jones and Lock, 1993; Romaní and Sabater, 2000b). Hence, biofilms growing under light reduce the uptake of dissolved organic compounds from the flowing water pool (Romaní *et al.*, 2004a).

Although many studies have reported strong algal-bacterial coupling in freshwater biofilms, bacteria are less likely to utilize algal-generated DOC under specific conditions: (1) under extreme light or nutrient limitation, (2) under high nutrient and/or labile DOC availability in the flowing water.

Low light availability and/or extremely oligotrophic conditions may produce the development of a thin biofilm with low algal biomass and competition between algae and bacteria for nutrients may occur. Even though, results from Rier and Stevenson (2002) and Carr *et al.* (2005) point that there was no evidence of algae being negatively affected by competition with bacteria for nutrients. Streams with light limiting conditions (like forested streams), often have heterotrophic biofilms that rely on allochthonous organic matter rather than algal-derived extracellular organic compounds (Findlay *et al.*, 1993). In these light-limited environments, some algae are capable of heterotrophically metabolizing a diverse range of organic carbon sources. The proximal relationship between heterotrophically metabolizing algae and bacteria may also lead to competition for organic substrates. However, Tuchman *et al.* (2006) found that bacteria were able to oxidize a more diverse array of organic substrates including carboxylic acids and large polymers, while diatoms appeared to more readily utilize the complex carbohydrates. By oxidizing different organic substrates than bacteria, heterotrophically metabolizing diatoms may be reducing direct competition and enhancing coexistence with bacteria.

The degree of coupling between algae and bacteria in biofilms could decrease in high nutrient conditions, algae becoming less reliant on bacterial mineralization of nutrients and bacteria becoming less stressed for photosynthetically derived extracellular organic carbon (Rier and Stevenson, 2002; Scott *et al.*, 2008). The concentration of readily available organic matter compounds in the flowing water can also weaken the algal-bacterial coupling (Ylla *et al.*, 2009). Sobczak (1996) observed that labile DOC in the water column may be controlling bacteria throughout biofilm colonization, with epilithic algae becoming increasingly important as a DOC source in late stages of biofilm development. A thin biofilm should rely primarily on nutrients from the water column (allochthonous DOC), whereas a thick, mature biofilm may rely on both allochthonous and autochthonous carbon.

The algal bacterial coupling mainly refers to the positive effect of algae on bacteria by providing algal exudates, while algae can also benefit from bacteria. Algae require special vitamins (B vitamins) for their growth which are basically synthesized by heterotrophic bacteria (Cole, 1982; Lock, 1993). Autotrophic organisms are also favoured by the CO_2 released during heterotrophic respiration (Cole, 1982). Moreover, bacteria may also release extracellular enzymes that degrade complex organic molecules not transportable across cell membranes (Sinsabaugh and Linkins, 1988), making both organic and inorganic carbon (N and P) available to algae (Klug, 2005). Heterotrophic bacteria may also rely on the use of O_2 released during algal photosynthesis (Kuhl *et al.*, 1996). Thus, this algal–bacterial relationship can be therefore defined as mutualism or cooperation within the biofilm (Kreft, 2004).

Bacterial–fungal interactions

In freshwater biofilms bacteria and fungi may interact in the microbial decomposition of plant organic matter since both are key agents involved in litter breakdown and mineralization (Baldy *et al.*, 1995). In aquatic ecosystems, this fungal–bacterial interaction will be especially relevant in biofilms developing on sand sediments and leaf litter were recalcitrant and plant degrading material is available. In contrast, in epilithic biofilms (developing on rocks and cobbles) fungi are often very low represented (Artigas *et al.*, 2009) and fungal–bacterial interactions may be less relevant

Interaction between bacteria and fungi is connected to competition for substrate (Mille-Lindblom and Tranvik, 2003; Mille-Lindblom *et al.*, 2006.) modulating the expression of extracellular enzyme activities (Romaní *et al.*, 2006b). Aquatic fungi are able to decompose recalcitrant and complex polymers such as those found in plant material. They produce a wide range of enzymes including pectinases, hemicellulases, cellulases, and chitinases (Suberkropp, 1992; Abdel-Raheem and Shearer, 2002) and lignin-degrading enzymes (Fisher *et al.*, 1983; Bucher *et al.*, 2004). In bacteria, production of cellulolytic and xylanolytic (hemicellulolytic) enzymes has been reported to also occur (Tanaka, 1993; Sala and Güde, 2004). However, it is generally assumed that bacteria mainly decompose polysaccharides and polymeric compounds after the previous decomposition of high molecular and/or lignified compounds by fungi. Both synergistic and antagonistic interaction is described for the interaction between bacteria and fungi in plant organic matter degradation (Bengtsson, 1992; Gulis and Suberkropp, 2003a). The greater capacity of the fungi to decompose plant polymers such as lignin, cellulose, and hemicellulose (greater phenol oxidase, cellobiohydrolase and β-xylosidase) might provide bacteria with simple polysaccharides. Gulis and Suberkropp (2003b) postulated that bacteria in leaf biofilms depend on the activity of fungi; the macerated leaf tissue (resulting from fungal action) and fungal hyphae provide increased surface area for bacterial colonization (Suberkropp and Klug, 1980). Moreover, bacteria can utilize organic compounds made available by partial enzymatic degradation of the plant material or compounds released from lysis of dead fungal mycelia (Gulis and Suberkropp, 2003b). Bacteria, in contrast, appear to be more efficient than fungi in the use of simple polysaccharides and peptides (β-glucosidase and leucine-aminopeptidase activities), inhibiting rapid fungal growth (Wohl and McArthur, 2001; Romaní *et al.*, 2006b). In summary, fungal and bacterial colonization dynamics and their extracellular enzyme activities, affect the rates of organic matter decomposition.

Other significant antagonistic relationships between fungi and bacteria are well known such as the release of fungicides or chitinolytic enzymes by bacteria (Lloyd and Lockwood, 1966; Weller, 1988), decomposing the fungal cell wall. Conversely, the most familiar example of fungal suppression of bacteria is the production of various antibiotic substances.

Potential effects of Protozoa on biofilm extracellular enzyme activity

Even though protozoa should be considered as potentially very important enzyme producers (Vrba *et al.*, 2004), few studies demonstrating organic matter use by protozoa have been done. The expression of extracellular enzymes involved in peptidoglycan and chitin degradation, such as β-*N*-acetylhexosaminidases and β-*N*-acetylglucosaminidase have been described for protozoa (Vrba *et al.*, 1993; Vrba, 2004). Some studies, report that protozoa may acquire the availability to express the enzyme activities from the digested bacteria (González *et al.*, 1993; Vrba *et al.*, 1993).

In biofilms, direct enzyme activity measurements in protozoa have not been performed although indirect effects of protozoa might modulate the biofilm enzyme expression and capacity to use organic matter use, mainly due to their predation on bacteria. Bacteria are consumed mostly by protists such as flagellates and ciliates (Bott and Kaplan, 1990). Protozoa predation in biofilms might regulate bacterial populations (Norf *et al.*, 2007; Wey *et al.*, 2008), increase cell viability (Joubert *et al.*, 2006) and determine changes in bacterial community structure (Matz and Kjelleberg, 2005). Thus, bacterial enzyme activities in biofilms might change in response to the abundance, composition and activity of protozoan communities.

Conclusions, trends and perspectives

The measuring of extracellular enzyme activities in aquatic biofilms has provided relevant results in the understanding of organic matter cycling in freshwater ecosystems. Enzyme activities provide sensible metabolic tools which respond to changes in water nutrient availability as well as pollution. At the same time, microbial interactions and decomposition capabilities within the biofilm are also described from enzyme activities. However, the measurement and interpretation of extracellular enzyme activities in aquatic biofilms still offers significant challenges for microbial ecologists in order to better our knowledge of biofilm functioning and its relationship to aquatic ecosystem functioning. In this sense, we could define three specific challenging research lines: (1) the biofilm stoichiometry approach, (2) the relationships between biofilm 'enzyme diversity' and biofilm microbial diversity, and (3) the separation of the different biofilm microbial groups, i.e. who is the most responsible for the enzyme activity?

One of the objectives microbial ecologists pursue when measuring extracellular enzyme activities in aquatic biofilms is to generate better knowledge of the quality of organic matter being actively degraded by biofilm microorganisms and thus being available in the flowing water. However, biofilm organic matter content and recycling within the biofilm (modulated by the whole biofilm complexity: microbial interactions, biofilm thickness, density, EPS content, community composition) usually makes the interpretation and comparison of extracellular measurements in freshwater biofilms difficult. The stoichiometric approach, which summarizes the whole biofilm characteristics of carbon, nitrogen and phosphorus content and their respective molar relationships, has been shown to be a powerful tool to follow this objective. Although the whole biofilm complexity is in part lost with the biofilm stoichiometry approach, it permits comparable analyses of enzyme capabilities and efficiencies from different environments (Sinsabaugh *et al.* 2010). Further analysis in this sense (including diverse biofilms and aquatic environments) would probably provide relevant general trends.

The measurement of extracellular enzyme activities is frequently linked to their independent capabilities, using the specific substrate each enzyme decomposes; whole analysis of the enzyme capabilities is much less frequent. The dissolved organic matter pool in the environment includes organic molecules from different origins, often including complex and polymeric compounds, needing a diverse pool of enzymes for their decomposition. Therefore, a greater 'enzyme diversity' would be beneficial for the microbial community. Even though the diversity of extracellular enzyme activities actually measured are limited by the substrates commercially available for their analysis, this 'enzyme diversity' approach can be further developed. Also, the complementary use of molecular tools, such as the analysis of the diversity of functional genes encoding for specific enzymes might highlight the relationship between diversity and function in biofilm microbial communities. This 'enzyme multifunctionality' approach might improve our knowledge of the links between microbial processes occurring at the biofilm level and aquatic ecosystem functioning.

Probably the more difficult challenge to improve our knowledge of enzymes in biofilms is the specific role of the different microbial groups (algae, bacteria, fungi, protozoa) for the whole biofilm enzyme activity. The structural complexity of the biofilm hampers our ability to directly study the relations between their components. The separation of the different trophic groups of microorganisms supposes the realization of laboratory experiments attempting to sort out the different biofilm compartments through the utilization of 'pure' microbial cultures grown in artificial substrata (Bruckner and Kroth, 2009) and/or the utilization of restrictive conditions for one or other groups like the photoautotrophs (splitting out the biofilms with an active photosynthetic community from those totally heterotrophic, Romaní and Sabater, 1999b), also maintaining the biofilm three-dimensional structure as intact as possible. Significant results can be also obtained from intensive analysis of biofilm formation sequences, when the microbial interactions and the respective role of the colonizing organisms might be highlighted (i.e. Díaz *et al.*, 2011). Specific experimental designs should be developed to allow a better understanding of such biofilm interactions and their implications for the whole biofilm metabolism.

Acknowledgement

The writing of this chapter was supported by the Spanish Ministry of Science and Education (CGL2008-05618-C02-01/BOS).

References

Abdel-Raheem, A., and Shearer, C.A. (2002). Extracellular enzyme production by freshwater ascomycetes. Fungal Divers *11*, 1–19.

Ainsworth, A.M., and Goulder, R. (2000). The effects of sewage-works effluent on riverine extracellular aminopeptidase activity and microbial leucine assimilation. Water Res. *34*, 2551–2557.

Allen, A.P., and Gillooly, J.F. (2009). Towards an integration of ecological stoichiometry and the metabolic theory of ecology to better understand nutrient cycling. Ecol. Lett. *12*, 369–384.

Allison, S.D., and Vitousek, P.M. (2005). Responses of extracellular enzymes to simple and complex nutrient inputs. Soil Biol. Biochem. *37*, 937–944.

Arnosti, C. (2003). Microbial extracellular enzymes and their role in dissolved organic matter cycling. In Aquatic Ecosystems. Interactivity of Dissolved Organic Matter, Findlay, S.E.G., and Sinsabaugh, R.L., eds. (Academic Press, San Diego), pp. 315–342.

Artigas, J., Romaní, A.M., and Sabater, S. (2008). Relating nutrient molar ratios to organic matter utilization in a forested stream. Fund. Appl. Limnol. *173*, 255–264.

Artigas, J., Romaní, A.M., Gaudes, A., Muñoz, I., and Sabater, S. (2009). Organic matter availability structures microbial biomass and activity in a Mediterranean river. Freshwat. Biol. *54*, 2025–2036.

Atlas, R.M., and Bartha, R. (1987). Biogeochemical cycling: carbon, hydrogen, and oxygen. In Microbial Ecology, Fundamentals and Applications. Second Edition, Atlas, R.M., and Bartha, R., eds. (San Francisco, The Benjamin/Cummings Publishing Company, Inc.), pp. 307–332.

Baldy, V., Gessner, M.O., and Chauvet, E. (1995). Bacteria, fungi and the breakdown of leaf litter in a large river. Oikos *74*, 93–102.

Barranguet, C., VanBeusekom, S.A.M., Veuger, B., Neu, T.R., Manders, E.M.M., Sinke, J.J., and Admiraal, W. (2004). Studying undisturbed autotrophic biofilms: still a technical challenge. Aquat. Microb. Ecol. *34*, 1–9.

Battin, T.J, Kaplan, L.A., Newbold, J.D., and Hansen, C.M.E. (2003). Contributions of microbial biofilms to ecosystem processes in stream mesocosms. Nature *426*, 439–442.

Bengtsson, G. (1992). Interactions between fungi, bacteria and beech leaves in a stream mesocosm. Oecologia *89*, 542–549.

Bertilsson, S., and Jones, J.B.J. (2003). Supply of dissolved organic matter to aquatic ecosystems: authocthonous sources. In Aquatic Ecosystems. Interactivity of Dissolved Organic Matter, Findlay, S.E.G., and Sinsabaugh, R.L., eds. (Academic Press, San Diego), pp. 4–24.

Bjoernsen, P.K. (1988). Phytoplankton exudation of organic matter: Why do healthy cells do it? Limnol. Oceanogr. 33, 151–154.

Bögershausen, A. and Marxsen, J. (1995). Die Enzymspezifität von Methylumbelliferyl-Verbindungen. In Messung Extrazellulärer Enzymaktivitäten im Grundwasser. Deutscher Verband für Wasserwirtschaft und Kulturbau, Bögershausen, A., Hendel, B., Marxsen, J., and Preuss, G. (e.V., DVWK-Fachausschuss Grundwasserbiologie, Bonn), pp. 1–77.

Bonnineau, C., Guasch, H., Proia, L., Ricart, M., Geisinger, A., Romaní, A.M., and Sabater, S. (2010). Fluvial biofilms: a pertinent tool to assess b-blockers toxicity. Aquat. Toxicol. *96*, 225–233.

Bott, T.L., and Kaplan L.A. (1990). Potential for protozoan grazing of bacteria in streambed sediments. J. N. Am. Benthol. Soc. *9*, 336–445.

Bott, T.L., Kaplan, L.A., and Kuserk, F.T. (1984). Benthic bacterial biomass supported by streamwater dissolved organic matter. Microb. Ecol. *10*, 335–344.

Boyer, J.N., and Groffman, P.M. (1996). Bioavailability of water extractable organic carbon fractions in forest and agricultural soil profiles. Soil Biol. Biochem. *28*, 783–790.

Bucher, V.V.C., Hyde, K.D., Pointing, S.B., and Reddy C.A. (2004). Production of wood decay enzymes, mass loss and lignin solubilization in wood by marine ascomycetes and their anamorphs. Fungal Divers *15*, 1–14.

Burns, R.G. (1983). Extracellular enzyme–substrate interactions in soil. In Microbes in their Natural Environments, Slater, J.H., Whittenbury, R., and Wimpenny, J.W.T., eds. (Symposium 34. The Society for General Microbiology, Cambridge University Press, Cambridge), pp. 249–298.

Campbell, C.D., Chapman, S.J., Cameron, C.M., Davidson, M.S., and Potts, J.M. (2003). A rapid microtiter plate method to measure carbon dioxide evolved from carbon substrate amendments so as to determine the physiological profiles of soil microbial communities by using whole soil. Appl. Environ. Microbiol. *69*, 3593–3599.

Cardinale, B.J., Srivastava, D.S., Duffy, J.E., Wright, J.P., Downing, A.L., Sankaran, M., and Jouseau, C. (2006). Effects of biodiversity on the functioning of trophic groups and ecosystems. Nature *443*, 989–992.

Carr, G.M., Morin, A., and Chambers, P.A. (2005). Bacteria and algae in stream periphyton along a nutrient gradient. Freshwat. Biol. *50*, 1337–1350.

Chapell, K.R., and Goulder, R. (1992). Epilithic extracellular enzyme activity in acid and calcareous headstreams. Arch. Hydrobiol. *125*, 129–148.

Chappell, K.R., and Goulder, R. (1994). Enzymes as river pollutants and the response of native epilithic extracellular-enzyme activity. Environ. Poll. *86*, 161–169.

Chapell, K.R., and Goulder, R. (1995). A between-river comparison of extracellular-enzyme activity. Microb. Ecol. *29*, 1–17.

Chróst, R.J. (1989). Characterization and significance of β-glucosidase activity in lake water. Limnol. Oceanogr. *34*, 660–672.

Chróst, R.J. (1990). Microbial ectoenzymes in aquatic environments. In Aquatic Microbial Ecology: Biochemical and Molecular Approaches, Overbeck, J., and Chróst, R.J., eds. (Springer-Verlag, New York), pp. 47–78.

Chróst, R.J. (1991). Environmental control of the synthesis and activity of aquatic microbial ectoenzymes. In Microbial Enzymes in Aquatic Environments, Chróst, J., ed. (Brock/Springer Verlag), pp. 29–59.

Chróst, R.J., and Krambeck, H.J. (1986). Fluorescence correction for measurements of enzyme activity in natural waters using methylumbelliferyl-substrates. Arch. Hydrobiol. *106*, 79–90.

Chróst, R.J., and Rai, H. (1993). Ectoenzyme activity and bacterial secondary production in nutrient-impoverished and nutrient-enriched freshwater mesocosms. Microb. Ecol. *25*, 131–150.

Confer, D.R., and Logan, B.E. (1997). Molecular weight distribution of hydrolysis products during biodegradation of model macromolecules in suspended and biofilm cultures I. Bovine serum albumin. Wat. Res. *31*, 2127–2136.

Conrad, A., Cadoret, A., and Block, J.C. (2000). Extracellular enzymatic activities in activated sludge flocs. In Proceedings of the International Specialist Conference: Extracellular Polymeric Substances: The Construction Material of Biofilms, Mülheim, Flemming, H.C., and Leis, A., eds (IWA Publishing, London).

Cole, J.J. (1982). Interactions between bacteria and algae in aquatic ecosystems. Ann. Rev. Ecol. and Syst. *13*, 291–314.

Cottrell, M.T., and Kirchman, D.L. (2000). Natural assemblages of marine Proteobacteria and members of the Cytophaga-Flavobacter cluster consuming low- and high-molecular-weight dissolved organic matter. Appl. Environ. Microbiol. *66*, 1692–1697.

Cummins, K.W., Sedell, J.R., Swanson, F.J., Minshall, G.W., Fisher, S.G., Cushing, C.E., Petersen, R.C., and Vannote, R.L. (1983). Organic matter budgets for stream ecosystems: problems in their evaluation. In Stream Ecology: Application and Testing of General Ecological Theory, Barnes, J.R., and Minshall, G.W., eds. (Plenum Press, New York), pp. 299–353.

Davis, S.H., and Goulder, R. (1993). Deterioration in bacteriological quality of water through fish farms. J. Appl. Bacteriol. *74*, 336–339.

Deshpande, V., and Eriksson, K.-E. (1988). 1,4-β-Glucosidases of Sporotrichum pulverulentum. In Methods in Enzymology Volume 160 PartA Cellulose and Hemicellulose, Wood, W.A., and Kellog, S.T., eds. (New York, Academic Press, Inc.), pp. 415–424.

Díaz Villanueva, V., Font, J., Schwartz, T., and Romaní, A.M. (2011). Biofilm formation at warming temperature: acceleration of microbial colonization and microbial interactive effects. Biofouling *27*, 59–71.

Dodds, W.K. (2003). The role of periphyton in phosphorus retention in a shallow freshwater aquatic ecosystem. J. Phycol. *39*, 840–849.

Douterelo, I., Goulder, R., and Lillie, M. (2010). Soil microbial community response to land-management and depth, related to the degradation of organic matter in English wetlands: Implications for the *in situ* preservation of archaeological remains. Appl. Soil Ecol. *44*, 219–227.

Espeland, E.M., Francoeur, S.N., and Wetzel, R.G. (2001). Influence of algal photosynthesis on biofilm bacterial production and associated glucosidase and xylosidase activities. Microb. Ecol. *42*, 524–530.

Espeland, E.M., and Wetzel, R.G. (2001). Complexation, stabilization, and UV photolysis of extracellular and surface-bound glucosidase and alkaline phosphatase: implications for biofilm microbiota. Microb. Ecol. *42*, 572–585.

Findlay, S., and Sinsabaugh, R.L. (2003). Response of hyporeic biofilm metabolism and community structure to nitrogen amendments. Aquat. Microb. Ecol. 33, 127–136.

Findlay, S., Strayer, D., Goumbala, C., and Gould, K. (1993). Metabolism of streamwater dissolved organic carbon in the shallow hyporheic zone. Limnol. Oceanogr. *38*, 1493–1499.

Findlay, S., Quinn, J.N., Hickey, C.W., Burrell, G., and Downes, M. (2001). Effects of land use and riparian flowpath on delivery of dissolved organic carbon to streams. Limnol. Oceanogr. *46*, 345–355.

Fisher, P.J., Davey, R.A., and Webster, J. (1983). Degradation of lignin by aquatic and aero-aquatic hyphomycetes. Trans. Brit. Mycol. Soc. *80*, 166–168.

Fisher, S.G., and Likens, G.E. (1973). Energy flow in Bear Brook, New Hampshire: an integrative approach to stream ecosystem metabolism. Ecology *43*, 421–439.

Fogg, G.E. (1966). The extracellular products of algae. Oceanogr. Mar. Biol. Ann. Rev. *4*, 195–212.

Francoeur, S.N., and Wetzel, R.G. (2003). Regulation of periphytic leucine-aminopeptidase activity. Aquat. Microb. Ecol. *31*, 249–258.

Francoeur, S.N., Biggs, B.J.F., Smith, R.A., and Lowe, R.L. (1999). Nutrient limitation of algal biomass accrual in streams: seasonal patterns and a comparison of methods. J. N. Am. Benthol. Soc. *18*, 242–260.

Francoeur, S.N., Schaecher, M., Neely, R.K., and Kuehn, K.A. (2006). Periphytic photosynthetic stimulation of extracellular enzyme activity in aquatic microbial communities associated with decaying Typha litter. Microb. Ecol. *52*, 662–669.

Freeman, C., and Lock, M. (1992). Recalcitrant high-molecular-weight material, an inhibitor of microbial metabolism in river biofilms. Appl. Environ. Microbiol. *58*, 2030–2033.

Freeman, C., and Lock M.A. (1995). The biofilm polysaccharide matrix: A buffer against changing organic substrate supply? Limnol. Oceanogr. *40*, 273–278.

Gajewski, A.J., and Chróst, R.J. (1995). Microbial enzyme activities and phytoplankton and bacterial production in the pelagial of the Great Mazurian Lakes (North-eastern Poland) during summer stratification. Ekol. Polska *43*, 245–265.

Garland, J.L., and Mills, A.L. (1991). Classification and characterization of heterotrophic microbial communities on the basis of patterns of community-level sole-carbon-source utilization. Appl. Environ. Microbiol. *57*, 2351–2359.

Garland, J.L. (1996). Analytical approaches to the characterization of samples of microbial communities using patterns of potential C source utilization. Soil Biol. Biochem. *28*, 213–221.

Geesey, G.G., Mutch, R., Costerton, J.W., and Green, R.B. (1978). Sessile bacteria: An important component of the microbial population in small mountain streams. Limnol. Oceanogr. *23*, 1214–1223.

Golladay, S.W., and Sinsabaugh, R.L. (1991). Biofilm development on leaf and wood surfaces in a boreal river. Freshwat. Biol. *25*, 437–450.

Gonzalez, J.M., Sherr, B.F., and Sherr, E.B. (1993). Digestive enzyme activity as a quantitative measure of protistan grazing: The acid lysozyme assay for bacterivory. Mar. Ecol. Prog. Ser. *100*, 197–206.

Graça, M.A.S. (1993). Patterns and processes in detritus-based stream systems. Limnologica *23*, 107–114.

Grover, J.P., and Chrzanowski, T.H. (2000). Seasonal patterns of substrate utilization by bacterioplankton: case studies in four temperate lakes of different latitudes. Aquat. Microb. Ecol. *23*, 41–54.

Gulis, V., and Suberkropp, K. (2003a). Interactions between stream fungi and bacteria associated with decomposing leaf litter at different levels of nutrient availability. Aquat. Microb. Ecol. *30*, 149–157.

Gulis, V., and Suberkropp, K. (2003b). Effect of inorganic nutrients on relative contributions of fungi and bacteria to carbon flow from submerged decomposing leaf litter. Microb. Ecol. *45*, 11–19.

Haack, T.K., and McFeters, G.A. (1982a). Microbial dynamics of an epilithic mat community in a high alpine stream. Appl. Environ. Microbiol. *43*, 702–707.

Haack, T.K., and McFeters, G.A. (1982b). Nutritional relationships among microorganisms in an epilithic biofilm community. Microb. Ecol. *8*, 115–126.

Harbott, E.L., and Grace, M.R. (2005). Extracellular enzyme response to bioavailability of dissolved organic C in streams of varying catchment urbanization. J. N. Am. Benthol. Soc. *24*, 588–601.

Healey, F.P., and Hendzel, L.L. (1979). Indicators of phosphorus and nitrogen deficiency in five algae in culture. J. Fish. Res. Board Can. *36*, 1364–1369.

Hepinstall, J.A., and Fuller, R.L. (1994). Periphyton reactions to different light and nutrient levels and the response of bacteria to these manipulations. Arch. Hydrobiol. *131*, 161–173.

Hollibaugh, J.T. (1994). Relationship between thymidine metabolism, bacterioplankton community metabolic capabilities and sources of organic matter. Microb. Ecol. *28*, 117–131.

Hopkinson, C.S., Buffam, I., Hobbie, J., Vallino, J., Perdue, M., Eversmeyer, B., Prahl, F., Covert, J., Hodson, R., Moran, M.A., Smith, E., Baross, J., Crump, B., Findlay, S., and Foreman, K. (1998). Terrestrial inputs of organic matter to coastal ecosystems: An intercomparison of chemical characteristics and bioavailability. Biogeochemistry *43*, 211–234.

Hoppe, H.-G. (1983). Significance of exoenzymatic activities in the ecology of brackish water: measurement by means of methylumbelliferyl-substrates. Mar. Ecol. Progr. Ser. *11*, 299–308.

Hoppe, H.-G., Kim, S.-J., and Gocke, K. (1988). Microbial decomposition in aquatic environments: combined process of extracellular enzyme activity and substrate uptake. Appl. Environ. Microbiol. *54*, 784–790.

Hoppe, H.-G., Ducklow, H., and Karrasch, B. (1993). Evidence for dependency of bacterial growth on enzymatic hydrolysis of particulate organic matter in the mesopelagic ocean. Mar. Ecol. Progr. Ser. *93*, 277–283.

Hutchinson, G.E. (1957). A Treatise on Limnology. (John Wiley & Sons, New York).

Insam, H. (1997). A new set of substrates proposed for community characterization in environmental samples. In Microbial Communities: Functional versus Structural Approaches, Insam, H., and Rangger, A., eds. (Berlin and Heidelberg, Springer-Verlag), pp. 259–260.

Jansson, M., Olsson, H., and Pettersson, K. (1988). Phosphatases; origin, characteristics and function in lakes. Hydrobiology *170*, 157–175.

Jellet, J.F., Li, W.K.W., Dickie, P.M., Boraie, A., and Kepkay, P.E. (1996). Metabolic activity of bacterioplankton communities assessed by flow cytometry and single carbon substrate utilization. Mar. Ecol. Prog. Ser. *136*, 213–225.

Jenkinson, H.F., and Lappin-Scott, H. (2001). Biofilms adhere to stay. Trends Microbiol. *9*, 9–10.

Joffre, R., Agren, G.I., Gillon, D., and Bosatta, E. (2001). Organic matter quality in ecological studies: theory meets experiment. Oikos 93, 451–458.

Jones, J.G. (1971). Studies on freshwater bacteria: factors which influence the population and its activity. J. Ecol. *59*, 593–613.

Jones, S.E., and Lock, M.A. (1989). Hydrolytic extracellular enzyme activity in heterotrophic biofilms from two contrasting streams. Freshwat. Biol. *22*, 289–296.

Jones, S.E., and Lock, M.A. (1991). Peptidase activity in river biofilms by product analysis. In Microbial Enzymes in Aquatic Environments, Chróst, R.J., ed. (Berlin and Heidelberg, Springer-Verlag), pp. 144–154.

Jones, S.E., and Lock, M.A. (1993). Seasonal determinations of extracellular hydrolytic activities in heterotrophic and mixed heterotrophic/autotrophic biofilms from two contrasting rivers. Hydrobiology *257*, 1–16.

Joubert, L.-M., Wolfaardt, G.M., and Botha, A. (2006). Microbial exopolymers link predator and prey in a model yeast biofilm system. Microb. Ecol. 52, 187–197.

Judd, K.E., Crump, B.C., and Kling, G.W. (2006). Variation in dissolved organic matter controls bacterial production and community composition. Ecology *87*, 2068–2079.

Kaplan, L.A., and Bott, T.L. (1982). Diel fluctuations of DOC generated by algae in a piedmont stream. Limnol. Oceanogr. *27*, 1091–1100.

Kaplan, L.A., and Bott, T.L. (1989). Diel fluctuations in bacterial activity on streambed substrata during vernal algal blooms: effects of temperature, water chemistry and habitat. Limnol. Oceanogr. *34*, 718–733.

Kaplan, L.A., and Newbold, J.D. (1995). Measurement of stream water biodegradable dissolved organic carbon with a plug-flow bioreactor. Wat. Res. *29*, 2696–2706.

Keyser, P., Kirk, T.K., and Zeikus J.G. (1978). Ligninolytic enzyme system of Phanerochaete chrysosporium: synthesized in the absence of lignin in response to nitrogen starvation. J. Bacteriol. *135*, 790–797.

King, G.M. (1986). Characterization of β-glucosidase activity in intertidal marine sediments. Appl. Environ. Microbiol. *51*, 373–380.

Kirchman D.L. (2003). The contribution of monomers and other low-molecular weight compounds to the flux of dissolved organic material in aquatic ecosystems. In Aquatic Ecosystems: Interactivity of Dissolved Organic Matter, Findlay, S.E., and Sinsabaugh, R.L., eds. (Academic Press, San Diego), pp. 217–241.

Klotz, R.L. (1992). Factors influencing alkaline phosphatase activity of stream epilithon. J. Freshwat. Ecol. 7, 233–242.

Klug, J.L. (2005). Bacterial response to dissolved organic matter affects resource availability for algae. Can. J. Fish. and Aquat. Sci. *62*, 472–481.

Koetsier, P., McArthur, J.V., and Leff, L.G. (1997). Spatial and temporal response of stream bacteria to sources of dissolved organic carbon in a blackwater stream system. Freshwat. Biol. *37*, 79–89.

Konopka, A., Oliver, L., and Turco, R.F. (1998). The use of carbon substrate utilization patterns in environmental and ecological microbiology. Microb. Ecol. *35*, 103–115.

Kuhl, M., Glud, R.N., Ploug, H., and Ramsing, N.B. (1996). Microenvironmental control of photosynthesis and photosynthesis-coupled respiration in an epilithic cyanobacterial biofilm. J. Phycol. *32*, 799–812.

Kreft, J.U. (2004). Biofilms promote altruism. Microbiology *150*, 2751–2760.

Kuserk, F.T., Kaplan, L.A., and Bott, T.L. (1984). *In situ* measures of dissolved organic carbon flux in a rural stream. Can. J. Fish. Aquat. Sci. *41*, 964–973.

Lachke, A.H. (1988). 1,4-β-D-Xylan Xylohydrolase of Sclerotium rolfsii. In Methods in Enzymology. Volume 160. Par A Cellulose and Hemicellulose, Wood, W.A., and Kellog, S.T., eds. (New York, Academic Press, Inc.), pp. 679–684.

Langenheder, S., Bulling, M.T., Solan, M., and Prosser, J.I.. (2010). Bacterial biodiversity-ecosystem functioning relations are modified by environmental complexity. Plos One 5, e10834.

Lewis, D.L., Hodson, R.E., and Hwang, H.-M. (1988). Kinetics of mixed microbial assemblages enhance removal of highly dilute organic substrates. Appl. Environ. Microbiol. *54*, 2054–2057.

Lloyd, A.B., and J.L. Lockwood (1996). Lysis of fungal hyphae in soil and its possible relation to autolysis. Phytopathology 56, 595–602.

Lock, M.A., Wallace, R.R., Costerton, J.W., Ventullo, R.M., and Charlton, S.E. (1984). River epilithon: toward a structural-functional model. Oikos 42, 10–22.

Lock, M.A., and Ford, T.E. (1986). Colloidal and dissolved organic carbon in undisturbed boreal forest catchments: a seasonal study of apparent molecular weight spectra. Freshwat. Biol. *16*, 187–195.

Lock, M.A. (1990). The dynamics of dissolved and particulate organic material over the substratum of water bodies. In The Biology of Particles in Aquatic Systems, Whitton, R.J., ed. (CRC Press, Inc., Boca Raton), pp. 117–144.

Lock, M.A. (1993). Attached microbial communities in rivers. In Aquatic Microbiology: An Ecological Approach, Ford, T.E., ed. (Boston, Blackwell Scientific Publications), pp. 113–138.

Majeau, J.A., Brar, S.K., and Tyagi, R.D. (2009). Laccases for removal of recalcitrant and emerging pollutants. Biores. Technol. *101*, 2331–2350.

Marxsen, J., and Schmidt, H.-H. (1993). Extracellular phosphatase activity in sediments of the Breitenbach, a Central European mountain stream. Hydrobiology 253, 207–216.

Marxsen, J., and Fiebig, D.M. (1993). Use of perfused cores for evaluating extracellular enzyme activity in stream-bed sediments. FEMS Microbiol. Ecol. *13*, 1–12.

Marxsen, J., Zoppini, A., and Wilczek, S. (2010). Microbial communities in streambed sediments from dessication. FEMS Microbiol. Ecol. *71*, 374–386.

Matz, C., and Kjelleberg, S. (2005). Off the hook-how bacteria survive protozoan grazing. Trends Microbiol. *13*, 302–307.

Meyer, J.L. (1994). The microbial loop in flowing waters. Microb. Ecol. *28*, 195–199.

Mille-Lindblom, C., and Tranvik, L.J. (2003). Antagonism between bacteria and fungi on decomposing aquatic plant litter. Microb. Ecol. *45*, 173–182.

Mille-Lindblom, C., Fischer, H., and Tranvik, L.J. (2006). Antagonism between bacteria and fungi: substrate competition and a possible trade-off between fungal growth and tolerance towards bacteria. Oikos *113*, 233–242.

Mohamed, M.N., Lawrence, J.R., and Robarts, R.D. (1998). Phosphorus limitation of heterotrophic biofilms from the Fraser River, British Columbia, and the effects of pulp mill effluent. Microb. Ecol. *36*, 121–130.

Montuelle, B., and Volat, B. (1998). Impact of wastewater treatment plant discharge on enzyme activity in freshwater sediments. Ecotoxicol. Environ. Saf. *40*, 154–159.

Münster, U., Einiö, P., and Nurminen, J. (1989). Evaluation of the measurements of extracellular enzyme activities in a polyhumic lake by means of studies with 4-methylumbelliferyl-substrates. Arch. Hydrobiol. *115*, 321–337.

Murray, R.E., Cooksey, K.E., and Priscu, J.C. (1986). Stimulation of bacterial DNA synthesis by algal exudates in attached algal-bacterial consortia. Appl. Environ. Microbiol. *52*, 1177–1182.

Nedoma, J., Strojsova, A., Vrba, J., Komárková, J., and Simek, K. (2003). Extracelluar phosphatase activity of natural plankton studied with ELF97 phosphate: fluorescence quantification and labelling kinetics. Environ. Microbial. *5*, 462–472.

Norf, H., Arndt, H., and Weitere, M. (2007). Impact of local temperature increase on the early development of biofilm-associated ciliate communities. Oecologia *151*, 341–350.

Nybroe, O., Jorgensen, P.E., and Henze, M. (1992). Enzyme activities in waste water and activated sludge. Wat. Res. *26*, 579–584.

Paulsson, M., Mànsson, V., and Blanck, H. (2002). Effects of zinc on the phosphorus availability to periphyton communities from the river Göta Álv. Aquat. Toxicol. *56*, 103–113.

Raunkjaer, K., Hvitved-Jacobsen, T., and Nielsen, P.H. (1994). Measurement of pools of protein, carbohydrate and lipid in domestic wastewater. Wat. Res. *28*, 251–262.

Reichardt, W., Overbeck, J., and Steubing, L. (1967). Free dissolved enzymes in lake waters. Nature *216*, 1345–1347.

Ricart, M., Barceló, D., Geiszinger, A., Guasch, H., López de Alda, M., Romaní, A.M., Vidal, G., Villagrasa, M., and Sabater, S. (2009). Effects of low concentrations of the phenylurea herbicide diuron on biofilm algae and bacteria. Chemosphere *76*, 1392–1401.

Rier, S.T., and Stevenson, R.J. (2002). Effects of light, dissolved organic carbon, and inorganic nutrients on the relationship between algae and heterotrophic bacteria in stream periphyton. Hydrobiology *489*, 179–184.

Rier, S.T., Kuehn, K.A., and Francoeur, S.N. (2007). Algal regulation of extracellular enzyme activity in stream microbial communities associated with inert substrata and detritus. J. N. Am. Benthol. Soc. *26*, 439–449.

Rogers, H.J. (1961). The dissimilation of high molecular weight organic substances. In The Bacteria, Gunsalus, I.C., and Stanier, R.Y., eds. (New York, Academic Press), pp. 261–318.

Romaní, A.M. (2000). Characterization of extracellular enzyme kinetics in two Mediterranean streams. Arch. Hydrobiol. *148*, 99–117.

Romaní, A.M. (2010). Freshwater biofilms. In Biofouling, Dürr, S., and Thomason, J.C., eds. (Oxford, Wiley-Blackwell), pp. 137–153.

Romaní, A.M., and Sabater, S. (1999). Effect of primary producers on the heterotrophic metabolism of a stream biofilm. Freshwat. Biol., *41*, 729–736.

Romaní, A.M., and Sabater, S. (2000a). Variability of heterotrophic activity in Mediterranean stream biofilms: A multivariate analysis of physical–chemical and biological factors. Aquat. Sci. *62*, 205–215

Romaní, A.M., and Marxsen, J. (2002). Extracellular enzyme activities in epilithic biofilms of the Breitenbach: microhabitat differences. Arch. Hydrobiol. *155*, 541–555.

Romaní, A.M., and Sabater, S. (2000b.) Influence of algal biomass on extracellular enzyme activity in river biofilms. Microb. Ecol. *41*, 16–24.

Romaní, A.M., and Sabater, S. (2001). Structure and activity of rock and sand biofilms in a Mediterranean stream. Ecology *82*, 3232–3245.

Romaní, A.M., Butturini, A., Sabater, F., and Sabater, S. (1998). Heterotrophic metabolism in a forest stream sediment: surface versus subsurface zones. Aquat. Microb. Ecol. *16*, 143–151.

Romaní, A.M., Guasch, H., Muñoz, I., Ruana, J., Vilalta, E., Schwartz, T., Emtiazi, F., and Sabater, S. (2004a) Biofilm structure and function and possible implications for riverine DOC dynamics. Microb. Ecol. 47, 316–328.

Romaní, A.M., Giorgi, A., Acuña, V., and Sabater, S. (2004b). The influence of substratum type and nutrient supply on biofilm organic matter utilization in streams. Limnol. Oceanogr. *49*, 1713–1721.

Romaní, A.M., Vázquez, E., and Butturini, A. (2006a). Microbial activity and size fractionation of dissolved organic carbon after drought in an intermittent stream: biogeochemical link across the stream–riparian interface. Microb. Ecol. *52*, 501.512.

Romaní, A.M., Fischer, H., Mille-Lindblom, C., and Tranvik, L.J. (2006b). Interactions of bacteria and fungi on decomposing litter: Differential extracellular enzyme activities. Ecology 87, 2559–2569.

Romaní, A.M., Fund, K., Artigas, J., Schwartz, T., Sabater, S., and Obst, U. (2008). Relevance of polymeric matrix enzymes during biofilm formation. Microb. Ecol. *56*, 427–436.

Ryan, D.R., Leukes, W.D., and Burton, S.G. (2005). Fungal bioremediation of phenolic wastewaters in air lift reactor. Biotechnol. Progr. *21*, 1068–1074.

Sabater, S., and Romaní, A.M. (1996). Metabolic changes associated with biofilm formation in an undisturbed Mediterranean stream. Hydrobiology *335*, 107–113.

Sala, M.M., Karner, M., Arin, L., and Marrasé, C. (2001). Measurement of ectoenzyme activities as an indication of inorganic nutrient imbalance in microbial communities. Aquat. Microb. Ecol. *23*, 310–311.

Sala, M.M., and Güde, H. (2004). Ectoenzymatic activities and heterotrophic bacteria decomposing detritus. Arch. Hydrobiol. *160*, 289–303.

Sala, M.M., Terrado, R., Lovejoy, C., Unrein, F., and Pedrós-Alió, C. (2008). Metabolic diversity of heterotrophic bacterioplankton over winter and spring in the coastal Arctic Ocean. Environ. Microbiol. *10*, 942–949.

Salomo, S., Münch, C., and Röske, I. (2009). Evaluation of the metabolic diversity of microbial communities in four different filter layers of a constructed wetland with vertical flow by BiologTM analysis. Wat. Res. *43*, 4569 - 4578.

Schade, M., and Lemmer, H. (2006). *In situ* enzyme activities of filamentous scum bacteria in municipal activated sludge wastewater treatment plants. Acta Hydrochim. Hydrobiol. *34*, 480–490.

Scott, J.T., Back, J.A., Taylor, J.M., and King, R.S. (2008). Does nutrient enrichment decouple algal-bacterial production in periphyton? J. N. Am. Benthol. Soc. *27*, 332–344.

Scott, J.T., Lang, D.A., King, R.S., and Doyle, R.D. (2009). Nitrogen fixation and phosphatase activity in periphyton growing on nutrient diffusing substrata: evidence for differential nutrient limitation in stream periphyton. J. N. Am. Benthol. Soc. *28*, 57–68.

Sell, A.F. (1994). Phytoplankton-excreted organic carbon. In Microbial Ecology of Lake Plubsee, Overbeck, J., and Chróst, R.J., eds. (Ecological Studies 105, Springer-Verlag), pp. 81–91.

Sekar, R., Nair, K.V.K., Rao, V.N.R., and Venugopalan, V.P. (2002). Nutrient dynamics and successional changes in a lentic freshwater biofilm. Freshwat. Biol. *47*, 1893–1907.

Serra, A., Guasch, H., Admiraal, W., Van der Geest, H.G., and Van Beusekom, S.A.M. (2009). Influence of phosphorus on copper sensitivity of fluvial periphyton: the role of chemical, physiological and community-related factors. Ecotoxicology *19*, 770–780.

Sinsabaugh, R.L. (2010). Phenol oxidase, peroxidase and organic matter dynamics of soil. Soil Biol. Biochem. *42*, 391–404.

Sinsabaugh, R.L., and Findlay, S. (1995). Microbial production, enzyme activity, and carbon turnover in surface sediments of the Hudson River estuary. Microb. Ecol. *30*, 127–141.

Sinsabaugh, R.L., and Follstad Shah, J.J. (2010). Integrating resource utilization and temperature in metabolic scaling of riverine bacterial production. Ecology *91*, 1455–1465.

Sinsabaugh, R.L., and Foreman, C.M. (2001). Activity profiles of bacterioplankton in a eutrophic river. Freshwat. Biol. *46*, 1239–1249.

Sinsabaugh, R.L., and Linkins, A.E. (1988). Exoenzyme activity associated with lotic epilithon. Freshwat. Biol. *20*, 249–261.

Sinsabaugh, R.L., and Moorhead, D.L. (1994). Resources allocation to extracellular enzyme production: a model for nitrogen and phosphorus control of litter decomposition. Soil Biol. Biochem. *26*, 1305–1311.

Sinsabaugh, R.L., Golladay, S.W., and Linkins, A.E. (1991a). Comparison of epilithic and epixylic biofilm development in a boreal river. Freshwat. Biol. *25*, 179–187.

Sinsabaugh, R.L., Repert, D., Weiland, T., Golladay, S.W., and Linkins, A.E. (1991b). Exoenzyme accumulation in epilithic biofilm. Hydrobiology 222, 29–37.

Sinsabaugh, R.L., Antibus, R.K., Linkins, A.E., McClaugherty, C.A., Rayburn, L., Repert, D., and Weiland, T. (1993). Wood decomposition: Nitrogen and phosphorus dynamics in relation to extracellular enzyme activity. Ecology *74*, 1585–1593.

Sinsabaugh, R.L., Gallo, M.E., Lauber, C., Waldrop, M.P., and Zak, D.R. (2005). Extracellular enzyme activities and soil organic matter dynamics for northern hardwood forests receiving simulated nitrogen deposition. Biogeochemistry *75*, 201–215.

Sinsabaugh, R.L., Hill, B.H., and Follstad Shah, J.J. (2009). Ecoenzymatic stoichiometry of microbial organic nutrient acquisition in soil and sediment. Nature *462*, 795–798.

Sinsabaugh, R.L., Van Horn, D.J., Follstad Shah, J.J., and Findlay, S. (2010). Ecoenzymatic sotichiometry in relation to productivity for freshwater biofilm and plankton communities. Microb. Ecol.,60, 885–893.

Sirová, D., Vrba, J., and Rejmánková, E. (2006). Extracellular enzyme activities in benthic cyanobacterial mats: comparison between nutrient-enriched and control sites in marshes of northern Belize. Aquat. Microb. Ecol. *44*, 11–20.

Smucker, N.J., DeForest, J.L., and Vis, M.L. (2009). Different methods and storage duration affect measurements of epilithic extracellular enzyme activities in lotic biofilms. Hydrobiology *636*, 153–162.

Sobczak, W.V. (1996). Epilithic bacterial responses to variations in algal biomass and labile dissolved organic carbon during biofilm colonization. J. N. Am. Benthol. Soc. *15*, 143–154.

Somville, M. (1984). Measurement and study of substrate specificiy of exoglucosidase activity in eutrophic water. Appl. Environ. Microbiol. *48*, 1181–1185.

Somville, M., and Billen, G. (1983). A method for determining exoproteolytic activity in natural waters. Limnol. Oceanogr. *28*, 190–193.

Stelzer, R.S., and Lamberti, G.A. (2001). Effects of N: P ratio and total nutrient concentration on stream periphyton community structure, biomass, and elemental composition. Limnol. Oceanogr. *46,* 356–367.

Sterner, R.W., and Elser, J.J. (2002). Ecological Stoichiometry: The Biology of Elements from Molecules to the Biosphere (Princeton Univ. Press).

Stoodley, P., Wilson, S., Hall-Stoodley, L., Boyle, J.D., Lappin-Scott, H.M., and Costerton, J.W. (2001). Growth and detachment of cell clusters from mature mixed-species biofilms. Appl. Environ. Microb. *67,* 5608–5613.

Suberkropp, K. (1992). Aquatic hyphomycete communities. In The Fungal Community: Its Organization and Role in the Ecosystem, Wicklow, D.T., and Carroll, G.C., eds. (Marcel Dekker, Inc., New York), pp. 729–747.

Suberkropp, K., and Klug, M.J. (1980). The maceration of deciduous leaf litter by aquatic hyphomycetes. Can. J. Bot.-Rev. Can. Bot. *58,* 1025–1031.

Tanaka, Y. (1993). Aerobic cellulolytic bacterial flora associated with decomposing Phragmites leaf litter in a seawater lake. Hydrobiology *263,* 145–154.

Thurmann, E.M. (1985). Organic Geochemistry of Natural Waters. (Nijhoff/Junk, Dordrecht).

Tuchman, N.C., Schollet, M.A., Rier, S.T., and Geddes, P. (2006). Differential heterotrophic utilization of organic compounds by diatoms and bacteria under light and dark conditions. Hydrobiology *561,* 167–177.

Turley, C. (1994). Controls of the microbial loop: nutrient limitation and enzyme production, location and control. Microb. Ecol. *28,* 287–289.

Van Ommen Kloeke, F., and Geesey, G.G. (1999). Localization and identification of populations of phosphatase-active bacterial cells associated with activated sludge flocs. Microb. Ecol. *38,* 210–214.

Vázquez, E., Romaní, A.M., Sabater, F., and Butturini, A. (2007). Effects of the dry-wet hydrological shift on dissolved organic carbon dynamics and fate across stream–riparian interface in a Mediterranean catchment. Ecosystems *10,* 239–251.

Vrba, J., Simek, K., Nedoma, J., and Hartman, P. (1993). 4-Methylumbelliferyl-β-N-acetylglucosaminide hydrolysis by a high-affinity enzyme, a putative marker of protozoan bacterivory. Appl. Environ. Microbiol. *59,* 3091–3101.

Vrba, J., Callieri, C., Bittl, T., Šimek, K., Bertoni, R., Filandr, P., Hartman, P., Hejzlar, J., Macek, M., and Nedoma, J. (2004). Are bacteria the major enzyme producers of extracellular glycolytic enzymes in aquatic environments? Int. Rev. Hydrobiol. *89,* 102–117.

Weller, D.M. (1988). Biological control of soilborne plant pathogens in the rhizosphere with bacteria. Ann. Rev. Phytopathol. *26,* 379–407.

Wetzel, R.G. (1991). Extracellular enzymatic interactions: storage, redistribution, and interspecific communication. In Microbial Enzymes in Aquatic Environments, Chróst, R.J., ed. (Springer-Verlag, New York), pp. 6–28.

Wey, J.K., Scherwass, A., Norf, H., Arndt, H., and Weitere, M. (2008). Effects of protozoan grazing within river biofilms under semi-natural conditions. Aquat. Microb. Ecol. *52,* 283–296.

Wohl, D.L., and McArthur, J.V. (2001). Aquatic Actinomycete–Fungal interactions and their effects on organic matter decomposition: a microcosm study. Microb. Ecol. *42,* 446–457.

Willey, J.D., Kieber, R.J., Eyman, M.S., and Avery, G.B. (2000). Rainwater dissolved organic carbon: Concentrations and global flux. Glob. Biogeochem. Cycl. *14,* 139–148.

Ylla, I., Borrego, C., Romaní, A.M., and Sabater, S. (2009). Availability of glucose and light modulates the structure and function of a microbiol biofilm. FEMS Microbiol. Ecol. *69,* 27–42.

Ylla, I., Sanpera-Calbet, I., Vázquez, E., Romaní, A.M., Muñoz, I., Butturini, A., and Sabater, S. (2010). Organic matter availability during pre- and post-drought periods in a Mediterranean stream. Hydrobiologia *657,* 217–232.

Energy from Slime? Power from Microbial Fuel Cells

10

Koichi Nishio, Atsushi Kouzuma, Souichiro Kato and Kazuya Watanabe

Abstract

Microbial fuel cells (MFCs) are devices that exploit microbial catabolic activities to generate electricity from a variety of starting materials, including complex organic waste and renewable biomass. The use of these energy sources provides MFCs with a great advantage over chemical fuel cells that utilize only purified reactive fuels (e.g. hydrogen). In an MFC bioreactor, microbes that respire using an anode with organics as electron donors grow preferentially, resulting in accelerated and increased current generation with time. The placement of an anode in either soil or sediment represents a simplified MFC system, known as a sediment MFC, which generates current as soil microbes utilize the anode as an electron acceptor. In addition, the irradiation of an MFC system results in the proliferation of photosynthetic microbes together with anode-respiring microbes, resulting in the syntrophic conversion of light energy into electricity. These examples demonstrate that the MFC system is based on a variety of fundamental and sustainable bioenergy processes, and we suggest that a deeper understanding of how microbes transfer electrons to anodes is essential for further developments of MFC systems.

Introduction

Beginning with the Industrial Revolution in the eighteenth century, human society has largely depended on fossil fuels. Petroleum is a representative fossil fuel, which occurs naturally in an easy-to-use and energy-intensive form, and is a starting material for a variety of chemical products. In 2006, petroleum and other major fossil fuels supported 86% of the total US energy supply (Energy Information Administration, 2007), indicating that the dependency on fossil fuels continues in the twenty-first century.

Recently, however, it has become widely accepted that the excess use of fossil fuels results in serious problems for human society, which include the future shortage of energy sources and global warming. As their name indicates, fossil fuels were mostly generated by long-term geochemical reactions with ancient plants and algae serving as original resources, and have gradually accumulated in underground reservoirs (Craig *et al.*, 2010). Although these geochemical reactions are also presently occurring, the consumption of fossil fuels rapidly exceeds these reactions, resulting in a sharp decrease in fossil fuel reserves. In addition, a large fraction of the petroleum stored in easily accessible reservoirs is considered to have been consumed, and engineers predict that the costs for extracting and producing petroleum will dramatically increase in the twenty-first century. Another significant problem results from the combustion of fossil fuels, as this process generates carbon dioxide that accumulates in the atmosphere, resulting in the green-house effect and global warming (Lashof and Ahuja, 1990). According to an assessment report in 2007 by the Intergovernmental Panel on Climate Change (IPCC), global surface temperature increased by 0.74 ± 0.18°C during the twentieth century (IPCC, 2007), which corresponds to an increase in the carbon dioxide concentration in the atmosphere from 0.028% to 0.037% (Colls, 2002). We therefore consider that the extraction and combustion of fossil fuels effectively serves to convert normally inaccessible carbon into carbon dioxide that is released into the atmosphere.

Under these circumstances, alternative energy sources to fossil fuels are strongly desired for human society in the twenty-first century, particularly those sources that are renewable and not associated with increasing levels of carbon dioxide in the atmosphere. Of primary importance should be solar energy due to its virtually endless supply; however, biomass also represents a renewable energy source with significant potential (Shinnar and Citro, 2006). Biomass includes all biologically synthesized chemicals that are the primary production of photosynthetic organisms, which use sunlight as the energy source and carbon dioxide as the raw material (i.e. biological primary production) (Field *et al.*, 2008). In addition to sunlight, geochemical energy is also available for biological primary production (Banfield and Nealson, 1997); this process involves chemolithotropic bacteria that utilize the chemical energy conserved in reduced inorganic chemicals to synthesize organic compounds from carbon dioxide. Notably, biomass is the most abundant carbonaceous material at the surface of our planet, and the total amount has been estimated to be 75 billion tonnes (Field *et al.*, 2008). Although the combustion of biomass generates carbon dioxide, this is considered to be a part of natural carbon cycling that occurs along with biological primary production. In this sense, the use of biomass as an energy source is regarded as a 'carbon-neutral process'. Humans have a long history of using biomass as an energy source, beginning from the burning of wood by ancient people for heating and cooking, and continuing to the present day, as we continue to develop technologies to more efficiently convert and use the chemical energy trapped in biomass. Recent technical achievements include the conversion of raw biomass into convenient forms of energy (e.g. liquid fuels and electricity), and the use of biomass that has traditionally been considered as difficult to utilize for energy (e.g. sources that contain a large amount of water).

Recently, biological processes are the focus of intensive research and development in the field of renewable biomass energy, and include the production of bi-ethanol (Grey *et al.*, 2006), bio-hydrogen (Levin *et al.*, 2004), and bio-electricity (Logan *et al.*, 2006; Watanabe, 2008). The recent increased interest in this field is probably because researchers have realized the advantages of biological processes over chemical processes, which include (i) their environmentally friendly nature, (ii) the fact that a wide variety of chemicals can be used as fuels, and (iii) their self-sustaining potential. It is also important to note that waste biomass from agricultural and industrial processes can also be used as an energy source to generate fuel gases, such as hydrogen and methane (Lettinga, 1995), and electricity (Watanabe, 2008) if microbial communities are used as mixed catalysts; such an energy source would be impossible for chemical processes.

In this chapter, we first introduce several biological processes for electricity generation. The microbial fuel cell (MFC) represents a fundamental technology that utilizes these processes, and includes several variations, such as sediment MFCs (Reimers *et al.*, 2001), plant MFCs (Kaku *et al.*, 2008), and microbial solar cells (MSCs) (Nishio *et al.*, 2010). A significant feature of these systems is that they function using living microbes rather than extracted enzymes as sustainable catalysts. In addition, it is also critical that the active microbes (mostly bacteria) self-develop electron paths from cells to the electrodes (anodes) that are used for respiration (i.e. anode respiration). As these bacteria typically form biofilms on anodes, where active anode respiration occurs, we will therefore explain electron transfer mechanisms that have been elucidated using model electricity-generating bacteria (i.e. members of the genera *Shewanella* and *Geobacter*).

Microbial fuel cells

As mentioned, MFCs are devices that exploit microbial catabolic activities to generate electricity from a variety of materials, including complex organic wastes and renewable sources of biomass, such as cellulose (Ishii *et al.*, 2008a; Ishii *et al.*, 2008b). In MFCs, microbes utilize organic compounds as energy and carbon sources. In order for microbes to generate energy for growth, organics are first decomposed, and the released chemical energy is then utilized (i.e. fermentation). In addition, high-energy electrons released from the decomposed organics are transferred to oxidized chemicals that serve as electron acceptors, such as molecular oxygen, to conserve electrochemical energy (i.e. respiration). In microbial cells, electrons released from organics are initially accepted by intercellular electron-shuttling compounds [e.g. nicotinamide adenine dinucleotide (NAD)], and are subsequently transferred to electron acceptors via respiratory electron-transport chains. If a mechanism by which electrons released from organics are transferred to an extracellular electrode exists during any step in the intercellular electron-transfer pathway, then microbial oxidation of organics can be coupled to electricity generation (i.e. an MFC). These mechanisms are termed 'extracellular electron-transfer (EET) mechanisms', and represent the fundamental biological process behind MFCs. Studies have revealed that the natural environment, particularly soils and sediments, harbours a variety of microbes that have extracellular electron-transfer capabilities (Lovley, 2006).

Fig. 10.1 illustrates two typical MFC structures, which includes a traditional double-chamber MFC (Fig. 10.1A) and a single-chamber MFC equipped with a membrane-type oxygen-diffusion cathode (termed an air cathode) (Watanabe, 2008) (Fig. 10.1B). As shown in this figure, electrons released by microbes are first captured by an anode and then transferred to a cathode according to the potential gradient between them. The transferred electrons are used in cathode chemical reactions, which also utilize protons released at the anodes that diffusively transfer to the cathodes [via a proton-transfer membrane (e.g. nafion) in a double-chamber MFC], thereby completing an electric circuit. Cathode reactions are dependent on the species of electron acceptor and catalyst. Although ambient oxygen is the most commonly used electron acceptor, other oxidants, such as ferricyanide, can also serve this role (Logan *et al.*, 2006). As a cathode catalyst, platinum is widely used for oxygen reduction. In a traditional double-chamber reactor, the oxygen-reduction reaction in the cathode chamber is typically the rate-limiting step that determines the electric output from an MFC, because the solubility of oxygen in water is low (approximately 0.25 mM), resulting in a shortage of reductant (Oh *et al.*, 2004). To circumvent this limitation, a number of MFC researches have used ferricyanide as the cathode reductant (Oh *et al.*, 2004). Although this approach resulted in large improvements in electric outputs, as ferricyanide cannot be recycled, the MFCs were non-sustainable during long-term operation. However, if a membrane-type air cathode is used (Liu and Logan, 2004), oxygen molecules can be sufficiently supplied for the cathode reaction, and it is no longer the rate-limiting step. Recently, a variation of the single-chamber MFC, termed a cassette-electrode MFC, has been reported (Shimoyama *et al.*, 2008). In that MFC configuration, anodes are assembled with air cathodes across

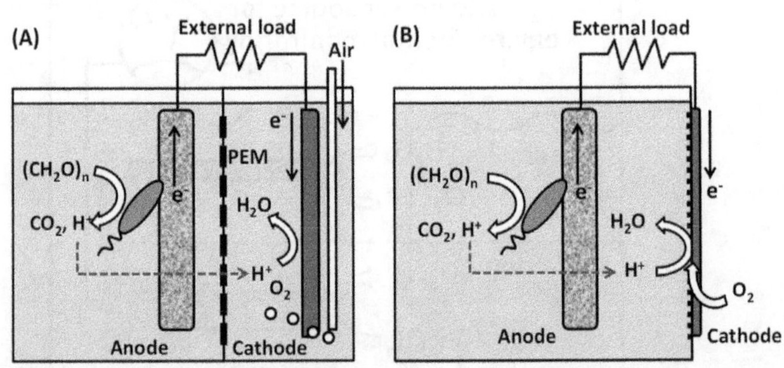

Figure 10.1 Typical microbial fuel cell (MFC) configurations. (A) A double-chamber MFC, in which a cathode chamber is bubbled with air to supply oxygen for the cathode reaction. PEM, proton-exchange membrane. (B) A single-chamber MFC equipped with an oxygen-diffusion membrane cathode.

separator membranes to form multiple cassette-electrode boxes, which are then placed adjacent to one another in a chamber to establish an MFC system.

It has been suggested that MFCs will have a number of possible future applications, including water treatment coupled to energy recovery, portable fuel cells, and biosensors. Among these potential applications, waste treatment is considered to be the most promising, and MFCs may be able to function as energy- and cost-saving options for waste treatment facilities (Watanabe, 2008). Despite this promising possibility, MFC technology has not yet been applied to practical waste treatment, as it is still an emerging technology, and more time is required for technical maturation. Another significant hurdle is that its process performance is still considered low compared with those of its competitors (e.g. methanogenic anaerobic digesters). On average, modern methanogenic anaerobic digesters treat organic wastes at efficiencies of ~8–20 kg chemical oxygen demand units per m^3 of reactor volume per day (Lettinga, 1995), equivalent to ~1200–3000 W/m^3. Given that waste-to-gas and gas-to-electricity conversion efficiencies are typically 80% and 40%, respectively, these methanogenic digesters are estimated to generate electricity at ~380–960 W/m^3. In contrast, a recent MFC study reported a power density of 130 W/m^3 using a cassette-electrode system supplied with a model organic waste comprising starch, peptone, and fish extract as the fuel (Shimoyama *et al.*, 2008). For MFCs to exceed performances of anaerobic digesters, at least a ~7-fold increase in power output is necessary. To date, extensive studies have been done to develop methods to increase power outputs from MFCs, and the results of these reports have been summarized in the literature (Watanabe, 2008).

Another adaptation of the MFC configuration is sediment MFCs (sMFCs) (Reimers *et al.*, 2001), in which anodes are submerged in freshwater or marine sediments, while cathodes are placed in the water phase above the sediment (Fig. 10.2). Since sediment and water are anaerobic and aerobic, respectively, potential gradients are generated between them. It is therefore possible to generate an electric current in sMFCs, if microbes possessing self-sustainable EET mechanisms are enriched around anodes (Lovley, 2006). To date, sMFCs have mostly been set in shallow marine coastal areas, since the high ionic strength in sea water is beneficial to electron transfer. Researchers aim at applying sediment MFCs as on-site power sources for environmental monitoring systems (Donovan *et al.*, 2008).

Microbial solar cells

Photosynthetic organisms utilize solar energy, a free and abundant energy source, to sustain growth. The amount of solar energy that strikes the

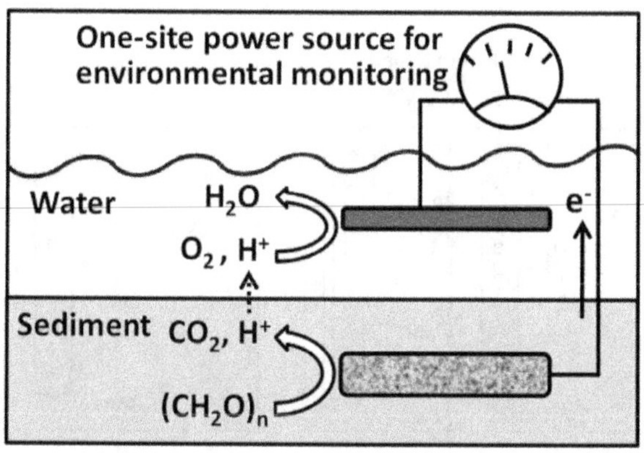

Figure 10.2 Sediment microbial fuel cell (MFC). An anode is submerged into sediment, while a cathode is set in the water above the sediment. In the sediment, microbes oxidize organics and discharge electrons to the anode.

Earth every hour (~4.3×10^{20} J) is approximately equal to the total amount of energy consumed by human society every year (Donohue and Cogdell, 2006). Hence, global energy needs can be substantially satisfied even if only a small fraction of the available solar energy can be captured. To date, extensive research has been performed to develop technologies, such as photovoltaics (Fahrénbruch and Bube, 1983), which convert sunlight into easy-to-use energy. Researchers have also focused their attention on the application of natural photosynthetic systems for energy conversion (Gust *et al.*, 1993; Das and Veziroglu, 2001). For instance, studies have been performed on photosynthetic microbes that generate hydrogen gas using sunlight as the sole energy input (Das and Veziroglu, 2001; Prince and Kheshgi, 2005).

In a number of investigations, attempts have been made to utilize photosynthetic microbes for converting light energy into electricity in devices termed microbial solar cells (MSCs). MSCs convert light energy into electricity with the aid of photosynthetic microorganisms. To date, several types of MSCs have been reported, including mediator-, electrocatalyst-, plant-, and community-type MSCs. An overview of these MSCs is given below.

Mediator-type MSCs

In a mediator-type MSC, living photosynthetic microbes convert solar energy into electric energy with the aid of artificially supplemented electron mediator compounds. Electron mediators, also referred to as electron shuttles, are organic molecules that can be reversibly oxidized and reduced, a property which confers their capacity to serve as electron carriers for multiple redox reactions (Watanabe *et al.*, 2009). In mediator-type MSCs, electron mediators enter microbial cells and are subsequently reduced upon taking electrons from photosynthetic and respiratory electron-transport chains. After diffusing out of microbial cells, the reduced mediators are oxidized at the surface of anodes through the release of electrons. Several investigations have measured electric outputs from MSCs inoculated with cyanobacteria (e.g. *Anabaena variabiliis* and *Synechococcus* spp.) in the presence of an electron mediator, such as 2-hydroxy-1,4-naphthoquinone (HNQ) and 2,6-dimethyl-1,4-benzoquinone (DMBQ), under light and dark conditions (Tanaka *et al.*, 1988; Tsujimura *et al.*, 2001; Yagishita *et al.*, 1993a,b, 1996, 1997, 1998, 1999). These studies suggest that MSCs generate electricity in association with photosynthesis and endogenous carbohydrate catabolism in light and catabolism only in the dark. Yagishita *et al.* (1997) reported an open-circuit voltage of 800 mV, a short-circuit current of 320 μA/cm (per anode area), and a maximum power density of 190 μW/cm^2 for an MSC inoculated with a *Synechococcus* strain and supplemented with HNQ. In this case, power-conversion efficiencies (PCE, obtained by dividing maximum power density by light power density) were estimated to be 3.3% or lower under sustained illumination. However, the use of unsustainable and, in many cases, toxic artificial electron mediators has prevented mediator-type MSCs from practical realization. It is, therefore, not surprising that this type of MSC is no longer popular in recent studies.

Electrocatalyst-type MSCs

In electrocatalyst-type MSCs, anode surfaces are doped with chemical catalysts that generate electricity by oxidizing metabolites produced by photosynthetic microbes. In a typical example, a platinum-doped anode was used to convert hydrogen photosynthetically produced by the green alga *Chlamydomonas reinhardtii* into electric current (Rosenbaum *et al.*, 2005). Several studies have also attempted the conversion of hydrogen into electricity in electrocatalyst-type MSCs, which have been supported by research related to hydrogen production by green algae (Prince and Kheshgi, 2005).

Gaffron and Rubin (1942) examined hydrogen metabolism of unicellular eukaryotic algae and found that after the anaerobic induction of algae for several hours in the dark, hydrogenase activity could be detected for a short time period under irradiation. Following this discovery, a great number of phototropic hydrogen-producing species, including eukaryotic (green algae) and prokaryotic (cyanobacteria and purple bacteria) microbes, have been identified. However, although photosynthetic hydrogen production is considered a promising method for the biological conversion of sunlight into fuels, this approach is currently still far from practical application. The major problem relates to the extreme sensitivity of hydrogenase enzymes towards molecular oxygen. Oxygen suppresses not only the expression of hydrogenase genes, but also irreversibly inactivates the enzymatic activity of hydrogenases (Ghirardi *et al.*, 1997). Efforts have been made to overcome this problem; for instance, *C. reinhardtii*, a model organism that expresses an Fe-hydrogenase enzyme shortly after anaerobic adaptation (Happe and Kaminski, 2002), was genetically engineered to produce hydrogenases with improved O_2 tolerance (Flynn *et al.*, 2002). Another strategy involves so-called indirect photobiolysis by temporarily separating the water-splitting and oxygen-evolving reactions from the hydrogen production process. Using this approach, sustained photobiological hydrogen production was achieved by transferring growing cells of *C. reinhardtii* into sulfur-deprived nutrient conditions (Melis, 2002). The lack of sulfur led to a rapid diminution of photosystem II activity, which is the oxygen-evolving part of the photosynthetic apparatus, and to a decrease in the oxygen-production rate below what was suitable for mitochondrial respiration (Melis *et al.*, 2000).

The above-described hydrogen production studies are fundamental for the further development of electrocatalyst-type MSCs. Rosenbaum *et al.* (2005) proposed an electrocatalysed MSC system, in which the green alga *C. reinhardtii* photosynthetically produced hydrogen that was then oxidized *in situ* to generate current. Notably, this system used a platinum electrocatalyst that was overlaid with conductive polymers, such as polyaniline and its fluorinated derivatives, which served to not only preserve the electrocatalyst from deactivation, but also improved the efficiency of hydrogen oxidation (Nießen *et al.*, 2004). This system also allowed electricity generation from the decomposition of complex carbohydrates by exploiting the catabolism of chemotropic microorganisms.

Community-type MSCs

In nature, photosynthetic microbes are typically found in communities with other microbes, including heterotrophs that utilize organic compounds produced by photosynthetic microbes (Larsen *et al.*, 2007). Syntrophic interactions are known to exist between photosynthetic microbes, such as cyanobacteria and microalgae, and heterotrophs, and often occur in microbial mats (Stal *et al.*, 1985). Such interactions have been applied to sediment MSCs to generate electric outputs (He *et al.*, 2009; Malik *et al.*, 2009). In one example, the irradiation of a sediment MFC resulted in the growth of photosynthetic microbes in the water phase and their subsequent deposition on the sediment, where they served as substrates for heterotrophic bacteria to generate electricity (He *et al.*, 2009). It is also possible to construct community-type MSC reactors without sediments. For instance, electric currents were generated using MSC reactors equipped with air cathodes and graphite-felt anodes and supplemented with green hot-spring microbial mats (Nishio *et al.*, 2010) (Fig. 10.3A). When these reactors were irradiated, an electric current was generated between the anode and cathode in accordance with the formation of green biofilm on the anode. Analysis of the biofilm community suggested that green algae and heterotrophic bacteria cooperatively converted light energy into electricity. This system effectively demonstrates the possibility for constructing sustainable biological light/electricity conversion systems using natural biofilm microbial communities.

Plant-type MSCs

Plant photosynthesis fixes carbon dioxide in the air to produce organic matter that serves as the primary source of carbon and energy for most organisms. Such organic matter (biomass) is also utilized as an energy source for human society. Recently, a novel concept has been proposed for on-site conversion of the energy conserved in photosynthesized organic matter into electricity (De Schamphelaire *et al.*, 2008; Kaku *et al.*, 2008; Strik *et al.*, 2008) (Fig. 10.3B). To cite one instance, Strik *et al.* (2008) developed a laboratory pot-culture system to generate electric power from organics exuded from the roots of reed mannagrass (*Glyceria maxima*). In their system, a graphite-felt anode was embedded in the pot soil, and a cation-exchange membrane was inserted between the pot anode and a neighbouring cathode compartment

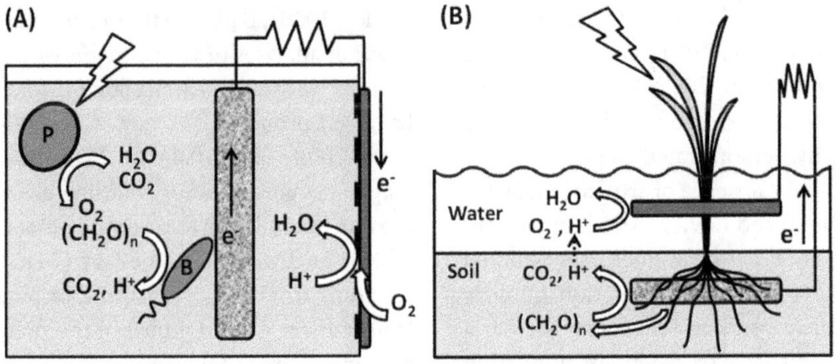

Figure 10.3 Microbial fuel cell (MFC -based biosolar cells. (A) A community-type MSC, in which photosynthetic microbes (P) supply organics to electricity-generating bacteria (B), resulting in light/electricity conversion. (B) A plant-type MSC, in which soil bacteria utilize organic compounds that are photosynthesized by plants and exuded from roots for anode respiration, resulting in light/electricity conversion.

to establish a plant-type MSC that produced a maximum power output of 67 mW/m^2. A similar laboratory pot experiment was also conducted with rice plants (De Schamphelaire *et al.*, 2008). Furthermore, on-site electricity generation in a rice paddy field has also been demonstrated (Kaku *et al.*, 2008). In that field trial, graphite-felt anodes and cathodes were set in the rice rhizosphere and flooded water, respectively, and it was observed that electricity (as high as 30 mW/m^2, normalized to the anode projection area) was generated in a sunlight-dependent manner (Kaku *et al.*, 2008; Takanezawa *et al.*, 2010). The results of this experiment suggest that the paddy-field electricity-generation system functions as an ecological solar cell that couples plant photosynthesis to the microbial conversion of organics into electricity (Kaku *et al.*, 2008; Takanezawa *et al.*, 2010). It is considered that such plant/microbe-interacting solar cells are widely applicable for electricity generation in plant-inhabiting wetlands.

Extracellular electron transfer by microbes

MFCs and MSCs are dependent on the ability of bacteria to transfer electrons to solid extracellular electron acceptors, i.e. anodes. For bacteria to grow, it is necessary to discharge electrons that are obtained through the decomposition and oxidization of organic compounds. In addition, if high-energy (low electrochemical potential) electrons are used to reduce compounds with high electrochemical potentials, such as oxygen, electrochemical energy can be conserved (i.e. as occurs in the process of respiration). In cases where water-soluble electron-acceptor compounds are not present, bacteria that are able to use solid anodes for respiration (i.e. anode respiration) outcompete and overgrow microbes without this capability, which represents the preferential condition for MFCs and MSCs. To improve the electric outputs of these systems, it is therefore necessary to better understand the molecular mechanisms of bacterial EET. For this purpose, extensive research has been performed on two model bacteria, namely, *Shewanella oneidensis* and *Geobacter sulfurreducens*. In this section, we summarize the existing knowledge on bacterial EET mechanisms for these two model organisms.

Geobacter

Members of the genus *Geobacter*, which is affiliated with the class *Deltaproteobacteria*, were originally studied as dissimilatory metal reducing bacteria with the potential to facilitate the bioremediation of heavy metal contaminants (Lovley *et al.*, 2004). In recent years, however, *Geoobacter* spp. have also attracted attention as current-generating bacteria in MFC systems. It has been reported that *Geobacter* species (and closely related bacteria such as *Desulfuromonas* species) are selectively enriched at anode surfaces of sediment MFCs (Tender *et al.*, 2002; Holmes *et al.*, 2004) and reactor-type MFCs (Jung and Regan, 2007; Lee *et al.*, 2008). In addition, pure cultures of *Geobacter* species have been clearly demonstrated to generate currents in MFCs in the absence of artificial mediator compounds (Bond and Lovley, 2003). In particular, *G. sulfurreducens* has been intensively studied owing to its ability for current generation, publicly available complete genome sequence (Methé *et al.*, 2003), and existence of genetic-manipulation systems (Coppi *et al.*, 2001).

EET mechanisms of *Geobacter* were first studied under metal-reducing conditions. To date, two mechanisms have been proposed for bacterial EET, namely, indirect electron transfer via extracellular mediator compounds and direct electron transfer through outer-membrane proteins [e.g. c-type cytochromes (c-Cyts)] (Lovley, 2006). In the case of *Geobacter* spp., direct electron transfer is considered to be the major active pathway used to reduce insoluble metals, as they do not naturally secrete mediator compounds. This conclusion is based on experiments that demonstrated *Geobacter* was unable to reduce insoluble iron coated within porous beads, which effectively prevented cells from direct contact with the insoluble iron, but iron was reduced after mediator compounds were artificially added to the bacterial/bead suspension (Nevin and Lovley, 2000). Based on transcriptomic and gene-deletion analyses of *G. sulfurreducens*, an electron transfer pathway to extracellular insoluble metals has been proposed (Lovley *et al.*, 2004; Lovley, 2006, 2008) (Fig. 10.4). In this pathway, c-Cyts play important roles for shuttling electrons from the inner membrane (IM) to the outer membrane (OM); an IM-associated c-Cyt (MacA) accepts electrons from the cytoplasmic respiratory electron transfer chain (probably the quinine pool) and transports them to the OM pathway via soluble, periplasmic c-Cyts, including PpcA. Electrons are subsequently transported to OM c–Cyt complexes (mainly composed of OmcB, OmcE, and OmcS) that are considered to directly transfer electrons to extracellular insoluble metals.

Initially, MFCs with pure cultures of *G. sulfurreducens* achieved relatively low current densities (typically below 300 mA/m^2), mainly owing to instrumental and design limitations. Under low-current conditions, *G. sulfurreducens* formed thin, monolayer biofilms on electrodes (Bond and Lovley, 2003). The removal of planktonic cells and water-soluble compounds from *G. sulfurreducens* MFCs by exchanging spent medium in the anodic chamber with fresh medium did not affect the current density, indicating that direct electron transfer was the major pathway for current generation, as is the case for metal reduction by *Geobacter* strains (Bond and Lovley, 2003).

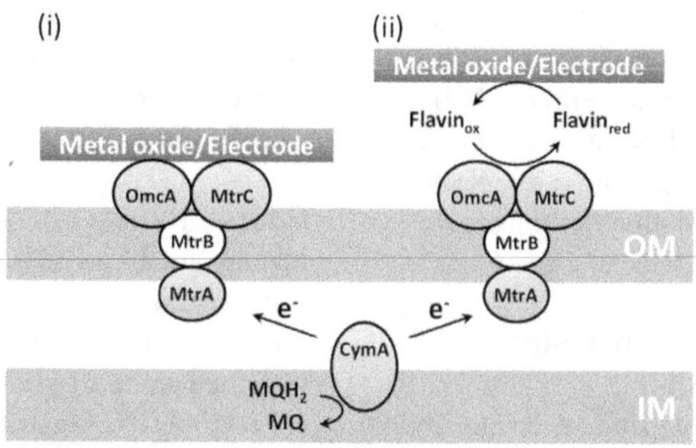

Figure 10.4 Proposed extracellular electron-transfer (EET) pathways in *Geobacter sulfurreducens*. (i) An EET pathway for metal reduction and current generation in thin biofilms. (ii) An EET pathway for current generation in thick biofilms. OM, outer membrane; IM, inner membrane; MQH$_2$, reduced form of menaquinone; MQ, oxidized form of menaquinone. Oval shapes depict various c-type cytochromes.

This concept was also supported by electrochemical experiments, as cyclic voltammetry and impedance spectroscopy of electrode-attached *G. sulfurreducens* revealed that electron transfer was mediated by catalysts adsorbed on electrodes (Marsili *et al.*, 2008b). In addition, transcriptomic analyses revealed that two OM c-Cyts (OmcS and OmcE) that are required for metal reduction were highly expressed in *G. sulfurreducens* cells attached to MFC anodes (Holmes *et al.*, 2004). Although deletion of the *omcS* gene resulted in a decrease in current density, indicating that OmcS is essential for current generation in low power-density MFCs, current generation by an *omcE* mutant was initially suppressed, but current-generating ability was restored after prolonged cultivation. Furthermore, OmcB, an OM c-Cyt essential for metal reduction, was also moderately expressed in MFCs, however, its deletion did not affect current generation. Taken together, these results indicate that although the EET pathway used for current generation in MFCs partially overlaps with the metal-reduction pathway, they are clearly not identical.

The improvement of MFC devices resulted in significant increases in current densities for *G. sulfurreducens* MFCs (~4.5 A per m^2); these values are comparable to current densities reported for efficient MFC reactors containing microbial communities (Ishii *et al.*, 2008c; Nevin *et al.*, 2008). In these high current-density MFCs, *G. sulfurreducens* produced thick biofilms (50 μm or thicker) on electrodes (Reguera *et al.*, 2006; Nevin *et al.*, 2008). Analyses of *G. sulfurreducens* biofilms at different growth stages in MFCs revealed that biofilm thickness and current density were well correlated. Microscopic analyses disclosed that bacterial cells were viable and metabolically active at all sites within the biofilm, suggesting that bacterial cells distant from electrodes and not directly attached to electrodes can also utilize electrodes for respiration (Reguera *et al.*, 2006; Franks *et al.*, 2009). As discussed above, since a mediator-dependent indirect electron transfer pathway does not naturally function in *G. sulfurreducens* MFCs, it has been proposed that *G. sulfurreducens* is able to transfer electrons to anodes across biofilms via conductive matrices. This speculation is also supported by electrochemical analyses (Marsili *et al.*, 2008b; Richter *et al.*, 2009) and simulation studies (Picioreanu *et al.*, 2007), although the exact molecular mechanisms operating within conductive biofilms have not yet been elucidated.

Possible candidates for mediating the movement of electrons across *G. sulfurreducens* biofilms are conductive pili (also called as nanowires), which were first discovered by atomic force microscopy as possible conduits for long-range electron transfer from cells to distant insoluble Fe(III) oxides (Reguera *et al.*, 2005). These analyses demonstrated that *G. sulfurreducens* abundantly produced pili and highly expressed the *pilA* gene (encoding a structural component of pili) when it respired using insoluble metals. Significantly, its iron-oxide reducing ability was impaired by the deletion of *pilA*, and transcriptomic analyses of *G. sulfurreducens* cells in thick biofilms under both current-generating and fumarate-reducing conditions revealed that *pilA* was highly expressed only under the former (Nevin *et al.*, 2009). Deletion mutagenesis of *pilA* resulted in decreases in current density and in biofilm thickness in MFCs, suggesting that conductive pili are important for long-range electron transfer in high-power MFCs (Nevin *et al.*, 2009, Richter *et al.*, 2009). Despite this evidence, the roles of pili for long-range electron transfer are still under debate, as electric conductivity has only been measured in the longitudinal direction of a pilus (Reguera *et al.*, 2005), and no information is available for the conductivity in the horizontal direction. In addition, since the main component of pili (PilA) is a redox inert protein, redox active components are presumably required in pili to exhibit conductivity. Presently, c-Cyts are the primary candidates for redox active agents in pili. In fact, immunological analyses have detected OmcS along *G. sulfurreducens* pili (Leang *et al.*, 2010); however, this observation is insufficient to explain conductivity in the horizontal direction, since apparent distances between OmcS proteins were shown to be much greater than 2 nm, which represents the maximum distance for effective electron transfer between cytochromes. Furthermore, the deletion of the *pilA* gene seemed to affect biofilm formation itself (Reguera *et al.*, 2007), and in some cases caused mislocalization of cytochromes (Juarez *et al.*, 2009).

Another probable explanation for long-range electron transfer is that extracellular matrices are inherently electrically conductive. Extracellular matrices in microbial biofilms are mainly composed of polysaccharides, proteins, and DNA, and generally act as insulators (Dheilly *et al.*, 2008). In conductive biofilms, c-Cyts secreted by *G. sulfurreducens* may be able to confer the property of electric conductance. Transcriptomic analyses demonstrated that OmcZ, an outer-surface c-Cyts with unknown function, is abundantly expressed by *G. sulfurreducens* cells in biofilms exhibiting high-current densities (Nevin *et al.*, 2009), while biochemical and immunogold-localization analyses revealed that OmcZ proteins are dispersed in the extracellular matrix (Inoue *et al.*, 2010). The deletion of *omcZ* inhibited thick biofilm formation and current generation in MFCs, whereas the mutant had no impairment for biofilm growth under fumarate-reducing conditions (Nevin *et al.*, 2009, Richter *et al.*, 2009). Electrochemical analyses of the *omcZ*-deletion mutant demonstrated that OmcZ is important for electron transfer in the bulk of biofilm (Richter *et al.*, 2009). Taken together, these observations indicate that OmcZ is an important component that confers conductivity to *G. sulfurreducens* biofilms, while other c-Cyts, such as OmcS and OmcB, may also engage in this role (Richter *et al.*, 2009).

In this section, our knowledge of EET mechanisms of *G. sulfurreducens* under metal-reducing, and low and high current-generating conditions is summarized. Although c-Cyts are fundamentally important in all of these mechanisms, it is noteworthy that *G. sulfurreducens* appears to use different c-Cyts for different situations. Since *Geobacter* strains are known to harbour numerous c-Cyt genes in their genome (e.g. *G. sulfurreducens* encodes greater than 100 putative cytochromes) (Methé *et al.*, 2003; Butler *et al.*, 2009), it is also possible that certain cytochromes have overlapping functions, and it is highly probable that *Geobacter* possesses other as yet unknown mechanisms for EET.

Shewanella

Shewanella species, which are affiliated with the class *Gammaproteobacteria*, are widely distributed in nature, including marine, freshwater, sedimentary, and soil environments (Fredrickson *et al.*, 2008). Members of this genus have attracted considerable attention in recent years in terms of their respiratory versatility and potential applicability to biotechnological processes, such as bioremediation and MFCs. *S. oneidensis* MR-1 is the most extensively studied strain in the genus *Shewanella* because of its annotated genome sequence (Heidelberg *et al.*, 2002), genetic accessibility, and capability to utilize a wide range of electron acceptors (Myers and Nealson, 1988). This bacterium can respire various organic and inorganic substrates, including oxygen, fumarate, nitrate, nitrite, thiosulfate, elemental sulfur, trimethylamine N-oxide, dimethyl sulfoxide (DMSO), and anthraquinone-2,6-disulfonate, as well as both soluble and solid metals such as iron, manganese, uranium, chromium, cobalt, technetium, and vanadium (Nealson and Saffarini, 1994; Liu *et al.*, 2002; Carpentier *et al.*, 2003). In addition, MR-1 can transfer electrons to an anode and generate current in MFCs without exogenously added mediator (Kim *et al.*, 1999).

As mentioned, solid metal/anode respiration requires specific mechanisms to transfer electrons from intracellular electron sources (e.g. NADH) to insoluble and extracellular electron acceptors. Similar to *G. sulfurreducens*, extensive studies have been performed to understand EET mechanisms in *S. oneidensis* MR-1. In this strain, two main mechanisms of EET have been proposed: (i) direct EET through the physical contact of bacterial cells via OM cytochromes (Shi *et al.*, 2007), and (ii) indirect EET via self-secreted electron shuttles, such as flavins (Marsili *et al.*, 2008a; Von Canstein *et al.*, 2008). Although the detailed mechanisms underlying these EET processes are not completely understood, genetic and biochemical investigations have identified five primary protein components (i.e. CymA, MtrA, MtrB, MtrC, and OmcA) that constitute a pathway for electron transfer from the IM respiratory chain (including the quinone/quinol pool) to the surface of the OM (the Mtr pathway; Fig. 10.5).

CymA (SO_4591) is a tetraheme c-Cyt that shows homology to NapC/NirT-family quinol dehydrogenases. Like other members of the NapC/NirT family, CymA consists of a short N-terminal region that is anchored in the IM and

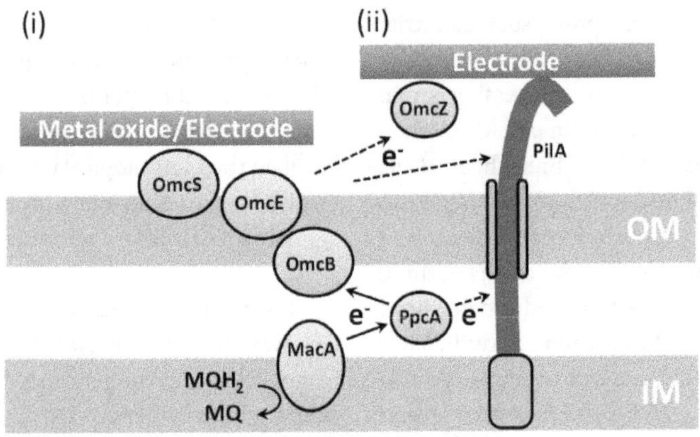

Figure 10.5 Proposed extracellular electron transfer (EET) pathways in *Shewanella oneidensis* (Mtr pathway). (i) A direct EET pathway. (ii) An indirect EET pathway mediated by electron shuttles (flavins). OM, outer membrane; IM, inner membrane; MQH_2, reduced form of menaquinone; MQ, oxidized form of menaquinone. Oval shapes depict various c-type cytochromes.

a long C-terminal region that protrudes into the periplasm and contains four haem-binding sites (Myers and Myers, 1997a). A deletion mutant of the *cymA* gene has impaired ability to reduce not only extracellular electron acceptors, such as Fe(III)/Mn(IV) oxides and MFC anodes, but also soluble substrates, such as fumarate, nitrate, and nitrite (Myers and Myers, 1997a; 2000; Schwalb *et al.*, 2003; Lies *et al.*, 2005; Bretschger *et al.*, 2007). Therefore, CymA is considered to mediate electron transfer from the quinone/quinol pool to terminal reductases for the reduction of these soluble compounds, and is thought to function in close association with MtrA, a periplasmic protein that delivers electrons to OM proteins in the EET pathway.

MtrA (SO_1777) is a soluble decaheme c-Cyt that is localized in the periplasm (Pitts *et al.*, 2003). A mutant lacking MtrA displays decreased ability to reduce soluble/insoluble metals and generate current in MFCs, although its ability to reduce fumarate, nitrate, nitrite, DMSO, TMAO, thiosulfate, and sulfide remains unchanged, demonstrating its specific role for metal/anode respiration (Beliaev *et al.*, 2001). Although the exact function of MtrA is currently unknown, this protein is considered to play a critical role in the transfer of electrons to OM proteins, such as MtrB, based on its periplasmic location and association with CymA.

Of the five primary components required for metal reduction and EET, MtrB (SO_1776) is the only protein that is not a c-Cyt (Beliaev and Saffarini, 1998; Lies *et al.*, 2005). MtrB contains a signal peptide necessary for protein secretion and is therefore predicted to be transported across the IM. Analyses of membrane fractions have also indicated that MtrB is present in OM fractions (Beliaev and Saffarini, 1998). Although the exact role of MtrB remains to be determined, evidence suggests that it is required for the proper localization and insertion of MtrC and OmcA into the OM (Myers and Myers, 2002).

OmcA and MtrC are OM c-Cyts that contain ten haem-binding sites and a lipoprotein consensus sequence (Myers and Myers, 2003; 2004). These c-Cyts are translocated across the OM by the type-II protein-secretion system, where they are exposed to the extracellular environment (Shi *et al.*, 2008). OmcA and MtrC are known to associate with each other and form a stable complex with 2:1 stoichiometry (Shi *et al.*, 2006), while MtrC also interacts with MtrB and MtrA to form a complex with a 1:1:1 stoichiometry (Ross *et al.*, 2007). Single-deletion mutants of either *mtrC* or *omcA*, and a *mtrC*/*omcA* double-deletion mutant show decreased current generation in MFCs and reduction rates for insoluble metal oxides, including Mn(IV) and Fe(III) oxides, although these mutants have normal abilities to reduce soluble electron acceptors, such as nitrite and nitrate (Beliaev *et al.*, 2001; Bretschger *et al.*, 2007). However, an *mtrC*-overexpressing strain generated 35% more current in an MFC than wild-type *S. oneidensis* MR-1 (Bretschger *et al.*, 2007). These biochemical and genetic analyses clearly indicate the importance of OmcA and MtrC in the EET pathway. The functions of these two c-Cyts appear to be partially redundant, since the overproduction of MtrC can restore the ability of an *omcA*-deletion mutant to reduce MnO_2 (Myers and Myers, 2003b). However, several studies have also described functional differences between OmcA and MtrC. For example, Lower *et al* (2007) reported that OmcA shows a higher affinity towards haematite (α-Fe_2O_3) than MtrC. In addition, it has been suggested that OmcA is required for attachment and/or biofilm formation on carbon electrodes in MFC systems, whereas MtrC acts as the primary terminal reductase transferring electrons to electrodes (Coursolle *et al.*, 2010).

Direct EET to solid substances by *Shewanella* strains has been proposed based on the fact that the purified c-Cyts OmcA and MtrC strongly bind to and reduce crystalline Fe(III) oxide haematite (Xiong *et al.*, 2006; Lower *et al.*, 2007). However, evidence for indirect EET have also been provided by Nevin and Lovley (2002) and Lies *et al.* (2005), who demonstrated that *Shewanella* can reduce Fe(III) oxides at a distance without direct contact. In addition, von Canstein *et al.* (2008) and Marsili *et al.* (2008) reported that *Shewanella* secretes flavins, including riboflavin and flavin mononucleotide (FMN), that function as electron shuttles (mediators) for EET. Coursolle *et al.* (2010) also reported that *S. oneidensis* MR-1 transfers electrons from the cell interior to flavins using the Mtr pathway, with the majority (~90%) of this activity occurring at the OM using OmcA and MtrC. As both direct and indirect pathways appear active in this bacterium, it is still under debate which is the dominant route for EET by *Shewanella*. Recent studies have indicated that soluble flavins have critical roles in electron transfer to both metals and electrodes in *S. oneidensis* MR-1. For instance, Marsili *et al.* (2008a) reported that the accumulation of flavins in *Shewanella* biofilms increased the rate of electron transfer to an electrode by at least 370%. Significantly, kinetic analyses of purified OmcA and MtrC have shown that the direct reduction of insoluble metal oxides by these cytochromes proceeded too slowly to explain the physiological rates of electron transfer, while the reaction rates of these enzymes were greatly increased in the presence of flavins (Ross *et al.*, 2009).

Four genes encoding components of the Mtr pathway (*omcA*, *mtrC*, *mtrA*, and *mtrB*) constitute a gene cluster in the genome of *S. oneidensis* MR-1, although no evidence exists that these genes form an operon and are simultaneously expressed. Transcriptomic and other gene-expression experiments suggest that the transcription of these mtr genes (including *cymA*) is activated under anaerobic conditions, similar to other genes involved in anaerobic respiration (Myers and Myers, 1997b; Beliaev *et al.*, 2005). In strain MR-1, a cyclic AMP (cAMP) receptor protein (CRP) is involved in the regulation of the mtr and other anaerobic respiratory genes (Saffarini *et al.*, 2003), whereas in many other bacteria, anaerobic growth is controlled by the redox-sensing protein FNR (fumarate nitrate-reduction regulator; Kiley and Beinert, 1998). As *crp*-deletion mutants of *S. oneidensis* MR-1 are deficient in the anaerobic reduction of Fe(III), Mn(IV), fumarate, nitrate, and DMSO, CRP appears to activate genes involved in the respiration of these electron acceptors (Saffarini *et al.*, 2003). However, mechanisms for CRP activation under anaerobic conditions remain unknown, because CRP lacks obvious redox-sensing domains and may not respond to changes in oxygen concentrations like FNR. The detection of fumarate-reductase activity in aerobic cultures of MR-1 supplemented with cAMP suggests that the control of CRP activity under anaerobic conditions is probably directly linked to adenylate cyclase activity (Saffarini *et al.*, 2003). Charania *et al.* (2009) reported that although MR-1 has three adenylate cyclases, a membrane-bound class-III adenylate cyclase (CyaC) plays the central role in CRP-dependent anaerobic gene expression.

Although the primary components of the Mtr pathway and the specific electron shuttles that play central roles in EET have been identified, other (extra)cellular components are also involved in this complex catabolic process. Gorby *et al.* (2006) reported that under O_2-limited conditions, *S. oneidensis* MR-1 produces extracellular pilus-like structures (nanowires) that can conduct a tunnelling current and be imaged by scanning tunneling microscopy (STM). Nanowires have been implicated in the reduction of silica-hydrous ferric oxide (Si-HFO) and generation of electrical currents in MFCs. MtrC and OmcA have been proposed to have functional roles in nanowire structure, as *mtrC*/*omcA*-deletion mutants produce nanowire structures that are unable to be imaged by STM, and have impaired abilities to reduce Si-HFO and generate current in MFCs (Gorby *et al.*, 2006).

Cell-surface polysaccharides (CSP) of *S. oneidensis* MR-1 also affect the EET process. Kouzuma *et al.* (2010) reported that a mutant lacking CSP generated 1.5-fold higher current than wild-type MR-1 in MFCs. Compared with wild-type MR-1, the CSP-deficient mutant exhibited an enhanced ability to form biofilms on graphite anodes, and cells of the mutant formed large aggregates in solution. In addition, the cell surface of the mutant was found to be more hydrophobic and electrochemically negative than the wild-type surface, suggesting that CSP modify cell-surface physicochemical properties, resulting in altered behaviours in MFCs (Kouzuma *et al.*, 2010). Although *Shewanella* cells produce flavins as soluble electron shuttles, EET mediated by these shuttles is limited by their diffusion when cells are located at a distance from solid electron acceptors, such as anodes. In addition, flavins are reported to adsorb onto carbon electrodes and bacterial cells, particularly in matured biofilms (Marsili *et al.* 2008). Therefore, cell adhesion and biofilm formation on anodes probably represents an important event to efficiently generate currents in MFCs even for bacteria, such as *Shewanella* strains, that are capable of indirect EET using electron shuttles.

As described above, EET by *Shewanella* strains is largely dependent on self-secreted electron shuttles. It is therefore probable that EET by *Shewanella* and *Geobacter* are significantly different, suggesting that EET mechanisms vary from bacteria to bacteria. To date, a number of current-generating bacteria have been isolated from MFCs, and further studies on these species are necessary to understand the diversity and ubiquity of bacterial EET mechanisms.

Conclusions

In this chapter, the MFC system and its variations are outlined along with the latest knowledge on microbial electricity-generation mechanisms. As described, MFC systems have not yet been put into practical use, as further technical developments are necessary, particularly those for improving process efficiencies and power outputs. To achieve this goal, a deeper understanding of microbial electricity-generation mechanisms (i.e. EET mechanisms) is absolutely required. We also expect that such investigations will help elucidate unique features of functional microbial biofilms and open new areas of study in the field of biofilm research.

References

Banfield, J.F., and Nealson, K.H. (1997). Geomicrobiology: Interactions Between Microbes and Minerals. (The Mineralogical Society of America, Washington DC., USA).

Beliaev, A.S., and Saffarini, D.A. (1998). *Shewanella putrefaciens mtrB* encodes an outer membrane protein required for Fe(III) and Mn(IV) reduction. J. Bacteriol. *180*, 6292–6297.

Beliaev, A.S., Klingeman, D.M., Klappenbach, J.A., Wu, L., Romine, M.F., Tiedje, J.M., Nealson, K.H., Fredrickson, J.K., and Zhou, J. (2005). Global transcriptome analysis of *Shewanella oneidensis* MR-1 exposed to different terminal electron acceptors. J. Bacteriol. *187*, 7138–7145.

Beliaev, A.S., Saffarini, D.A., McLaughlin, J.L., and Hunnicutt, D. (2001). MtrC, an outer membrane decahaem *c* cytochrome required for metal reduction in *Shewanella putrefaciens* MR-1. Mol. Microbiol. 39, 722–730.

Bond, D.R., and Lovley, D.R. (2003). Electricity production by *Geobacter sulfurreducens* attached to electrodes. Appl. Environ. Microbiol. *69*, 1548–1555.

Bretschger, O., Obraztsova, A., Sturm, C.A., Seop Chang, I., Gorby, Y.A., Reed, S.B., Culley, D.E., Reardon, C.L.., Barua, S., Romine, M.F., *et al.* (2007). Current production and metal oxide reduction by *Shewanella oneidensis* MR-1 wild type and mutants. Appl. Environ. Microbiol. 73, 7033–7012.

Butler, J.E., Young, N.D., and Lovley, D.R. (2009). Evolution from a respiratory ancestor to fill syntrophic and fermentative niches: comparative genomics of six *Geobacteraceae* species. BMC Genomics *10*, 103.

Carpentier, W., Sandra, K., De Smet, I., Brigé, A., De Smet, L., and Van Beeumen, J. (2003). Microbial reduction and precipitation of vanadium by *Shewanella oneidensis*. Appl. Environ. Microbiol. *69*, 3636–3639.

Charania, M.A., Brockman, K.L., Zhang, Y., Banerjee, A., Pinchuk, G.E., Fredrickson, J.K., Beliaev, A.S., and Saffarini, D.A. (2009). Involvement of a membrane-bound class III adenylate cyclase in regulation of anaerobic respiration in *Shewanella oneidensis* MR-1. J. Bacteriol. *191*, 4298–4306.

Colls, J. (2002). Air Pollution (Spon Press, London, UK).

Coppi, M.V., Leang, C., Sandler, S.J., and Lovley, D.R. (2001). Development of a genetic system for *Geobacter sulfurreducens*. Appl. Environ. Microbiol. *67*, 3180–3187.

Coursolle, D., Baron, D.B., Bond, D.R., and Gralnick, J.A. (2010). The Mtr respiratory pathway is essential for reducing flavins and electrodes in *Shewanella oneidensis*. J. Bacteriol. *192*, 467–474.

Craig, J.R., Vaughan, D.J, and Skinner, B.J. (2010). Earth Resources and the Environment, 4th edition. (Prentice Hall, New Jersey, USA).

Das, D., and Veziroglu, T. (2001). Hydrogen production by biological processes: a survey of literature. Int. J. Hydrogen Energy *26*, 13–28.

De Schamphelaire, L., van den Bossche, L., Dang, H.S., Höfte, M., Boon, N., Rabaey, K., and Verstraete, W. (2008). Microbial fuel cells generating electricity from rhizodeposits of rice plants. Environ. Sci. Technol. *42*, 3053–3058.

Dheilly, A., Linossier, I., Darchen, A., Hadjiev, D., Corbel, C., and Alonso, V. (2008). Monitoring of microbial adhesion and biofilm growth using electrochemical impedancemetry. Appl. Microbiol. Biotechnol. *79*, 157–164.

Donohue, T.J., and Cogdell, R.J. (2006). Microorganisms and clean energy. Nat. Rev. Microbiol. *4*, 800.

Donovan, C., Dewan, A., Heo, D., and Beyenal, H. (2008). Batteryless, wireless sensor powered by a sediment microbial fuel cell. Environ. Sci. Technol. *42*, 8591–8596.

Fahrénbruch, A.L., and Bube, R.H. (1983). Fundamentals of Solar Cells: Photovoltaic Solar Energy Conversion. (Academic Press, New York).

Energy Information Administration. (2007). Annual Energy Outlook 2007 with predictions to 2030. Report N. DOE/EIA-060383(2007), Department of Energy, Washington D.C.

Field, C.B., Campbell, J.E., and Lobell, D.B. (2008). Biomass energy: the scale of the potential resource. Trends Ecol. Evol. *23*, 65–72.

Flynn, T., Ghirardi, M.L., and Seibert, M. (2002). Accumulation of O_2-tolerant phenotypes in H_2-producing strains of *Chlamydomonas reinhardtii* by sequential applications of chemical mutagenesis and selection. Int. J. Hydrogen Energy *27*, 1421–1430.

Franks, A.E., Nevin, K.P., Glaven, R.H., and Lovley, D.R. (2009). Microtoming coupled to microarray analysis to evaluate the spatial metabolic status of *Geobacter sulfurreducens* biofilms. ISME J. *4*, 509–519.

Fredrickson, J.K., Romine, M.F., Beliaev, A.S., Auchtung, J.M., Driscoll, M.E., Gardner, T.S., Nealson, K.H., Osterman, A.L., Pinchuk, G., Reed, J.L., *et al.* (2008). Towards environmental systems biology of *Shewanella*. Nat. Rev. Microbiol. *6*, 592–603.

Gaffron, H., and Rubin, J. (1942). Fermentative and photochemical production of hydrogen in algae. J. Gen. Physiol., *26*, 219–240.

Ghirardi, M.L., Togasaki, R.K., and Seibert, M. (1997). Oxygen sensitivity of algal H_2 production. Appl. Biochem. Biotechnol. *63*, 141–151.

Gorby, Y.A., Yanina, S., McLean, J.S., Rosso, K.M., Moyles, D., Dohnalkova, A., Beveridge, T.J., Chang, I.S., Kim, B.H., Kim, K.S., *et al.* (2006). Electrically conductive bacterial nanowires produced by *Shewanella oneidensis* strain MR-1 and other microorganisms. Proc. Natl. Acad. Sci. U.S.A. *103*, 11358–11363.

Gray, K.A., Zhao, L., and Emptage, M. (2006). Bioethanol. Curr. Opin. Chem. Biol. *10*, 141–146.

Gust, D., Moore, T.A., and Moore, L.A. (1993). Molecular mimicry of photosynthetic energy and electron transfer. Accounts. Chem. Res. *26*, 198–205.

Happe, T., and Kaminski, A. (2002). Differential regulation of the Fe-hydrogenase during anaerobic adaptation in the green alga *Chlamydomonas reinhardtii*. Eur. J. Biochem. *269*, 1022–1032.

He, Z., Kan, J., Mansfeld, F., Angenent, L.T., and Nealson, K.H. (2009). Self-sustained phototrophic microbial fuel cells based on the synergistic cooperation between photosynthetic microorganisms and heterotrophic bacteria. Environ. Sci. Technol. *43*, 1648–1654.

Heidelberg, J.F., Paulsen, I.T., Nelson, K.E., Gaidos, E.J., Nelson, W.C., Read, T.D., Eisen, J.A., Seshadri, R., Ward, N., Methe, B., *et al.* (2002). Genome sequence of the dissimilatory metal ion-reducing bacterium *Shewanella oneidensis*. Nat. Biotechnol. *20*, 1118–1123.

Holmes, D.E., Bond, D.R., O'Neil, R.A., Reimers, C.E., Tender, L.R., and Lovley, D.R. (2004). Microbial communities associated with electrodes harvesting electricity from a variety of aquatic sediments. Microb. Ecol. *48*, 178–190.

Holmes, D.E., Chaudhuri, S.K., Nevin, K.P., Mehta, T., Methé, B.A., Liu, A., Ward, J.E., Woodard, T.L., Webster, J., and Lovley, D.R. (2006). Microarray and genetic analysis of electron transfer to electrodes in *Geobacter sulfurreducens*. Environ. Microbiol. *8*, 1805–1815.

Inoue, K., Qian, X., Morgado, L., Kim, B.C., Mester, T., Izallalen, M., Salgueiro, C.A., and Lovley, D.R. (2010). Purification and characterization of OmcZ, an outer-surface, octaheme c-type cytochrome essential for optimal current production by *Geobacter sulfurreducens*. Appl. Environ. Microbiol. *76*, 3999–4007.

IPCC. (2007). Summary for Policymakers. Climate Change 2007: The Physical Science Basis. Contribution of Working Group I to the Fourth Assessment Report of the Intergovernmental Panel on Climate Change. (ICPP, Geneva, Switzerland).

Ishii, S., Shimoyama, T., Hotta, Y., and Watanabe, K. (2008a). Characterization of a filamentous biofilm community established in a cellulose-fed microbial fuel cell. BMC Microbiol. *8*, 6.

Ishii, S., Hotta, Y., and Watanabe, K. (2008b). Methanogenesis versus electrogenesis: morphological and phylogenetic comparisons of microbial communities. Biosci. Biotechnol. Biochem. *72*, 286–294.

Ishii, S., Watanabe, K., Yabuki, S., Logan, B.E., and Sekiguchi, Y. (2008c). Comparison of electrode reduction activities of *Geobacter sulfurreducens* and an enriched consortium in an air-cathode microbial fuel cell. Appl. Environ. Microbiol. *74*, 7348–7355.

Juárez, K., Kim, B.C., Nevin, K., Olvera, L., Reguera, G., Lovley, D.R., and Methé, B.A. (2009). PilR, a transcriptional regulator for pilin and other genes required for Fe(III) reduction in *Geobacter sulfurreducens*. J. Mol. Microbiol. Biotechnol. *16*, 146–158.

Jung, S., and Regan, J.M. (2007). Comparison of anode bacterial communities and performance in microbial fuel cells with different electron donors. Appl. Microbiol. Biotechnol. *77*, 393–402.

Kaku, N., Yonezawa, N., Kodama, Y., and Watanabe, K. (2008). Plant/microbe cooperation for electricity generation in a rice paddy field. Appl. Microbiol. Biotechnol. *79*, 43–49.

Kiley, P.J., and Beinert, H. (1998). Oxygen sensing by the global regulator, FNR: the role of the iron–sulfur cluster. FEMS Microbiol. Rev. *22*, 341–352.

Kim, B.H., Kim, H.J., Moon, M.S., and Park, D.H. (1999). Direct electrode reaction of Fe(III)-reducing bacterium *Shewanella putrefaciens*. J. Microbiol. Biotechnol. *9*, 127–131.

Kouzuma, A., Meng, X.Y., Kimura, N., Hashimoto, K., and Watanabe, K. (2010). Disruption of the putative cell surface polysaccharide biosynthesis gene SO3177 in *Shewanella oneidensis* MR-1 enhances adhesion to electrodes and current generation in microbial fuel cells. Appl. Environ. Microbiol. *76*, 4151–4157.

Larsen, K.S., Ibrom, A., Beier, C., Jonasson, S., and Michelsen, A. (2007). Ecosystem respiration depends strongly on photosynthesis in a temperate heath. Biogeochem. *85*, 201–213.

Lashof, D.A., and Ahuja, D.R. (1990). Relative contributions of greenhouse gas emissions to global warming. Nature *344*, 529 –531.

Leang, C., Qian, X., Mester, T., and Lovley, D.R. (2010). Alignment of the *c*-type cytochrome OmcS along pili of *Geobacter sulfurreducens*. Appl. Environ. Microbiol. *76*, 4080–4084.

Lee, H.S., Parameswaran, P., Kato-Marcus, A., Torres, C.I., and Rittmann, B.E. (2008). Evaluation of energy-conversion efficiencies in microbial fuel cells (MFCs) utilizing fermentable and non-fermentable substrates. Water. Res. *42*, 1501–1510.

Lettinga, G. (1995). Anaerobic digestion and wastewater treatment systems. Antonie Van Leeuwenhoek *67*, 3–28.

Levin, D.B., Pitt, L., and Love, M. (2004). Biohydrogen production: prospects and limitations to practical application. Int. J. Hydrogen Energy *29*, 173–185.

Lies, D.P., Hernandez, M.E., Kappler, A., Mielke, R.E., Gralnick, J.A., and Newman, D.K. (2005). *Shewanella oneidensis* MR-1 uses overlapping pathways for iron reduction at a distance and by direct contact under conditions relevant for biofilms. Appl. Environ. Microbiol. *71*, 4414–4426.

Liu, C., Gorby, Y.A., Zachara, J.M., Fredrickson, J.K., and Brown, C.F. (2002). Reduction kinetics of Fe(III), Co(III), U(VI), Cr(VI), and Tc(VII) in cultures of dissimilatory metal-reducing bacteria. Biotechnol. Bioeng. *80*, 637–649.

Liu, H., and Logan, B.E. (2004). Electricity generation using an air-cathode single chamber microbial fuel cell in the presence and absence of a proton exchange membrane. Environ. Sci. Technol. *38*, 4040–4066.

Logan, B.E., Hamelers, B., Rozendal, R., Schröder, U., Keller, J., Freguia, S., Aelterman, P., Verstraete, W., and Rabaey, K. (2006). Microbial fuel cells: methodology and technology. Environ. Sci. Technol. *40*, 5181–5192.

Lovley, D.R., Holmes, D.E., and Nevin, K.P. (2004). Dissimilatory Fe(III) and Mn(IV) reduction. Adv. Microb. Physiol. *49*, 219–286.

Lovley, D.R. (2006). Bug juice: harvesting electricity with microorganisms. Nat. Rev. Microbiol. *4*, 497–508.

Lovley, D.R. (2008). The microbe electric: conversion of organic matter to electricity. Curr. Opin. Biotechnol. *19*, 564–571.

Lower, B.H., Shi, L., Yongsunthon, R., Droubay, T.C., McCready, D.E., and Lower, S.K. (2007). Specific bonds between an iron oxide surface and outer membrane cytochromes MtrC and OmcA from *Shewanella oneidensis* MR-1. J. Bacteriol. *189*, 4944–4952.

Malik, S., Drott, E., Grisdela, P., Lee, J., Lee, C., Lowy, D.A., Gray, S., and Tender, L.M. (2009). A self-assembling self-repairing microbial photoelectrochemical solar cell. Energy. Environ. Sci. 2, 292–298.

Marsili, E., Baron, D.B., Shikhare, I.D., Coursolle, D., Gralnick, J.A., and Bond, D.R. (2008a). *Shewanella* secretes flavins that mediate extracellular electron transfer. Proc. Natl. Acad. Sci. U.S.A. *105*, 3968–3973.

Marsili, E., Rollefson, J.B., Baron, D.B., Hozalski, R.M., and Bond, D.R. (2008b). Microbial biofilm voltammetry: direct electrochemical characterization of catalytic electrode-attached biofilms. Appl. Environ. Microbiol. *74*, 7329–7337.

Melis, A. (2002). Green alga hydrogen production: progress, challenges and prospects. Int. J. Hydrogen Energy 27, 1217–1228.

Melis, A., Zhang, L., Forestier, M., Ghirardi, M.L., and Seibert, M. (2000). Sustained photobiological hydrogen gas production upon reversible inactivation of oxygen evolution in the green alga *Chlamydomonas reinhardtii*. Plant Physiol. *122*, 127–135.

Methe, B.A., Nelson, K.E., Eisen, J.A., Paulsen, L.T., Nelson, W., Heidelberg, J.F., Wu, D., Wu, M., Ward, N., Beanan, M.J., *et al.* (2003). Genome of *Geobacter sulfurreducens:* Metal reduction in subsurface environments. Science *302*, 1967–1969.

Myers, C.R., and Myers, J.M. (1997a). Cloning and sequence of *cymA*, a gene encoding a tetraheme cytochrome *c* required for reduction of iron(III), fumarate, and nitrate by *Shewanella putrefaciens* MR-1. J. Bacteriol. *179*, 1143–1152.

Myers, C.R., and Myers, J.M. (1997b). Outer membrane cytochromes of *Shewanella putrefaciens* MR-1: spectral analysis, and purification of the 83-kDa *c*-type cytochrome. Biochim. Biophys. Acta. *1326*, 307–318.

Myers, C.R., and Myers, J.M. (2002). MtrB is required for proper incorporation of the cytochromes OmcA and OmcB into the outer membrane of *Shewanella putrefaciens* MR-1. Appl. Environ. Microbiol. *68*, 5585–5594.

Myers, C.R., and Myers, J.M. (2003a). Cell surface exposure of the outer membrane cytochromes of *Shewanella oneidensis* MR-1. Lett. Appl. Microbiol. *37*, 254–258.

Myers, C.R., and Myers, J.M. (2004). The outer membrane cytochromes of *Shewanella oneidensis* MR-1 are lipoproteins. Lett. Appl. Microbiol. *39*, 466–470.

Myers, J.M., and Myers, C.R. (2000). Role of the tetraheme cytochrome CymA in anaerobic electron transport in cells of *Shewanella putrefaciens* MR-1 with normal levels of menaquinone. J. Bacteriol. *182*, 67–75.

Myers, J.M., and Myers, C.R. (2003b). Overlapping role of the outer membrane cytochromes of *Shewanella oneidensis* MR-1 in the reduction of manganese(IV) oxide. Lett. Appl. Microbiol. *37*, 21–25.

Myers, C.R., and Nealson, K.H. (1988). Bacterial manganese reduction and growth with manganese oxide as the sole electron acceptor. Science *240*, 1319–1321.

Nealson, K.H., and Saffarini, D. (1994). Iron and manganese in anaerobic respiration: environmental significance, physiology, and regulation. Annu. Rev. Microbiol. *48*, 311–343.

Nevin, K.P., and Lovley, D.R. (2000). Lack of production of electron-shuttling compounds or solubilization of Fe(III) during reduction of insoluble Fe(III) oxide by *Geobacter*. Appl. Environ. Microbiol. *66*, 2248–2251.

Nevin, K.P., and Lovley, D.R. (2002). Mechanisms for Fe(III) oxide reduction in sedimentary environments. Geomicrobiol. J. *19*, 141–159.

Nevin, K.P., Richter, H., Covalla, S.F., Johnson, J.P., Woodard, T.L., Orloff, A.L., Jia, H., Zhang, M., and Lovley, D.R. (2008). Power output and columbic efficiencies from biofilms of *Geobacter sulfurreducens* comparable to mixed community microbial fuel cells. Environ. Microbiol. *10*, 2505–2514.

Nevin, K.P., Kim, B.C., Glaven, R.H., Johnson, J.P., Woodard, T.L., Methé, B.A., Didonato, R.J., Covalla, S.F., Franks, A.E., Liu, A., and Lovley,

D.R. (2009). Anode biofilm transcriptomics reveals outer surface components essential for high density current production in *Geobacter sulfurreducens* fuel cells. PLoS One *20*, e5628.

Nießen, J., Schröder, U., Rosenbaum, M., and Scholz, F. (2004). Fluorinated polyanilines as superior materials for electrocatalytic anodes in bacterial fuel cells. Electrochem. Commun. *6*, 571–575.

Nishio, K., Hashimoto, K., and Watanabe, K. (2010). Light/electricity conversion by a self-organized photosynthetic biofilm in a single-chamber reactor. Appl. Microbiol. Biotechnol. *86*, 957–964.

Oh, S., Min, B., and Logan, B.E. (2004). Cathode performance as a factor in electricity generation in microbial fuel cells. Environ. Sci. Technol. *38*, 4900–4904.

Picioreanu, C., Head, I.M., Katuri, K.P., van Loosdrecht, M.C., and Scott, K. (2007). A computational model for biofilm-based microbial fuel cells. Water Res. *41*, 2921–2940.

Pitts, K.E, Dobbin, P.S., Reyes-Ramirez, F., Thomson, A.J., Richardson, D.J., and Seward, H.E. (2003). Characterization of the *Shewanella oneidensis* MR-1 decaheme cytochrome MtrA: expression in *Escherichia coli* confers the ability to reduce soluble Fe(III) chelates. J. Biol. Chem. *278*, 27758–27765.

Prince, R.C., and Kheshgi, H.S. (2005). The photobiological production of hydrogen: potential efficiency and effectiveness as a renewable fuel. Crit. Rev. Microbiol. *31*, 19–31.

Reguera, G., McCarthy, K.D., Mehta, T., Nicoll, J.S., Tuominen, M.T., and Lovley, D.R. (2005). Extracellular electron transfer via microbial nanowires. Nature 435, 1098–1101.

Reguera, G., Nevin, K.P., Nicoll, J.S., Covalla, S.F., Woodard, T.L., and Lovley, D.R. (2006). Biofilm and nanowire production leads to increased current in *Geobacter sulfurreducens* fuel cells. Appl. Environ. Microbiol. 72, 7345–7348.

Reguera, G., Pollina, R.B., Nicoll, J.S., and Lovley, D.R. (2007). Possible nonconductive role of *Geobacter sulfurreducens* pilus nanowires in biofilm formation. J. Bacteriol. *189*, 2125–2127.

Reimers, C.E., Tender, L.M., Fertig, S., and Wang, W. (2001). Harvesting energy from the marine sediment–water interface. Environ. Sci. Technol. *35*, 192–195.

Richter, H., Nevin, K.P., Jia, H.F., Lowy, D.A., Lovley, D.R., and Tender, L.M. (2009). Cyclic voltammetry of biofilms of wild type and mutant *Geobacter sulfurreducens* on fuel cell anodes indicates possible roles of OmcB, OmcZ, type IV pili, and protons in extracellular electron transfer. Energ. Environ. Sci. *2*, 506–516.

Rosenbaum, M., Schröder, U., and Scholz, F. (2005). Utilizing the green alga *Chlamydomonas reinhardtii* for microbial electricity generation: a living solar cell. Appl. Microbiol. Biotechnol. *68*, 753–756.

Ross, D.E., Ruebush, S.S., Brantley, S.L., Hartshorne, R.S., Clarke, T.A., Richardson, D.J., and Tien, M. (2007). Characterization of protein–protein interactions involved in iron reduction by *Shewanella oneidensis* MR-1. Appl. Environ. Microbiol. *73*, 5797–5808.

Ross, D.E., Brantley, S.L., and Tien, M. (2009). Kinetic characterization of OmcA and MtrC, terminal reductases involved in respiratory electron transfer for dissimilatory iron reduction in *Shewanella oneidensis* MR-1. Appl. Environ. Microbiol. *75*, 5218–5226.

Saffarini, D.A., Schultz, R., and Beliaev, A. (2003). Involvement of cyclic AMP (cAMP) and cAMP receptor protein in anaerobic respiration of *Shewanella oneidensis*. J. Bacteriol. *185*, 3668–3671.

Schwalb, C., Chapman, S.K., and Reid, G.A. (2003). The tetraheme cytochrome CymA is required for anaerobic respiration with dimethyl sulfoxide and nitrite in *Shewanella oneidensis*. Biochemistry *42*, 9491–9497.

Shi, L., Chen, B., Wang, Z., Elias, D.A., Mayer, M.U., Gorby, Y.A., Ni, S., Lower, B.H., Kennedy, D.W., Wunschel, D.S., *et al.* (2006). Isolation of a high-affinity functional protein complex between OmcA and MtrC: Two outer membrane decaheme *c*-type cytochromes of *Shewanella oneidensis* MR-1. J. Bacteriol. *188*, 4705–4714.

Shi, L., Squier, T.C., Zachara, J.M., and Fredrickson, J.K. (2007). Respiration of metal (hydr)oxides by *Shewanella* and *Geobacter*: a key role for multihaem *c*-type cytochromes. Mol. Microbiol. *65*, 12–20.

Shi, L., Deng, S., Marshall, M.J., Wang, Z., Kennedy, D.W., Dohnalkova, A.C., Mottaz, H.M., Hill, E.A., Gorby, Y.A., Beliaev, A.S., *et al.* (2008). Direct involvement of type II secretion system in extracellular translocation of *Shewanella oneidensis* outer membrane cytochromes MtrC and OmcA. J. Bacteriol. *190*, 5512–5516.

Shimoyama, T., Komukai, S., Yamazawa, A., Ueno, Y., Logan, B.E., and Watanabe, K. (2008). Electricity generation from model organic wastewater in a cassette-electrode microbial fuel cell. Appl. Microbiol. Biotechnol. *80*, 325–330.

Shinnar, R., and Citro, F. (2006). A road map to U.S. decarbonization. Science *313*, 1243–1244.

Stal, L.J., van Gemerden, H., Krumbein, W. E. (1985). Structure and development of a benthic marine microbial mat. FEMS Microbiol. Ecol. *31*, 111–125.

Strik, D.P.B.T.B., Hamelers, H.V.M., Snel, J.F.H., and Buisman, C.J.N. (2008). Green electricity production with living plants and bacteria in a fuel cell. Int. J. Energy Res. 32, 870–876.

Takanezawa, K., Nishio, K., Kato, S., Hashimoto, K., and Watanabe, K. (2010). Factors affecting electric output from rice-paddy microbial fuel cell. Biosci. Biotechnol. Biochem. *74*, 1271–1273

Tanaka, K., Kashiwagi, N., Ogawa, T. (1988). Effects of light on the electrical output of bioelectrochemical fuel-cells containing *Anabaena variabilis* M-2: mechanism of the post-illumination burst. J. Chem. Technol. Biotechnol. *42*, 235–240.

Tender, L.M., Reimers, C.E., Stecher, H.A., Holmes, D.E., Bond, D.R., Lowy, D.A., Pilobello, K., Fertig, S.J., and Lovley, D.R. (2002). Harnessing microbially generated power on the seafloor. Nat. Biotechnol. *20*, 821–825.

Tsujimura, S., Wadano, A., Kano, K., and Ikeda, T. (2001). Photosynthetic bioelectrochemical cell utilizing cyanobacteria and water-generating oxidase. Enzyme Microb. Technol. *29*, 225–231.

Von Canstein, H., Ogawa, J., Shimizu, S., and Lloyd, J.R. (2008). Secretion of flavins by *Shewanella* species and their role in extracellular electron transfer. Appl. Environ. Microbiol. *74*, 615–623.

Watanabe, K. (2008). Recent developments in microbial fuel cell technologies for sustainable bioenergy. J. Biosci. Bioeng. *106*, 528–536.

Watanabe, K., Manefield, M., Lee, M., and Kouzuma, A. (2009). Electron shuttles in biotechnology. Curr. Opin. Biotechnol. *20*, 633–641.

Xiong, Y., Shi, L., Chen, B., Mayer, M.U., Lower, B.H., Londer, Y., Bose, S., Hochella, M.F., Fredrickson, J.K., and Squier, T.C. (2006). High-affinity binding and direct electron transfer to solid metals by the *Shewanella oneidensis* MR-1 outer membrane *c*-type cytochrome OmcA. J. Am. Chem. Soc. *128*, 13978–13979.

Yagishita, T., Horigome, T., and Tanaka, K. (1993a). Biofuel-cells containing photosynthetic microorganisms. J. Electrochem. Soc. Jpn. *61*, 687–688.

Yagishita, T., Horigome, T., and Tanaka, K. (1993b). Effects of light, CO_2, and inhibitors on the current output of biofuel cells containing the photosynthetic organism *Synechococcus* sp. J. Chem. Technol. Biotechnol. *56*, 393–399.

Yagishita, T., Sawayama, S., Tsukahara, K., and Ogi, T. (1996). Photosynthetic bio-fuel cells using cyanobacteria. Renewable Energy *9*, 958–961.

Yagishita, T., Sawayama, S., Tsukahara, K., and Ogi, T. (1997). Effects of intensity of incident light and concentrations of *Synechococcus* sp. and 2-hydroxy-1,4-naphthoquinone on the current output of photosynthetic electrochemical cell. Solar Energy *61*, 347–353.

Yagishita, T., Sawayama, S., Tsukahara, K., and Ogi, T. (1998). Performance of photosynthetic electrochemical cells using immobilized *Anabaena variabilis* M-3 in discharge/culture cycles. J. Ferment. Bioeng. *85*, 546–549.

Yagishita, T., Sawayama, S., Tsukahara, K., and Ogi, T. (1999). Effects of glucose addition and light on current outputs in photosynthetic electrochemical cells using *Synechocystis* sp. PCC6714. J. Biosci. Bioeng. *88*, 210–214.

11 Catalytic Biofilms: A Powerful Concept for Future Bioprocesses

Rainer Gross, Andreas Schmid and Katja Buehler

Abstract

Biofilms are mainly known for causing problems in medical and industrial settings owing to their persistence towards treatment with bactericides, including antibiotics. However, in the area of bioremediation they are widely recognized for their ability to degrade hazardous or organic compounds to CO_2 and biomass. Biofilms represent a highly interesting biological concept since they unite important characteristics such as the ability of self-immobilization and increased robustness to various physical, chemical and biological stressors, which make them exceedingly attractive for productive catalysis.

The following review provides a detailed survey of biofilm applications for productive biocatalysis on laboratory-, pilot-, and industrial scales, regarding fermentation as well as biotransformation reactions. It discusses technological as well as biological challenges of biofilm driven catalysis, presenting developments in the field of biofilm reactor technology and the latest findings in understanding biofilm dynamics. Biocatalysis related issues like genetic stability, evolution, uncontrolled growth as well as detachment, contamination risks, monitoring of biomass, EPS, chemical and biological heterogeneity are considered.

Introduction

For decades, biofilms have been successfully exploited for bioremediation purposes. Microbial communities have been used in 'biofilters', to convert problematic substances to CO_2 and biomass. During the 1980s a number of publications emerged, describing the use of biofilms as naturally immobilized biocatalysts for the continuous fermentation of bulk chemicals such as acetic acid (vinegar) and ethanol (Crueger *et al.*, 1990; Qureshi *et al.*, 2005). The two main advantages of biofilm applications are (i) the possibility to produce large volumes of the desired end products due to the high cell densities prevailing in these communities and (ii) their 'robustness', allowing for continued production over several months, which are unrivalled by stirred tank processes with planktonic cells (Weuster-Botz *et al.*, 1993; Kunduru *et al.*, 1996; Qureshi *et al.*, 2004; van Groenestijn *et al.*, 2009). On the downside, a number of disadvantages have to be tackled, such as insufficient growth control which can lead to excessive growth and eventually to reactor obstruction (Qureshi *et al.*, 2005; Syron *et al.*, 2008). Furthermore, mass transfer limitations can create chemical and physiological heterogeneities, lowering the reactor performance (Stewart, 2003; Stewart *et al.*, 2008).

Despite these hurdles, recent studies have begun to explore the potential of biofilms for the synthesis of complex fine chemicals (Li *et al.*, 2006; Gross *et al.*, 2007), reactions which often suffer from the classical bottlenecks of insufficient process stability and substrate/product toxicity leading to low productivities and short process durations in planktonic bioreactor cultures (Bommarius *et al.*, 2004; Woodley, 2006a,b). New developments in biofilm based technologies are profiting from existing knowledge in the field of bioremediation, especially regarding reactor design and operation. This chapter introduces established biocatalytic concepts and points out the prevailing bottlenecks in chemical biocatalysis, for which some biofilm-based processes might provide a solution. Biofilm biology in the context of biocatalysis is discussed, and the challenges and unsolved problems, which arise from biofilm-based technologies is presented.

Whole-cell bioprocesses – state of the art of science and technology

Industrial bioprocesses utilize the enormous catalytic potential of organisms, either employing isolated enzymes or whole (predominantly microbial) cells, for the production of a wide variety of products ranging from high-priced complex pharmaceuticals with several chiral centres, to commodity chemicals such as amino acids, and low-value bulk chemicals like alcohols and organic acids (Patel, 2006; Buchholz *et al.*, 2008; Woodley, 2008; Wohlgemuth, 2009). This chapter exclusively discusses processes applying whole cells as biocatalysts.

Currently, there are about 150 reported biocatalysis processes running on an industrial scale with annual production ranging from 1 to several 100,000 tons, and employing various process concepts (Schmid *et al.*, 2001; Straathof *et al.*, 2002; Bommarius *et al.*, 2004; Liese *et al.*, 2006; Woodley, 2006a,b). These are based either on living cells or isolated enzymes, depending on the nature of the catalyst, the target reaction, and the substrate and product. The advantages and disadvantages of various biocatalysis-based approaches are summarized in Table 11.1. Of the 134 processes reviewed by Straathof *et al.* (2002), 62% utilize whole cells as biocatalysts. Whole cells are often preferred over isolated enzymes, as the necessary enzyme purification significantly adds to the overall process costs. They are mainly applied in processes involving instable, multicomponent, and membrane bound enzymes, or multistep and/or cofactor requiring reactions (Gbewonyo *et al.*, 1991; Murdock *et al.*, 1993; Buckland, 1999; Doig *et al.*, 2002; Zhao *et al.*, 2003; Walton *et al.*, 2004). This group comprises some enzyme classes with considerable potential for industrial applications (Faber, 2004), which often catalyse complex synthetically useful reactions such as stereoselective carbon–carbon bond formations (transferases) or selective redox reactions (oxidoreductases) (Woodley, 2006a; Pollard *et al.*, 2007). In whole cell biocatalysts, cellular metabolism provides a cheap cofactor regeneration system and continuous enzyme synthesis, while the cell membrane protects the enzymes from denaturing extracellular influences.

An elegant approach utilizes whole cells for multistep catalysis, where either existing pathways, modified pathways, or *de novo* pathways are used in a 'cell factory' (Woodley, 2006b). Thus the expense of multiple protein separation and a complex *in vitro* reaction system with various catalysts working at different reaction conditions is avoided. A successful example of multistep biocatalysis by whole cells is the synthesis of vitamin B_{12} involving 12 recombinant enzymes working together sequentially (Scott 2003). Such systems resemble 'bags stuffed with catalysts' (Ishige *et al.*, 2005), but most importantly require the host's metabolism.

In general, two main strategies in whole cell biocatalysis may be distinguished, namely fermentation and biotransformation (Fig. 11.1).

During a fermentative process, the product of choice is derived directly from the carbon source, while in a biotransformation reaction the product is derived via the conversion of an additional substrate, which cannot be utilized as a carbon source for cell growth (Stephanopoulos *et al.*, 1998). Hence, fermentative processes necessitate growing cells, while non-growing cells are commonly applied for biotransformations (Woodley, 2006a,b). In the latter case, either resting but metabolically active cells or dead (permeabilized) cells can be used. Approaches based on resting cells profit from a low degree of competing reactions, better process control and easier product recovery (Walton *et al.*, 2004).

Downstream processing is one of the bottlenecks in biocatalysis, and is especially challenging, when product titres are low. One possibility to increase the final product yield is biocatalyst immobilization, which allows high cell densities in a continuous mode of operation to produce

Table 11.1 Advantages and disadvantages of different bioprocess approaches

Nature of catalyst	Isolated enzymes	Whole cells
Pros	Easier control of kinetics to achieve high space–time yields, high mass transfer rates possible, no side reactions, thus lower downstream costs	Cellular envelope provides a protected environment, higher stability of biocatalyst, 'cell factory' allows multistep conversions, cellular co-factor regeneration, metabolic pathways, use of cheap and renewable precursors in fermentative processes
Cons	Not feasible for several enzyme types (membrane bound, multicomponent, cofactor-dependent), or multistep reactions, lower catalyst stability, higher preparation (upstream) costs	Side reactions and/or product degradation may occur, higher downstream costs, substrate and product transfer across cell membrane, high oxygen requirements, toxic effects of substrates, products, and/or solvents on host
Catalyst application	**Free biocatalyst**	**Immobilized biocatalyst**
Pros	Less or no mass transfer limitation, lower upstream costs	Multiple use of catalysts, high product/catalyst ratios, higher catalyst densities allow higher volumetric productivities, inexpensive downstream processing, increased biocatalyst stability
Cons	Loss of catalyst during continuous operation, higher downstream costs	Preparation costs (upstream), mass transfer limitations may occur, loss of catalytic activity may occur
Mode of operation	**Repeated batch/fed batch**	**Continuous**
Pros	Less expensive equipment, relatively simple reactor configuration, high cell densities achievable without immobilization, uncoupled downstream processing	Long-term process without interruptive reactor downtimes, reactions under steady state conditions, ease of operation, optimized energy consumption, constant product quality, smaller reactor volumes
Cons	Recurrent reactor downtimes increase maintenance costs, lower space-time yields	Expensive equipment, automated instrumentation and process control, integrated downstream processing technology necessary, availability of stable biocatalysts, high volumetric productivities need catalyst retention/immobilization

high space time yields [usually given as grams of product produced per unit reactor volume (L) and running time (h)], by reusing the immobilized catalyst. Furthermore, growth and bioconversion can be separated and thus the reaction conditions may be optimally set, disregarding growth (Woodley, 2006a).

On the reaction level, immobilization implies a change from a homogeneous to a heterogeneous format and increases the complexity of the reaction system (Buchholz *et al.*, 2005). Kinetics and yields change from suspended to immobilized biocatalysts, so that for V'_{max} (the maximum catalytic rate that can be achieved by a particular enzyme) and k'_M (the substrate concentration at which $\frac{1}{2}V'_{max}$ is achieved), the active cell fraction has to be identified. The reaction rate is not only determined by the catalytic properties but also by the mass transfer to, from, and inside the immobilized biocatalysts, and by the micro- and nano-environment (pH, ionic strength on a micrometre and nanometre scale) (Buchholz *et al.*, 2005). Furthermore, immobilization requires additional preparation steps and immobilization chemicals, which add to the overall costs. The main areas for the application of immobilized microorganisms at a large scale are complex transformations that require the regeneration

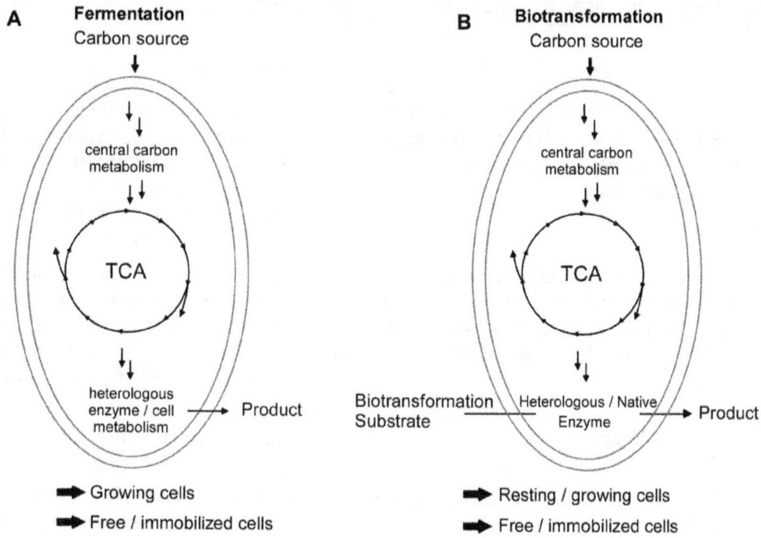

Figure 11.1 Different application concepts in whole cell biocatalysis. Organisms may either be applied in fermentations, where the product of choice is derived directly from the carbon source used for cell growth (A), or in biotransformations where an additional biotransformation substrate is supplied in addition to the carbon source (B).

of enzymes and cofactors and/or cell growth. These include wastewater treatment, notably with anaerobic cultures, and exhaust gas purification (Fujita *et al.*, 1991; Qureshi *et al.*, 2005; Kornaros *et al.*, 2006).

The major technique for cell immobilization is the entrapment of the organisms into a 'hydrogel', such as calcium alginate or χ-carrageenan (reviewed in Smidsrød *et al.*, 1990, and Salter *et al.*, 1991). However, in whole cell catalysis, immobilization seems to be less frequently used compared with isolated enzyme catalysis (Straathof *et al.*, 2002). This might be due to drawbacks including the lack of a universal entrapment technology, intricate preparation steps adding to the overall costs, mass transfer limitations in the immobilization matrix, and possibly the reduced viability of entrapped organisms (Webb *et al.*, 1996; Bickerstaff, 1997). In addition, many hydrogel matrixes are instable in the presence of organic solvents, or disintegrate due to abrasion (Smidsrød *et al.*, 1990). Other techniques use polymers in different configurations (e.g. beads, fibres or membranes) to immobilize living cells either by entrapment or crosslinking (Salter *et al.* 1991). In recent years, a new technology emerged, which utilizes hybrid organic–inorganic materials, so-called sol-gels for the entrapment of whole cells (Avnir *et al.*, 2006; Desimone *et al.*, 2009). Adsorption and adhesion (also flocculation and/or aggregation) can occur naturally without the help of any immobilizing chemicals and may be more economical to use than more engineered, artificial systems. In fact, the colonization of surfaces is an ubiquitous microbial phenomenon and a very successful strategy of survival for bacteria in natural settings (Costerton *et al.*, 1987). Biofilms have been applied for many years for the treatment of wastewaters, off-gases from industrial plants, and in the detoxification of contaminated soils (Nicolella *et al.*, 2000b; Peacock *et al.*, 2004; Qureshi *et al.*, 2005; Singh *et al.*, 2006). In these areas a wealth of knowledge and experience in the handling of biofilms has been collected. Industrially established biofilm-based production processes operating at large scale are scarce, except for the food sector, where acetic acid (vinegar) and ethanol (beer) are still commonly produced with traditional biofilm based processes.

An early report about the production of acetic acid using *Acetobacter* spp. attached to wood chips was published in 1847 (Knapp, 1847, cited in Buchholz *et al.*, 2005) and even today 60-m^3 reactors filled with curled beechwood shavings are used as so-called trickling generators with ethanol and oxygen as substrates (Crueger *et al.*, 1990). The use of immobilized yeast cells has been widely used for beer production and the influence of immobilized cell technology on the production of flavour-active compounds, such as higher alcohols, esters, and vicinal diketones has been investigated (Buchholz *et al.*, 2005; Willaert *et al.*, 2006). Using cells bound to replaceable carrier beads of various materials with a high specific surface area (such as porous glass, diatomaceous earth, wood chips, or silicon carbide rods), the continuous production of beer becomes financially and technically feasible compared with batch fermentation with suspended yeast cells (Willaert *et al.*, 2006).

Several studies reported further beneficial aspects of biofilms for industrial applications. Increased robustness against adverse conditions such as toxic chemicals, dessication, extreme temperatures and pH has been observed (Flemming, 1993; Costerton *et al.*, 1995). In particular, observations of enhanced tolerance to ethanol for immobilized yeast (Ciesarova *et al.*, 1998; Jirku, 1999) and *E. coli* (Zhou *et al.*, 2008) compared with the freely suspended counterparts make biofilms interesting alternative biocatalysts for a much broader range of uses in industrial biotechnology.

Another approach to deal with low product titres is the integration of *in situ* product extraction into the process, e.g. by the addition of an organic liquid phase. This is also commonly applied to meet challenges regarding substrate and product toxicity, and hydrophobicity. The two-phase approach is based on adding a water immiscible organic phase to the aqueous phase reaction system. It may simultaneously serve as a source of substrates and as a sink for the generated products depending on their chemical characteristics. Substrate concentrations can be maintained below inhibitory levels while still being bioavailable and products are continuously extracted into the second phase, thereby also preventing inhibition and toxicity (Woodley and Lilly, 1990; Leon *et al.*, 1998; Heipieper *et al.*, 2007). Furthermore, the continuous extraction of the product from the reaction phase shifts the chemical equilibrium, driving the reaction to completion (de Bont, 1998; Leon *et al.*, 1998; Heipieper *et al.*, 2007). On the downside, the viability of the respective host organisms might be affected due to solvent stress, depending on the nature of the organic phase (Vansonsbeek *et al.*, 1993; Heipieper *et al.*, 1994). Therefore the two-liquid phase approach has so far mostly been applied to isolated enzymes, mainly lipases, rather than to whole-cell catalysed reactions (Liese *et al.*, 2006).

Overall, insufficient biocatalyst stability is still one of the major drawbacks for bioprocess implementation and the main reason why most bioprocesses continue to be run in batch or fed-batch mode in industry. Continuous processes are aspired, because the catalyst costs per kilogram product can be reduced significantly below the price achieved with conventional systems (Wöltinger *et al.*, 2005). Utilizing the advantages of biofilm catalysts which include self-immobilization, greater robustness, solvent tolerance, and long-term activity compared with planktonic cells, we may overcome limitations typical of biocatalysis and allow continuous processing combined with high process stabilities.

Catalytic biofilms: expanding the toolbox of biocatalysis

In order to extensively utilize the catalytic power of biofilms for productive synthesis it is necessary to understand their biology. Biotechnologists can learn a lot about these microbial communities from studies conducted in the medical area, although the focus here is on the prevention, rather than on the utilization of biofilms. Biofilm research has begun to reveal the various protective functions a matrix embedded lifestyle provides for the respective microbes, and it is obvious that those resistance mechanisms can be exploited to stabilize biocatalytic processes. In fact, self-immobilized biocatalysts at high cell densities, which are protected against harsh environmental reaction reconditions are the ideal candidates for biocatalysis as they would solve one of the main challenges, namely limited process stability. They could enable long-term continuous operation at high volumetric productivities. This section will therefore focus on some areas of biofilm biology, which are interesting from a biotechnological point of view.

Biofilms are sessile microbial communities living as multicellular aggregates attached to any kind of surface at the interface of solid/liquid or liquid/gaseous media. There are a number of abiotic and biological factors influencing the biofilm performance in the context of biocatalysis (Fig. 11.2). Compared with their planktonic counterparts biofilms are known to exhibit altered specific gene expression and growth rates. The organisms are embedded in a matrix of so called extracellular polymeric substances (EPS), which they excrete in the course of biofilm formation. This substance works as a kind of glue keeping the whole structure together and giving the biofilm its three dimensional structure. It is also one of the most important factors regarding the robustness of biofilms (Donlan, 2002; Monds and O'Toole, 2009; Moons *et al.*, 2009). The EPS comprises mainly polysaccharides, proteins and extracellular DNA which form highly hydrated hydrogel matrices accounting roughly for 50–90% of the total biomass (Wingender *et al.*, 1999; Branda *et al.*, 2005).

The robustness against harsh conditions has successfully enabled microbial biofilms to prevail not only in nearly all natural environments but also in industrial settings with various implications, as well as in the human health sector (Hall-Stoodley *et al.*, 2004; Anderson and O'Toole, 2008). A crucial problem for biofilm combat is their recalcitrance against treatment with biocides, including antibiotics, which can be increased up to 1000-fold compared with their planktonic counterparts (Nickel *et al.*, 1985; Costerton and Stewart, 2001; O'Toole and Stewart, 2005; Bryers, 2008). In addition to being resistant to antibiotics, pathogenic bacterial biofilms responsible for the majority of hospital-acquired infections (e.g. *Staphylococcus aureus* and *Pseudomonas aeruginosa*) have been observed to evade the immune system by resisting phagocytosis (see Chapter 2), making infection control and treatment extremely difficult (Leid *et al.*, 2002).

In the industrial sector, all problems related to undesired biofilm growth are described as 'biofouling' (Flemming, 1993, 2002; Paulus, 2005). When biofilms develop on ship hulls, in industrial pipe systems and other kind of fluid flow systems, they increase frictional resistance, which may lead to substantial pressure drops and/or increase in energy consumption (Flemming, 2002). In water cooling systems they decrease heat energy transfer while increasing fluid frictional resistance, and may accelerate rates of metallic corrosion (Characklis *et al.*, 1990; Videla *et al.*, 1992). In addition, biofilms provide a habitat for potentially pathogenic microorganisms, which leads to severe problems in drinking water distribution systems posing health hazards to the consumer (Flemming, 2002; Paulus, 2005).

The outstanding robustness of biofilms is responsible for many biofilm related problems but on the other hand may provide the biggest advantage of biofilm biocatalysts, for example allowing the conversion of highly toxic substances in the presence of organic solvents. However, the underlying mechanisms responsible for the high robustness of biofilms are still a matter of intensive research. Several studies report induced resistance of biofilms when exposed to certain kinds of environmental stresses. There are known genetic

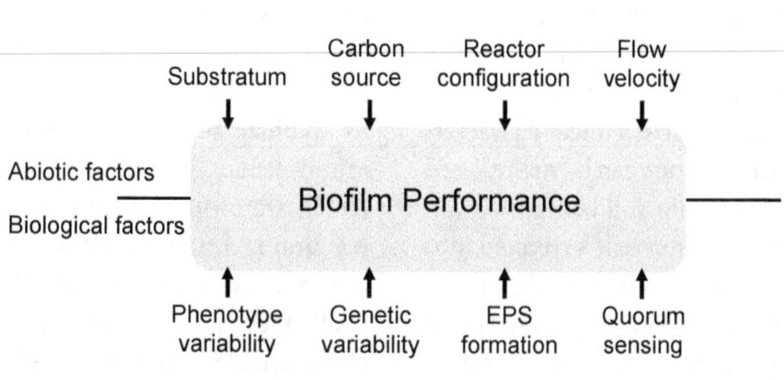

Figure 11.2 Abiotic and biological factors influencing biofilm performance in the context of biocatalysis.

networks (such as quorum sensing and c-di-GMP mediated regulation, later described), which regulate responses to anaerobiosis, starvation, and other stresses, including oxidative, osmotic, and pH induced stresses, as well as antimicrobial and toxic metal challenges (Harrison *et al.*, 2007; Stewart and Franklin, 2008). For example, *Pseudomonas aeruginosa* biofilms are reported to cope with oxidative stress caused by hydrogen peroxide through upregulation of the enzyme catalase, which functions to catalyse the degradation of hydrogen peroxide, and superoxide dismutase, which catalyses the dismutation of superoxide; both processes are reported to be controlled by quorum sensing (Hassett *et al.*, 1999). Another intriguing example is the increased drug tolerance of biofilm-associated organisms after exposure to sublethal concentrations of antibiotics (O'Toole and Stewart, 2005). Antibiotics are reported to activate regulatory pathways, leading to a profound effect on biofilm matrix production and the development of bacterial subpopulations with increased drug tolerance. In *Pseudomonas aeruginosa* biofilms one of the key mechanisms to overcome antibiotic challenge is upregulation of genes encoding for β-lactamase and multidrug efflux pumps after antibiotic exposure (Anderson and O'Toole, 2008). For tobramycin-treated *Pseudomonas aeruginosa* biofilms upregulation of the stress response chaperones *groES* and *dnaK* has been reported (Whiteley *et al.*, 2001). On the other hand there are studies reporting no difference in the expression pattern of the efflux pumps in biofilm bacteria compared with planktonic cells (De Kievit *et al.*, 2001). Another important issue regarding antibiotic resistance is horizontal gene transfer of the corresponding resistance genes by conjugation or transformation. The close proximity of cells in a biofilm seems to have a positive impact on horizontal gene transfer and a 1000-fold increase in the conjugation rate has been reported (Hausner *et al.*, 1999; Molin *et al.*, 2003), which facilitates the distribution of antibiotic resistance genes (Fux *et al.*, 2005).

Apart from active resistance mechanisms, there is another rather simple way of surviving biocides. Most biocides typically act upon rapidly growing bacteria, as an uptake of the compounds is necessary for toxicity. Thus slow or non-growing microorganisms would be protected from lethal effects. Several studies have shown a correlation between oxygen limitation, metabolic activity and protection from antibiotic killing in biofilms (Anderson and O'Toole, 2008). Oxygen and nutrient deprivation result in a decrease in bacterial metabolic activity and cessation of bacterial growth. By decreasing the growth rate and altering metabolic activity biofilm cells can enter a stationary phase. In this phase, the activity of RpoS, the stationary phase sigma factor, regulates the expression of stress response factors (note: sigma factors are prokaryotic transcription initiation factors that are activated in response to different environmental conditions). Microarray analysis of *Escherichia coli* biofilms revealed the upregulation of nearly 50% of all rpoS-regulated genes, while a rpoS deletion mutant failed to form biofilms at all (Schembri *et al.*, 2003).

A resistance mechanism, which is connected to metabolic quiescence, is 'persistence'. So called persisters are microbial cells with a much higher tolerance towards antibiotics, toxins, and metal cations, and thus represent multidrug tolerant cells (Lewis, 2007). They are able to repopulate a previously biofilm-covered surface after exposure to otherwise lethal concentrations of a bactericide. This specialized cell type is observed in many microbial organisms, including Gram-positive and Gram-negative bacteria as well as in yeast (Kussell *et al.*, 2005; Anderson and O'Toole, 2008). Both planktonic and biofilm growing organisms produce persister cells, although in biofilms they are found at a much higher frequency and amount to approximately 0.1–10% of a biofilm population (Harrison *et al.*, 2007; Lewis, 2007). Several studies report that besides the quiescent metabolism, they differ from normal cells by expression of genes associated with stress tolerance, such as chaperones, the multiple antibiotic (*marRAB*) regulon and cold-shock proteins (Shah *et al.*, 2006; Harrison *et al.*, 2007).

The extrapolymeric substances (EPS) excreted by the biofilm forming cells, represents another protection mechanism by acting as a physical diffusion and chemical reactive barrier. However, EPS is a rather ambivalent topic in the context of biofilm utilization, as apart from its protective function excessive EPS production poses a problem due to clogging or diffusion limitation, which will be discussed in detail later. Owing to its complex and varying composition including DNA, proteins, and chemically modified complex carbohydrates, the EPS is able to bind and sequester toxic metal ions, disinfectants and various antibiotics via often complex chemical reactions, which can be summarized under the term biosorption (Harrison *et al.*, 2007). Toxic compounds fail to reach potentially susceptible microbes in the deep core of the biofilm, as those are protected by the EPS. It has also been observed that a few antimicrobial agents actually stimulate EPS production with the result of increased resistance. In *Pseudomonas aeruginosa* biofilms, alginate expression was highly upregulated by subinhibitory treatment with the beta-lactam antibiotic imipenem (Bagge *et al.*, 2004). Alginate is an anionic exopolymer with the ability to trap cationic antimicrobial agents, such as aminoglycosides like tobramycin and other cationic quaternary ammonium compounds (Anderson and O'Toole, 2008). Furthermore, anionic extracellular polymeric substances bind and sequester toxic cationic heavy metals. *Pseudomonas aeruginosa* biofilms bind and sequester Fe, Cu, La, and Mn species in amounts that are significantly higher than what planktonic cells can tolerate (Harrison *et al.*, 2007). However, this protection mechanism cannot be generalized and greatly depends on the nature of the biofilm organisms as well as that of the antimicrobial agent and the environment.

Biofilm formation is influenced by many factors, external, such as experimental conditions, as well as internal, which are species dependent. There are examples of external factors influencing the expression of a particular bacterial factor, as described for *Pseudomonas aeruginosa.* For the initial attachment during the early phase of biofilm formation this organism requires a single polar flagellum, whose formation is carbon source dependent (Klausen *et al.*, 2003a). Furthermore, the architecture of *P. aeruginosa* biofilms varies with the kind of carbon source available. Growth on glucose leads to pronounced mushroom-like macrocolonies, whereas growth on citrate results in a flat biofilm with very little topology. This is due to catabolite repression as glucose represses a particular form of hypermotility, called twitching motility (O'Toole and Kolter, 1998). The formation of mushroom-like biofilms requires non-motile cells as a substratum for other cells to grow upon. In citrate-grown cells, twitching motility is not repressed, thereby preventing mushroom structure formation (Klausen *et al.*, 2003a,b). Biofilms are complex communities which consist of multiple species in a natural setting, but also in monoseptic films, which consist of one single species, highly specialized subpopulations will develop. Regulatory mechanisms for the different stages of attachment, maturation and dispersal have been intensively investigated for several years but still much remains to be elucidated. Owing to the high diversity of microorganisms capable of forming biofilms under various and also very extreme conditions, it is clear that regulation is complex, species dependent and influenced by many environmental factors. This is not very beneficial regarding process development, as each application needs to be tailor-made and even the scale-up could change biofilm behaviour significantly owing to different flow-conditions, etc. However, some common features concerning biofilm regulation on the molecular level have been found and are described in the following sections.

Quorum sensing

The phenotypic switch between the planktonic and the sessile phenotype is accompanied by altered gene expression during each stage of biofilm formation. These processes depend heavily on the biofilm species and the environmental conditions (Hall-Stoodley and Stoodley, 2002; Ren *et al.*, 2004; Patrauchan *et al.*, 2007). Furthermore, the regulation of gene expression varies widely across both spatial and temporal scales during the dynamic process of biofilm formation creating a variety of biofilm specific phenotypes (Stoodley *et al.*, 2002; Kirisits *et al.*, 2005; Koh *et al.*, 2007; Vlamakis *et al.*, 2008).

Quorum sensing (see Chapter 1) is a form of cell-to-cell communication, which involves producing, releasing, detecting, and responding to small hormone-like molecules termed autoinducers (Waters and Bassler, 2005); it plays an important role in biofilm development (Hammer *et al.*, 2003; Stanley and Lazazzera, 2004; Irie and Parsek, 2008). As the term quorum sensing implies, it needs a quorum (or minimum number) of cells to be effective. In Gram-negative bacteria, small molecules such as acyl homoserine lactones (AHLs) have been identified as the major players in this communication system (Parsek and Greenberg, 2000). In Gram-positive bacteria, autoinducing peptides (AIPs), which bind to receptor histidine kinases embedded within the cellular membranes, are commonly used for communication (Abraham, 2006). However, the only definitive quorum-sensing mechanism identified to date shared by both Gram-negative and Gram-positive bacteria involves autoinducer-2 (AI-2) as an effective signalling molecule (Richards and Melander, 2009). In *Pseudomonas aeruginosa*, numerous cellular functions are regulated by quorum sensing, including the production and secretion of siderophores, proteases, toxins and a type of surface motility called swarming (Kirisits and Parsek, 2006; Sakuragi and Kolter, 2007). There are also indications that EPS production is strongly regulated by quorum sensing (Branda *et al.*, 2005; Sakuragi and Kolter, 2007). However, as to be expected from such a complex system as biofilms, there are also examples for quorum sensing having no influence on biofilm formation (Parsek and Greenberg, 2005). It was suggested that this might be partly due to the global character of quorum sensing regulation, which regulates the expression of several hundreds of genes, depending upon the study, strain, and experimental conditions. Another hypothesis is based on the overall impact of the small signalling molecules, which might be strongly influenced by the architecture of the biofilm, flat or highly structured, the amount of EPS, the flow rate of the bulk liquid and mass transfer limitations (Parsek and Greenberg, 2005; Kirisits and Parsek, 2006).

c-di-GMP mediated regulation of biofilm formation

Another key player in the regulation of biofilm growth has been identified as bis-(3′–5′)-cyclic dimeric guanosine monophosphate (c-di-GMP). It is a universal intracellular messenger molecule in bacteria and has been shown to stimulate various biofilm-associated functions. In general, c-di-GMP is reported to stimulate the biosynthesis of adhesins and exopolysaccharide matrix substances in biofilms and to inhibit various forms of motility (Gjermansen *et al.*, 2006; Hengge, 2009; Ishihara *et al.*, 2009). Proteins with GGDEF and EAL domains are involved in the turnover of c-di-GMP. The GGDEF domain is responsible for the formation of this compound by diguanylate cyclase activity (Simm *et al.*, 2004, 2005), while the EAL domain was proposed to be responsible for its degradation through phosphodiesterase activity (Schmidt *et al.*, 2005). Manipulation of cellular c-di-GMP levels by overexpression of either GGDEF or EAL domain proteins strongly influenced biofilm growth. Enhanced synthesis of GGDEF domain proteins stimulated the synthesis of adhesins, biofilm matrix components, and interfered with motility, resulting in the formation of strong, thick biofilms. Overexpression of the EAL domain proteins gave rise to opposite phenotypes, hampering biofilm formation. Interestingly, GGDEF and EAL domains are not only ubiquitous in bacteria, but many species encode a surprisingly large number of these proteins. Moreover, c-di-GMP controls a plethora of other functions such as the virulence of animal and plant pathogens, progression through the cell-cycle, antibiotic production and other cellular functions, of which many still remain to be unveiled (Hengge, 2009).

These regulatory mechanisms may at one point provide a tool for biofilm formation control. Another fact to be considered is the formation of genetic variants during biofilm growth. Several studies have demonstrated that during the growth of bacteria in biofilms variant subpopulations emerge, which have altered phenotypic traits compared with the colonizing strain (Yildiz and Schoolnik, 1999; Haussler, 2004; Boles *et al.*, 2005; Kirisits *et al.*, 2005; Allegrucci and Sauer, 2007; Davies *et al.*, 2007; Hansen *et al.*, 2007a,b; Koh *et al.*, 2007). These phenomena have been observed in a wide range of bacterial species, both Gram-negative and Gram-positive. In most cases different morphotypes of the initially colonizing strain have been isolated from sheared fragments of the biofilm. The altered colony morphology phenotypes, are often described for *Pseudomonas* strains as rough small-colony variant (RSCV), small colony variant (SCV), or wrinkly spreaders (WS) (Drenkard *et al.*, 2002; Harrison *et al.*, 2007), or as smooth and wrinkled type cells (Fig. 11.3) (Gross *et al.*, 2010). For catalysis, morphotype variants are not critical as long as the biocatalytic activity is not altered by the change in morphology. So far there is only one study which looked into the correlation of biotransformation activity and morphology variants (Gross *et al.*, 2010). In this study a biofilm forming *Pseudomonas* sp. strain VLB120ΔC was utilized for the biotransformation of styrene to styrene oxide. During biofilm formation a biofilm specific wrinkled type variant developed (Fig. 11.3). However, this phenotypic variation had no influence on the biotransformation activity of the organism.

Boles *et al.* (2005) reported that *Pseudomonas aeruginosa* biofilms rapidly develop a greater diversity of heritable phenotypes, suggesting that genetic changes produced them. It was discovered that this diversity involved recombination functions and a mechanism that required the *recA* gene. The altered wrinkly phenotype showed 'hyperbiofilm formation' and increased biofilm-mediated resistance, demonstrating that the diversity enhanced the survival of biofilms that were subjected to a common environmental stress, in this study oxidative stress introduced by the addition of H_2O_2.

Drenkard and Ausubel (2002) identified a regulatory protein (PvrR) that controls the conversion between antibiotic-susceptible and antibiotic-resistant forms by inducing the formation of a rough colony phenotype (RSCV) with increased biofilm formation capability. In some cases, the cause of the phenotypic variation has been mapped to changes at specific genetic loci. For example, small rough colony variant cells, which are typically better at forming biofilms and less motile than their progenitors, are often recovered from aged biofilms of *Pseudomonas* spp. These variants were found to hyperexpress *psl* and *pel*, two of the three gene clusters encoding for polysaccharide production by *Pseudomonas aeruginosa* (Kirisits *et al.*, 2005). Other variants of *Pseudomonas putida* isolated from mixed-culture biofilms with *Acinetobacter* sp. were found to have mutations in the *wapH* homologue, which is involved in core lipopolysaccharide production (Hansen *et al.*, 2007a). Often these variants are correlated with higher multidrug resistance (Drenkard *et al.*, 2002) and multimetal resistance (Davies *et al.*, 2007; Hansen *et al.*, 2007a). It was assumed that such genetic variability occurs by random mutation processes, which in turn are enriched through natural selection in the local biofilm environment. Genetic variation and natural selection provide a slow but long-lasting

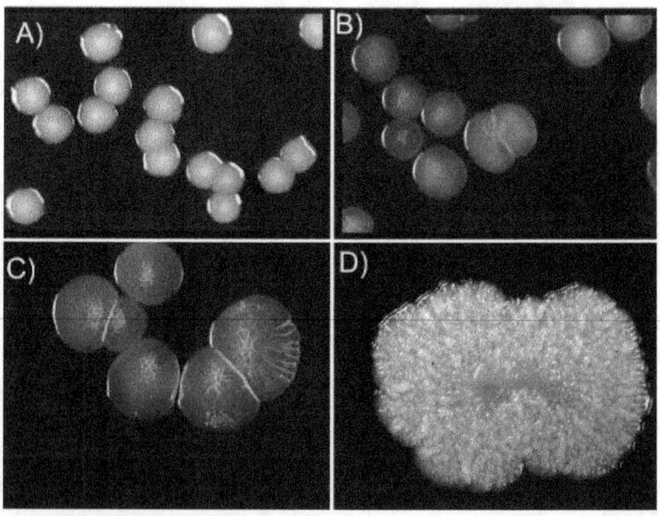

Figure 11.3 Development of a colony morphology variant isolated from *Pseudomonas* sp. strain VLB120ΔC biofilm. (A) Colonizing strain consisting of purely smooth type colonies. (B) Biofilm isolate, 20 hours incubated on LB agar, with early WR type of crater-like shape in the centre of the colonies. (C) Biofilm isolate, after around 30 hours incubation. (D) Fully developed WR type with distinguished three dimensional structures.

mechanism to generate physiological heterogeneity in biofilms. Furthermore, it has been proposed that genetic change and diversification of bacteria within a biofilm community provide an insurance policy against future antimicrobial challenges or environmental changes. Such a diversified population is predicted to be more robust than a single parent strain, as some phenotypes can survive environmental insults that others cannot. This idea is called the 'insurance hypothesis', meaning that random mutations or site-specific genetic rearrangements produce phenotypic diversity that ensures population survival (Drenkard *et al.*, 2002; Boles *et al.*, 2005).

In the next section, a survey on published biofilm processes gives a detailed overview of the productivities, scales, reactor types and organisms so far used in biofilm-based biocatalysis.

Current status of biofilm – technology in bioprocesses today

Although biofilms have been long exploited for the treatment of waste water, off-gas and contaminated soils (see Chapter 5), they are not yet established in industrial catalysis for the production of bulk and/or fine chemicals (Nicolella *et al.*, 2000b; Qureshi *et al.*, 2005). Only a few examples of productive catalysis using biofilms as biocatalysts are described in the literature and even fewer have been run on a large scale. This is despite the fact that a number of research studies at a laboratory scale have demonstrated the feasibility of biofilm reactors. These have mostly used single species biofilms for the conversion of renewable resources such as starch, sugars and glycerol into alcohols (e.g. ethanol, butanol, 2,3-butanediol, dihydroxyacetone) or organic acids (e.g. acetic, lactic, fumaric, succinic and citric acid) (Qureshi *et al.*, 2005). Why are biofilm-based processes not yet state of the art in biocatalysis? Bioprocesses have to be at least comparable to the productivities achieved with their chemical counterparts, to be attractive for industry. The requirements differ widely depending on the target compound. As the cost pressure increases from complex and new pharmaceuticals with no chemically derived counterparts, compared with fine chemicals or commodities, or to bulk chemicals with very low margins, the constraints for high volumetric productivities increases proportionally (Straathof *et al.*, 2002; Woodley, 2006b).

In a survey conducted by Straathof *et al.* (2002) it was found that successful process developments for the production of fine chemicals like β-lactams or amines involve, on average, a yield of 78%, a volumetric productivity of 15.5 g/l/h and a final product concentration of 108 g/l. For the production of pharmaceutical intermediates Pollard and Woodley (2007) report product concentrations of 50–100 g/l and an enzyme activity of 1000 g product per g enzyme or, in the case of whole cells, 15 g product per gram cell dry weight. However, depending on the complexity of the pharmaceutical such as the number of chiral centres, volumetric productivities as low as 0.1 g/l/h and final product titres of 1 g/l might still be economically sufficient (Straathof *et al.*, 2002).

Whether high volumetric productivities or space time yields and product titres can be achieved is mainly determined by the stability of the biocatalyst under the required process conditions. Process or operational stability is regarded as the key bottleneck in biocatalysis (Bommarius *et al.*, 2004). Considering process or operational stability, biofilm based catalysis has a clear advantage over traditional bioprocesses, as biofilms present a stable, continuously self-renewing catalytic system, which has been shown to work for months without replacement (Gross *et al.*, 2007). However, the final product titre in such processes is often very low, owing to the continuous flow and the rather low productivity of individual biofilm cells (as discussed in detail in the following section).

The examples shown in Table 11.2 illustrate the range of reactor configurations, surface materials and organisms (including recombinant and thermophilic strains) that have been considered, mostly on a laboratory scale. The table also shows that a number of products, such as solvents, fine chemicals, polymers and gases have been targeted. One of the oldest biofilm based processes is the well known production of vinegar (acetic acid) from ethanol using a mixed culture of acetic acid bacteria growing in a trickle bed reactor adsorbed to beech wood shavings as biocatalysts (Table 11.2, no. 1; Fig. 11.4). This process is run

Table 11.2 Examples of bulk chemicals biofilm fermentation (F) and biotransformation (B) processes (ordered according to scale)

	No.	Reactor	Organism	Surface material	Substrate	Product	Productivity (g/l h–1); (mmol/lh–1)	Reactor volume (L)	Reference
F	1	Trickle bed	Acetic acid bacteria	Beechwood shavings	Ethanol	Acetic acid (vinegar)	1.7; 28	60,000	Crueger and Crueger (1990)
F	2	Trickle bed	*Caldicellulosiruptor saccharolyticus*	Polyurethane foam	Sucrose	Hydrogen	0.044; 22	400	van Groenestijn *et al*. (2009)
F	3	Fluidized bed	*Zymomonas mobilis*	Macroporous glass beads	Hydrolysed starch	Ethanol (A)	13; 282	Cascade, 2 × 55	Weuster-Botz *et al*. (1993)
F	4	Packed bed	*Propionibacterium acidipropionici*	Spiral wound fibrous matrix	Lactate	Propionic acid	1.62; 21.8	2.9	Lewis and Yang (1992)
F	5	Rotary biofilm contactor	*Rhizopus oryzae*	Polysulfone plastic discs	Glucose	Fumaric acid	4.25; 36.6	2	Cao *et al*. (1996)
F	6	Packed bed bubble column	*Gluconobacter oxydans*	Silicone covered Ralu-rings	Glucose	Dihydroxyacetone	3.7; 41	2	Hekmat *et al*. (2003)
F	7	Packed bed	*Pediococcus acidilactici PO2*	Spiral wound fibrous matrix	Glucose	Pediocin	4.16 × 105 AU/l h–1	0.76	Cho *et al*. (1996)
F	8	Packed bed	*Lactococcus lactis*	Spiral wound fibrous matrix	Whey permeate	Nisin	1.0 × 107 AU/l h–1	0.36	Xia *et al*. (2005)
F	9	Hollow fibre membrane	Hydrogenotrophic and acetoclastic methanogens	Polysulfone membranes	Carbon dioxide	Methane	0.068; 4.25	0.33	Ju *et al*. (2008)
F	10	Packed bed	*Clostridium beijerinckii*	Clay brick pieces	Glucose	Acetone, butanol, ethanol (ABE)	16.1; 268	0.31	Qureshi *et al*. (2004)
F	11	Packed bed	*Lactobacillus casei*	Plastic composite supports	Glucose	Lactic acid	4.26; 47.3	0.3	Ho *et al*. (1997)

F	12	Packed bed	*Klebsilla pneumoniae*	Bonechar	Whey permeate	2,3-butanediol	11.7; 130	0.2	Maddox *et al.* (1988)
F	13	Packed bed	*Zymomonas mobilis*	Polypropylene	Glucose	Ethanol (B)	536; 11630	0.165	Kunduru and Pometto (1996)
F	14	Packed bed	*Zymomonas mobilis*	Anionic exchange resin beads	Glucose	Ethanol (C)	136; 2956	0.16	Krug and Daugulis (1983)
F	15	Packed bed	*Alcaligenes eutrophus*	DEAE-Sephadex A-25 beads	Glucose	Poly(3-hydroxybutyrate) (PHB)	0.027	0.031	Zhang *et al.* (2004)
B	16	Packed bed	*Zymomonas mobilis*	Glass beads	Benzaldehyde	Benzylalcohol	0.88; 8.1	0.02	Li *et al.* (2006)
B	17	Membrane aerated biofilm reactor	*Pseudomonas* sp. strain. VLB120ΔC	Ceramic membrane	Styrene	(*S*)-styrene oxide	1.2; 9.7	0.01	Halan *et al.* (2010)
B	18	Tubular membrane reactor	*Pseudomonas* sp. strain VLB120ΔC	Silicone membrane	Styrene	(*S*)-styrene oxide	0.67; 5.5	0.003	Gross *et al.* (2007)
F	19	Membrane gradostat biofilm reactor	*Phanerochaete chrysosporium*	Polysulfone membrane	Complex medium	Manganese Peroxidase	1.3 U/l/h	0.0025	Govender *et al.* (2010)

AU, arbitrary units.

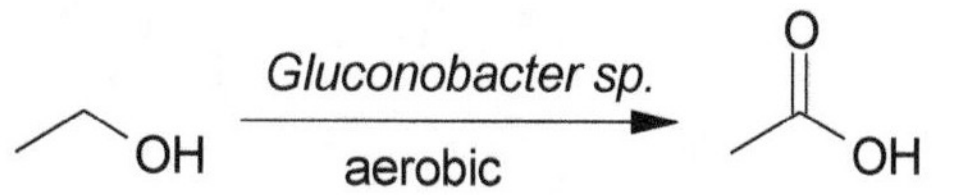

Figure 11.4 Production of acetic acid from ethanol by *Gluconobacter* sp. biofilms adsorbed to beechwood shavings in a trickling bed reactor.

on a 60 m^3 scale with productivities of 1.7 g/l/h (Crueger and Crueger, 1990). Although alternative routes with higher yields are available, the traditional process is still widely used, mainly due to the better flavour of the product (Rosche *et al.*, 2009).

A process for the production of hydrogen from sucrose running on a 400 l scale employs the organism *Caldicellulosiruptor saccharolyticus* as the biocatalyst (Table 11.2, no. 2; Fig. 11.5) (van Groenestijn *et al.* 2009). Also in this example a trickle bed reactor is used, in which the biofilm establishes itself on a polyurethane foam packing. This material is characterized by an exceptionally high surface area (in this case 1700 m^2/m^3), which is important for biofilm based processes as this increases the active surface area and thus the final product yield. For this process a product yield of 2.8 mol H_2/mol hexose converted was achieved.

Caldicellulosiruptor saccharolyticus is a thermophile, requiring a cultivation temperature of 73°C. The strain is anaerobic, which is beneficial for a biofilm-based process, as oxygen supply into biofilm reactors poses one of the main challenges of this technology. Another interesting aspect is that the system was run under non-axenic (microbiologically open) conditions and no contamination was observed in course of the fermentation. This may be attributed to the thermophilic character of *C. saccharolyticus*.

A continuous fluidized bed reactor has been realized exploiting the catalytic power of *Zymomonas mobilis* for the synthesis of ethanol from waste 'B-starch' (Table 11.2, no. 3; Fig. 11.6) (Weuster-Botz *et al.*, 1993)

The process utilizes *Zymomonas mobilis* immobilized on glass beads. Immobilization occurs by simple adsorption of the biomass onto the carrier. Up to the mid-1990s, publications often referred to self-immobilized cells or cells that grow as aggregates without the addition of any polymers or cross-linking chemicals instead of using the term biofilm. Utilizing *Zymomonas mobilis* biofilms, significantly increased the biomass yield in the reactor, which, combined with short residence times, allowed for an unsterilized input of B-starch without contaminating the reaction system as contaminants were simply outnumbered and washed out. By operating two 55-l reactors in a cascade, the system was evaluated on a semi technical scale. An ethanol production rate of 13 g/l/h was achieved over a period of 20 days. Operating the system with a non-sterile feed (or liquid input) is a big improvement comparing this approach to the one utilizing planktonic cultures, where the feed had to be heat treated before entering the reactor.

The highest productivity reported so far for ethanol production by *Zymomonas mobilis* biofilms was an outstanding 536 g/l/h based on the liquid volume in a polypropylene packed-bed reactor (Table 11.2, no. 13) (Kunduru and Pometto, 1996). In another study *Zymomonas mobilis* biofilms produced up to 136 g/l/h ethanol (Table 11.2, no. 14) based on total reactor volume (Krug and Daugulis, 1983). In a comparative continuous suspension culture only up to ~5 g/l/h were

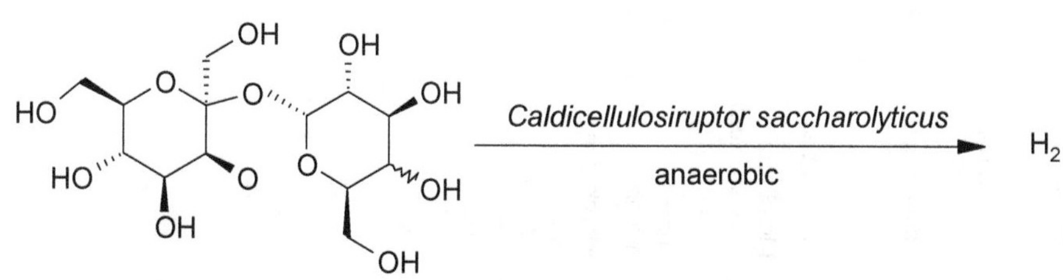

Figure 11.5 Hydrogen synthesis from sucrose utilizing anaerobic *Caldicellulosiruptor saccharolyticus* biofilms growing on polyurethane foam in a trickle bed reactor.

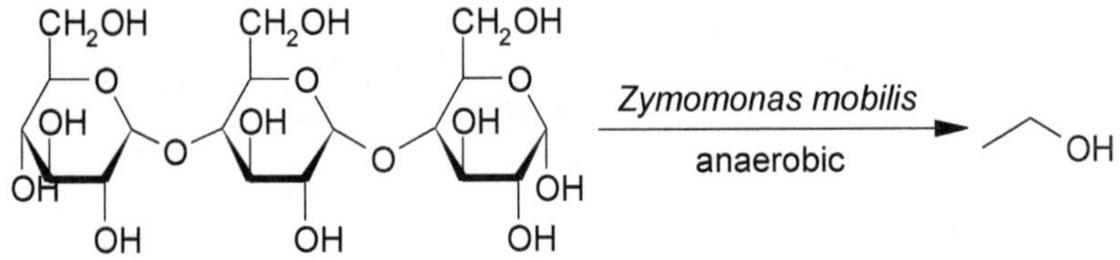

Figure 11.6 Ethanol synthesis from starch by anaerobic *Zymomonas mobilis* biofilms immobilized on glass beads in a fluidized bed reactor.

achieved (Kunduru and Pometto, 1996), which demonstrates that biofilm reactors have the potential for higher productivities than suspension reactors without cell retention.

A biofilm-based process for the synthesis of propionic acid was developed and run only at laboratory scale (Table 11.2, no. 4; Fig. 11.7) (Lewis and Yang, 1992). Propionic acid is an important chemical used in the production of cellulose plastics, herbicides and perfumes. It is also an effective mould inhibitor and its salts are widely used as food or feed preservatives. *Propionibacterium acidipropionici* adsorbed to spiral wound fibres in a conventional packed bed reactor was used as a biocatalyst to produce propionic acid from lactate (Fig. 11.7). At a high cell density of 37 g cell wet weight/l, product yields were 4-fold greater than batch systems utilizing planktonic growing *P. acidipropionici* and ran stable for 4 months without any clogging or contamination. In addition, the biocatalyst was not sensitive to a low nutrient and low pH-feed input.

A very elegant approach using biofilms as catalysts is the fermentative production of fumaric acid from glucose by the fungus *Rhizopus oryzae* (Table 11.2, no. 5; Fig. 11.8) (Cao *et al.*, 1996). Commercially, fumaric acid is so far derived exclusively from petroleum. Although various organisms produce this compound, yields are generally too low to be interesting for industrial application.

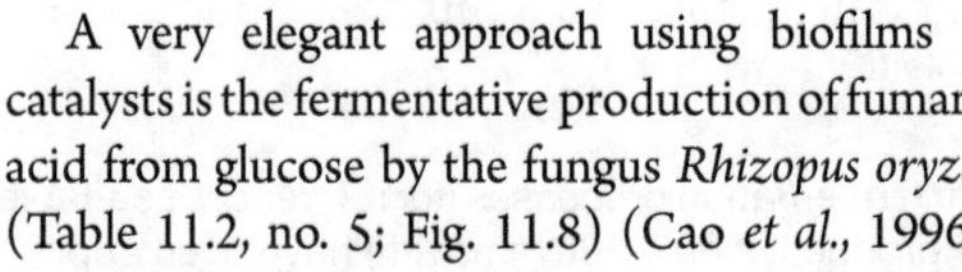

Figure 11.7 Synthesis of propionic acid from lactate by anaerobic *Propionibacterium acidipropionici* biofilms growing on spiral wound fibres in a packed bed reactor.

Rhizopus oryzae was adsorbed to polysulfone discs in a rotary biofilm contactor, which is a box like reactor, in which discs are mounted on a spindle that rotates the discs horizontally through the liquid/gas phases. The reactor was coupled to a continuous *in situ* product extraction by an anionic exchange column, which was beneficial for cell viability, preventing product inhibition, stabilizing the pH and enhancing reaction rates. An average yield of 4.25 g/l/h was achieved, which is the highest yield in a biological process for the synthesis of this compound. The system ran stable for 2 weeks without loss in activity. Nevertheless, this system was operated only at a laboratory scale in a 2-l reactor.

A packed bed bubble column reactor operated in a repeated feed-batch was utilized for the production of the fine chemical dihydroxyacetone by the acetic acid bacterium *Gluconobacter oxydans* (Table 11.2, no. 6; Fig. 11.9) (Hekmat *et al.*, 2003) attached to silicone coated Ralu rings. The coating was chosen because of its excellent oxygen permeability. In this set-up, a volumetric productivity of 3.7 g/l/h was reached, and the process was stable for 18 days. A comparable approach employing the same set-up with planktonic cells, showed three times lower productivities and a long lag phase of 4 hours.

A totally different approach was applied for the conversion of CO_2 to methane (Ju *et al.*, 2008). A hollow fibre module with 1000 fibres per module was utilized as the substratum. CO_2 was applied through this hollow fibre module as a carbon source. A mixed species biofilm originating from waste water plant sludge was used as a biocatalyst for the conversion. The whole system maintained stable operation for about 70 days

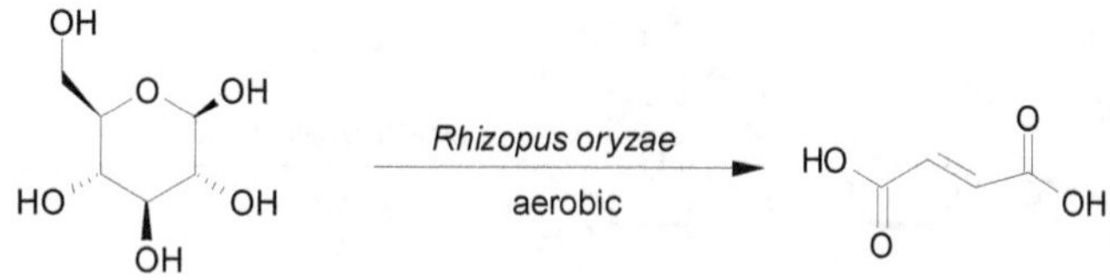

Figure 11.8 Synthesis of fumaric acid from glucose utilizing the fungus *Rhizopus oryzae* growing on polysulfone discs in a rotary biofilm contactor.

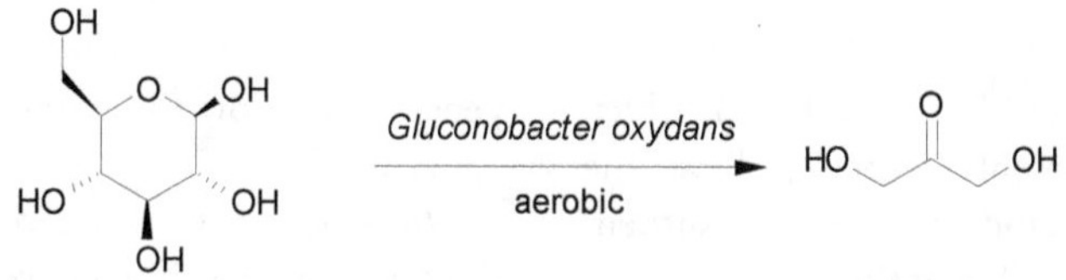

Figure 11.9 Synthesis of dihydroxyacetone from glucose utilizing *Gluconobacter oxydans* biofilms attached to silicone coated Ralu rings in a packed bed bubble column.

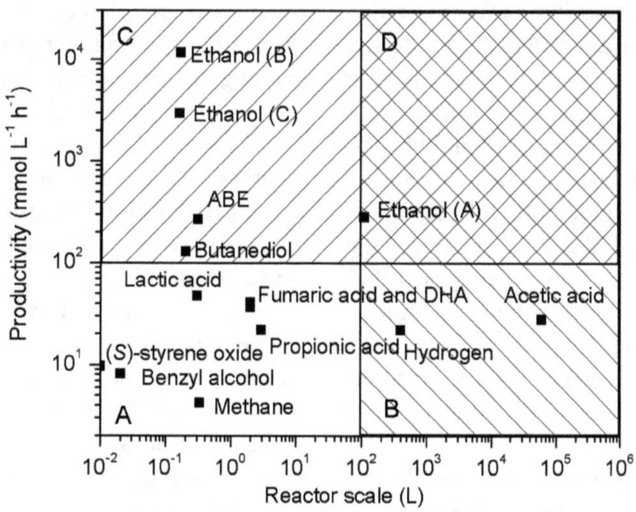

Figure 11.10 Productivities and scale comparison of biofilm fermentation processes (for references see Table 11.2). Borders for reactor scales (pilot scale ≥100 l) and for productivities (≥100 mmol/l/h) have been applied resulting in four process categories: (A) laboratory to technical scale with low to medium productivities, early developmental stage; (B) pilot to large scale processes with low to medium productivities; (C) laboratory scale processes with high productivities, high potentials; (D) large scale, high productivities, late developmental stage.

and no problems with membrane clogging were reported.

From these biofilm reaction concepts several interesting points can be observed in terms of developmental stage and application potential. In Fig. 11.10, the processes from Table 11.2 are plotted according to scale and productivities. Categories for large scale (≥100 l) and high productivity (≥100 mmol/l/h) reactors were defined, which allow characterization of the processes according to the potential scale and efficacy of biocatalysis operation.

Regarding process scale, only the first three out of the eighteen processes have been designed on a scale larger than 50 l, the others were pilot scale or even early development laboratory scale experiments. Trickle packed-bed configurations are obviously most suitable for reactors operating at larger scale as the investment costs as well as the amount of energy required for operation are low. Therefore, low cost bulk chemicals such as hydrogen, vinegar and ethanol can be produced in these reactor designs.

Outstanding productivities have been reported for anaerobic solventogenic fermentations with biofilms; however, only one process (ethanol A; Weuster-Botz *et al.*, 1993) falls into category D, being highly productive in a large-scale reactor. This process is already at a late stage of development and indisputably of high interest for industrial implementation. Irrespective of scale, the majority of the non-solvent producing biofilm fermentations report productivities in the range of 20 to 50 mmol/l/h (Table 11.2; Fig. 11.10).

Several studies reported that higher volumetric productivities could be achieved in biofilm reactors as compared with traditional batch reactors with planktonic cells, owing to the higher cell concentrations of over 70 g cell wet weight/l (Qureshi and Maddox, 1988). Unfortunately, for most of the processes described in Table 11.2, example biomass values were not given and thus a comparison with planktonic cell activities cannot be made.

All processes discussed so far are based on a fermentative approach utilizing obligate and/or facultative anaerobic organisms, which convert their carbon source to biomass, CO_2 and the product of choice. Examples for a biotransformation, which necessitates an additional biotransformation substrate besides the carbon source needed for growth (Fig. 11.1), are scarce. Pilot or even large scale examples are not published. This is possibly due to the fact that genetically modified organisms are usually utilized for biotransformation reactions, while all fermentative biofilm processes are so far based on wild type organisms. Most recently, biofilm processes involving toxic substrates and products have received attention. For the conversion of styrene to (*S*)-styrene oxide, a genetically modified *Pseudomonas* sp. strain VLB120ΔC biofilm was used (Table 11.2, no. 18; Fig. 11.11).

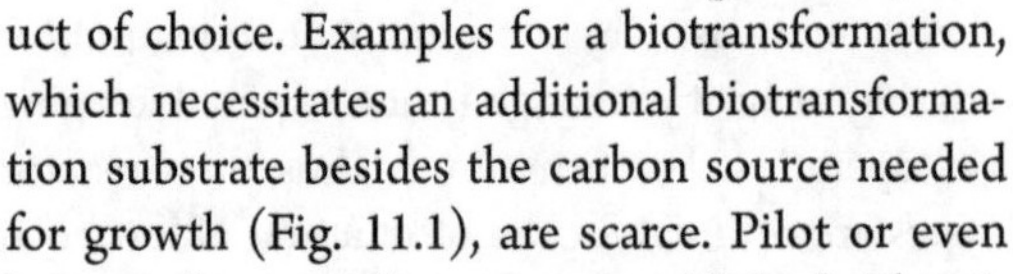

Figure 11.11 Conversion of styrene to *(S)*-styrene oxide utilizing genetically modified *Pseudomonas* sp. strain VLB120ΔC biofilm in a tubular reactor system.

This strain carried a gene deletion in the styrene degradation pathway, enabling it to accumulate styrene oxide while growing on an alternative carbon source. *Pseudomonas* sp. strain VLB120ΔC readily formed a biofilm on silicone tubing, which was used as substratum in a membrane attached biofilm reactor (MABR). In this example a two liquid phase approach was realized, as substrate and product are poorly water soluble and highly toxic to the cells. The biocatalyst grew inside the tubing, while on the outside a pure styrene phase was present. Styrene served a dual purpose: on the one hand it was the biotransformation substrate, entering the biofilm by diffusion through the membrane, while on the other hand it was the organic phase for *in situ* product recovery by extracting the product out of the biofilm. The organism showed a remarkable solvent tolerance and even grew submerged in styrene on an agar plate (Park *et al.*, 2007). Yields were comparable to the ones achieved with planktonic cells and the process remained operational for over 1600 hours, a considerable improvement over a typical batch operation for cultures grown in suspension that only lasts 10–50 hours (Park *et al.*, 2007). No clogging, contamination or genetic instability was observed. Also in case of *Zymomonas mobilis* converting highly toxic benzaldehyde to benzylalcohol an enhanced solvent tolerance of the biofilm was detected (Li *et al.*, 2006).

As high flow rates are applied in biofilm reactors in continuous mode, potentially toxic products are diluted below toxic levels. However, a detailed investigation of the influence of solvents on biofilm growing organisms is missing, so it is difficult to state how biofilms behave in general towards low log $P_{o/w}$ (partition coefficient between octanol and water) solvents, which are regarded as highly toxic for microbes. The toxicity of these compounds stems from their tendency to accumulate in the membranes of the microorganisms leading to alteration and disruption of their structure. This in turn causes a loss of function as a permeable barrier, a leakage of cellular compounds such as proteins, cofactors and ions which drive the proton motive force and eventually leads to cell death (Laane *et al.*, 1987; Sikkema *et al.*, 1995). Compounds with a log $P_{o/w}$ below one are considered to be too polar to dissolve into the membranes, while for compounds with a log $P_{o/w}$ above 4 the bioavailability in the aqueous solution is too low to affect the bacteria (Laane *et al.*, 1987). Although organic solvents with log $P_{o/w}$ values ranging from 1 to 4, are ideal with respect to their extraction abilities for many chemicals, this group is unfortunately not suitable for most whole cell biocatalysts (de Bont, 1998; Heipieper *et al.*, 2007). There are indications that biofilm growing organisms are better able to cope with solvent stress than their planktonic counterparts, and thus two liquid phase processes employing organic solvents, will probably become a main target for biofilm-based technologies.

A different field, where biofilms may become important is the synthesis of biofuels from lignocellulose (Wang and Chen, 2009). Current technologies of lignocellulose conversion are too costly owing to inefficient and complex processing steps in pre-treatment, saccharification (the breakdown of complex carbohydrates), and downstream processing. Employing biofilms might solve several bottlenecks in biofuel synthesis; biofilm-based approaches benefit from: (i) the high concentrations of cell-associated hydrolytic enzymes at the biofilm–substrate interface to increase reaction rates, (ii) a layered microbial structure in which multiple species may sequentially convert complex substrates, (iii) co-fermention of hexose and pentose as hydrolysates diffuse outward from the biofilm, and (iv) the possibility of fungal–bacterial symbioses that allow simultaneous delignification and saccharification. Through the application of biofilms, highly consolidated bioprocessing, including delignification, saccharification, fermentation, and separation may be possible in a single reactor. However, so far no process has been realized for biofilm-based biofuel production.

The diverse nature of biofilms and their range of applications have resulted in the development of several different reactor configurations. Reactor productivity, biofilm attachment and stability, the nature of the catalytic reaction and aeration requirements are important parameters to consider during the reactor design (Table 11.3 and Fig. 11.12) (Rosche *et al.*, 2009).

A number of reviews have extensively discussed the advantages and limitations of various biofilm reactors in fermentation and wastewater treatment (Nicolella *et al.*, 2000a,b; Qureshi *et al.*, 2005; Syron and Casey, 2008; Rosche *et al.*, 2009). Conventionally, biofilm reactors are distinguished into two main categories: static (or packed-bed) and dynamic reactor configurations (Table 11.3). These reactors have been operated for decades at full scale in wastewater treatment and some in fermentation processes and are considered a 'grown up' technology for which good design and scale-up guidelines are available (Nicolella *et al.*, 2000a).

Systems with static biofilms, as trickling filter reactors, have been used for over a century in wastewater treatment and for several decades for fermentations. They are characterized by small specific biofilm surface areas (typically less than 300 m^2 biofilm per m^3 reactor) available for substrate transport and reaction, and thus a limited reactor capacity. The oxygen mass transfer rate is typically less than 3 kg/m^3/day equalling 4 mmol/l/h (Nicolella *et al.* 2000a). Such rates are insufficient for most fermentation processes, as most of them report productivities over 20 mmol/l/h (Fig. 11.10). Advantages are that investment costs are low and the energy input for stirring or heat exchange is not required keeping the operational costs minimal. Thus, these reactor types are optimal for the production of low cost, bulk chemicals, which is underlined by the fact that the only large scale processes discussed above are operated in trickle bed reactors.

Higher productivities can be achieved with particle-based biofilm technologies, such as airlift and fluidized-bed reactors, owing to their large specific surface area, efficient mixing and higher mass transfer rates (Godia and Sola, 1995; Nicolella *et al.*, 2000a). In comparison to packed bed reactors, particle-based reactors can be operated with small particles without the risk of clogging, high liquid pressure drop or particle compression due to bed weight (Godia and Sola, 1995). In the case of

Table 11.3 Main parameters of the most common reactor configurations. Reprinted from Rosche *et al.* (2009) with permission from Elsevier

Process parameters	Trickle packed-bed reactor	Submerged packed-bed reactor	Airlift reactor	Fluidized-bed reactor
Oxygen supply	Medium–low	Low	Medium–high	Low
Shear stress	Low	Low	High	Medium–high
Biofilm establishment time	Weeks	Weeks	Weeks–month	Weeks–month
Energy consumption	Low	Low	Medium–high	High
m^2/m^3	Low–medium	Low–medium	Medium–high	Medium–high
Productivity/m^3	Low–medium	Low–medium	Medium–high	Medium–high

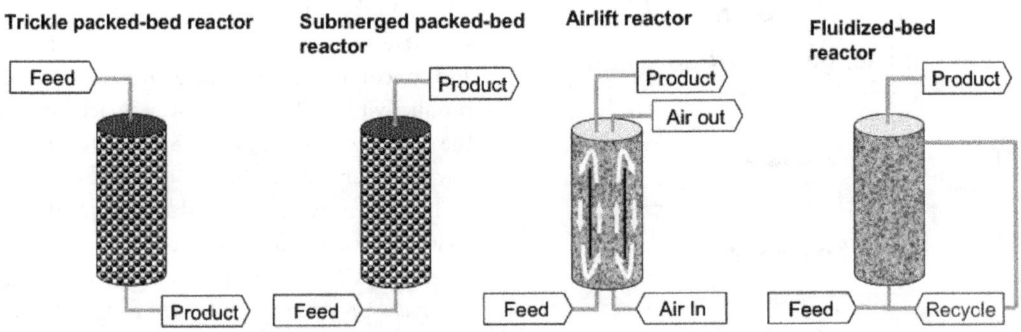

Figure 11.12 Basic configurations of the most common biofilm reactor systems. For reactor parameters, see Table 11.3. Reprinted from Rosche *et al.* (2009), with permission from Elsevier.

fluidized bed reactors, the operational regime can be chosen according to the reaction conditions. The flow rate can be decreased so that the reactor approaches a plug flow regime which is kinetically advantageous if the reaction is product inhibited. On the other hand with higher flow rates, complete mixing can be obtained, which is favourable for substrate limited conditions (Godia and Sola, 1995). However, harsh hydrodynamic conditions resulting in strong shear forces limit the application of these reactor types to strongly attached biofilms.

Another reactor type is the membrane attached biofilm reactor (MABR), which has been exclusively used for wastewater treatment so far (Fig. 11.13) (dos Santos *et al.*, 1995, 1997). This concept has been proven to be rather successful in terms of performance and process stability. The non-porous dense membranes consisting usually of silicone rubber, polyethylene and polypropylene serve as a support for the biofilm and are permeable for substrates like oxygen and small organic compounds (Brindle and Stephenson, 1996). Depending on the operation conditions, high mass transfer rates through the membranes have been achieved (Zhang *et al.*, 1998; Syron and Casey, 2008). In a recent patent application (2008/0305539 and 2008/0305540), Coskata (2008) reported the use of hydrophobic membranes for the biotransformation of syngas in biofilm reactors catalysed by biofilms (Hickey *et al.*, 2008).

A basic problem in all aerobic bioprocesses is oxygen supply which is considered as a main bottleneck in biofilm reactor design (Nicolella *et al.*, 2000a; Jesus *et al.*, 2009). Different methods for oxygen supply have been developed. Bubble aeration is a direct aeration technique using spargers, while bubble-free aeration employs thin membranes (e.g. silicone, polypropylene), which separate the liquid phase from a pressurized gas phase. Direct sparging bubble aeration is still the

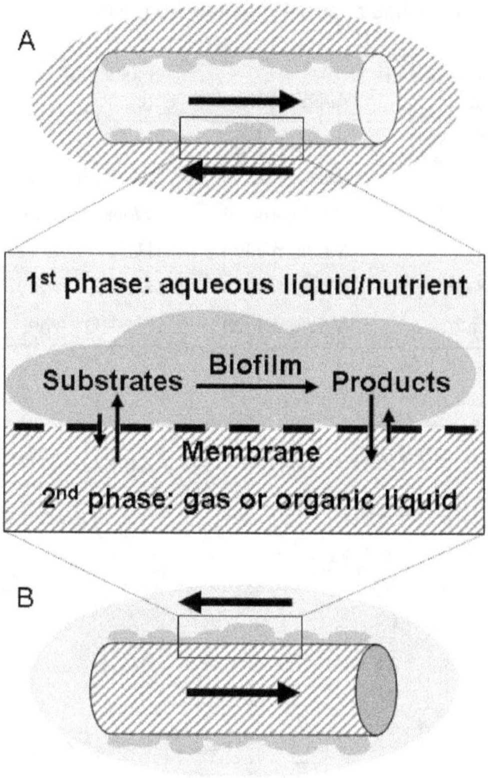

Figure 11.13 Concept of a tubular biofilm membrane reactor adapted to the two-liquid phase approach. (Close-up A) In this example the biofilm is growing inside a tubular silicone membrane in an aqueous nutrient solution. The biotransformation substrate is dissolved in an organic phase, which is situated outside of the tubing. The substrate is supplied to the biocatalyst via diffusion through the membrane, converted by the biofilm to the respective product and extracted back again into the organic phase. (Close-up B) In other set ups the biofilm is grown on the outside of the membrane, while passing oxygen, or other substrates/gases through the inner tube. Reprinted from Rosche *et al.* (2009), with permission from Elsevier.

most widely used technique at all application scales due to its simplicity and high oxygen mass transfer (Nehring *et al.*, 2004; Krahe, 2005; Syron and Casey, 2008). Ceramic spargers with homogeneous microporous distribution showed an enhanced performance in terms of oxygen transfer rates compared with bubble-free aeration systems (Nehring *et al.*, 2004). A significant advantage lies in the smaller required aeration surface area and chemical, thermal and mechanical stability (Sheldon and Small, 2005). Indirect aeration is less attractive because of the technical complexity, the high apparatus cost and difficulties in scale-up. Recently, Halan *et al.* (2010) reported on a concept combining MABRs and bubble free aeration, also employing *Pseudomonas* sp. strain VLB120ΔC as biocatalyst for the conversion of styrene to (*S*)-styrene oxide. A microporous ceramic unit was used as the surface for biofilm growth as well as an efficient oxygen transfer unit. Owing to this dual function, the reactor configuration could be significantly simplified by eliminating additional packing materials, as compared with traditional packed bed reactors. Silicone tubing filled with the biotransformation substrate styrene served for *in situ* substrate feed and concomitantly product recovery. A maximum productivity of 28 g/l aq/day was achieved and the system was stable for more than 30 days before it was actively terminated.

In a recent work, Govender *et al.* (2010) used an innovative biofilm reactor concept, a gradostat, for lignolytic enzyme production. The aerobic ligninase producing white rot fungus *Phanerochaete chrysosporium* was grown in the pores of an ultrafiltration capillary membrane through which nutrient medium was pumped, while from the shell side (external side) oxygen was provided to the biofilm. Nutrient gradients (hence the term gradostat) resulting from radial, outward (tube to shell side) biofilm development were exploited to provide both nutrient rich and poor zones. While this allowed for primary metabolism in the nutrient sufficient zone, secondary metabolism was induced by starvation in the nutrient poor zones. These nutrient-depleted zones constitute the active fraction of the biomass as starvation is the trigger for lignolytic enzyme production in this fungus (Ntwampe *et al.*, 2007; Govender *et al.*, 2010). By optimizing the composition of a complex-rich medium, an enzyme production phase of around 16 days could be obtained, with a peak volumetric productivity of manganese peroxidase of 1.3 units/l/h.

Challenges in biofilm mediated catalysis

Biofilms harbour a huge potential regarding their utilization as productive biocatalysts. However, there are a number of yet unsolved problems we are only beginning to understand, which have to be taken into close consideration when developing biofilm based bioprocesses. Table 11.4 summarizes advantages and challenges of biofilm based technologies.

Certainly, one major issue is the mass transfer of substrates and products and the consequent chemical heterogeneity generated inside the microbial community (Stewart, 2003). In particular, reactions requiring oxygen are strongly dependent on an efficient oxygen distribution in the biofilm. Diffusion limitation arises readily in biofilm systems because fluid flow is reduced and the diffusion distance is increased in the biofilm mode of growth. As biofilms mostly form elaborate structures consisting of distinct cell clusters of mushrooms or pillars and interstitial voids, investigation of transport phenomena is rather complex (O'Toole *et al.*, 2000). It has been suggested that these structures facilitate substrate delivery and product or waste removal (Beyenal and Lewandowski, 2001). Combining confocal scanning microscopy with microelectrode techniques, the relationship between structure and mass transfer was elucidated by de Beer and coworkers in mixed species biofilms of *Klebsiella pneumoniae, Pseudomonas aeroginosa* and *P. fluorescens* (de Beer *et al.*, 1994, 1997; de Beer and Stoodley, 1995). They showed that the distribution of solutes and oxygen strongly correlates with biofilm structures. In the voids the bulk liquid is moving allowing transport via diffusion and convection, while inside the cell clusters, the locally high cell densities and the presence of EPS arrest the flow of water. This limits the transport of solutes and oxygen to diffusion, which are orders of magnitude slower than transport by convection. The oxygen concentration in the voids was higher than in the clusters, inducing a diffusional flux from the voids into the clusters. However, this was dependent on the flow velocity: with increasing flow, convectional transport increased the oxygen concentration in the voids. Conversely, at very low velocities, the mass boundary layer is parallel to the substratum and no convective transport from the voids to the clusters occurred. Lower regions of the cell clusters stayed anoxic, depending on the total depth of the cluster.

A simple mathematical understanding of mass transfer of solutes in biofilms is given by Stewart (2003). Diffusion coefficients of various compounds in biofilms are provided and some rough estimation about the diffusional reduction [termed as ratio of the effective diffusion over diffusion in water (D_e/D_{aq})] was given:

The relative effective diffusivity depends on the biomass density in the biofilm and the physiochemical properties of the solute. D_e/D_{aq} in biofilm ranges from ca. 0.2 to 0.8 with a mean value of ca. 0.4. It was suggested to use a value of D_e/D_{aq} of 0.6 for light gases (e.g. oxygen, nitrous oxide, carbon dioxide, or methane) and a value of D_e/D_{aq} of 0.25 for most organic solutes (Stewart, 2003). Several conclusions could be drawn about diffusional processes in biofilms:

1 Diffusion is the predominant solute transport process within cell clusters.
2 The time scale for diffusive equilibration of a non-reacting solute will range from a fraction of a second to tens of minutes in most biofilm systems.
3 Diffusion limitation readily leads to gradients in the concentration of reacting solutes and hence to gradients in physiology.
4 Water channels can carry solutes into or out of the depths of a biofilm, but they do not guarantee access to the interior of cell clusters.

Oxygen diffusion is the best studied and most prominent example and its concentration is routinely measured using microelectrodes (Fig. 11.14) (Horn and Hempel, 1997; Wäsche and Hempel, 2000; Wäsche *et al.*, 2002). The penetration depth of oxygen into a biofilm is usually in a range of tens to a few hundreds of micrometres with an exponentially declining profile from the bulk liquid towards the deep layers of the biofilm. However, it has to be stated that it is not physical diffusional limitation that causes this oxygen gradient, as the diffusion rate of oxygen and other solutes of similar size inside a fully hydrated biofilm matrix is still about 60% of that in pure water. Oxygen actually fails to penetrate because it is actively respired by the cells in the upper layers of the biofilm (Stewart and Franklin, 2008).

Table 11.4 Advantages and challenges using microbial biofilms as biocatalysts compared with traditional planktonic biocatalysts

Advantages	Challenges
Self-immobilization of biocatalysts	–
Viable biocatalysts capable of adapting to changing conditions	–
High cell densities achievable enabling high volumetric productivities during continuous operation	Growth and detachment control to minimize risk of reactor obstruction; mass transfer limitations and gradients of oxygen, nutrients and other reaction compounds can lead to starvation and/or toxification of the biocatalyst
Resistance against toxic chemicals (depending on strain, conditions and toxic chemical)	Physiological heterogeneities resulting in a varying and often unknown fraction of active compared with inactive biocatalysts
Robust biocatalysts allow continuous biotransformations and fermentations over long periods (>months)	Constant pressure of natural evolution: accumulation of mutants, loss or gain of function over long process periods
Simple reactor configuration and ease of operation decrease capital and operating costs	A mathematical description of a biofilm is difficult owing to many unknown parameters

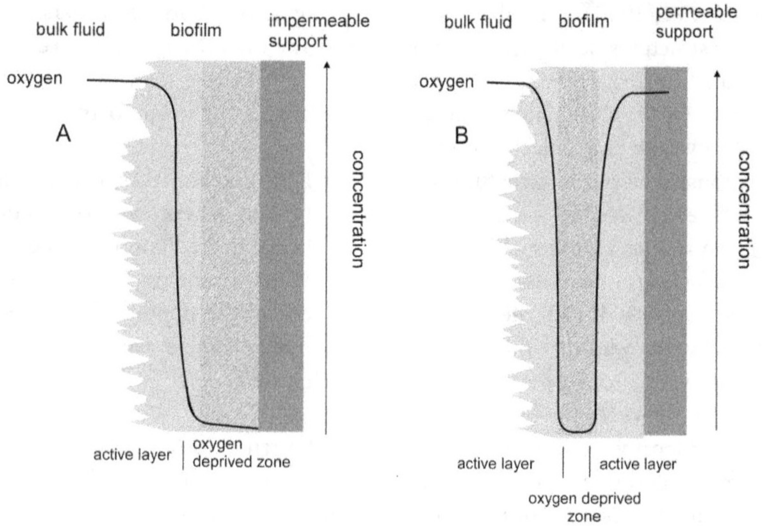

Figure 11.14 Schematic of oxygen concentration profiles in biofilms. (A) Biofilm growing on an oxygen impermeable support. Oxygen supply only occurs from one side of the biofilm. The oxygen deprived zone is localized directly above the support. (B) Biofilm growing on an oxygen permeable support as in the case of membrane attached biofilm reactors (MABR). Oxygen can enter the biofilm from two sides and the anaerobic layer is situated in the middle of the biofilm.

In contrast to oxygen or nutrients, a metabolic product is characterized by higher concentration in the biofilm interior and lower concentration outside the biofilm. Net transport of the product occurs along the concentration gradient and out of the biofilm. Other solutes are both produced and consumed in a biofilm. This situation arises with a metabolic intermediate or in mixed-species biofilms when the waste product of one species can serve as the substrate for another species. In such cases, concentration profiles exhibit a maximum within the biofilm (Stewart and Franklin, 2008). Apart from oxygen, microscale concentration profiles have been determined experimentally for a number of solutes, including nitrate, nitrite, ammonium, hydrogen sulfide, and carbon dioxide (Stewart and Franklin, 2008). Furthermore, microscale pH gradients have been observed and characterized (Hidalgo *et al.*, 2009). Hildalgo and coworkers developed a technique enabling functional tomographic (section-based) fluorescence imaging of pH microenvironments in biofilms by the use of small silica nanoparticle sensors (C dot sensors). The silica nanoparticles comprised a pH sensitive fluorescent dye in the silica shell and a pH insensitive reference fluorescent dye protected in the core of the particle. By that internally standardized design, quantitative pH measurements were possible, independent of changes in sensor concentration. With C dot sensors of 10 nm particle size a homogeneous staining throughout the biofilm was achieved. The advantage of using this technique is that it minimizes perturbation of the biofilm structure, which is a significant drawback of microelectrode-based techniques. Furthermore, silica is highly biocompatible, that is, non-toxic and not interacting with biofilm components as some of the free fluorescent dyes are, which may alter the sensitivities and perturb the resulting images. Using this method, the authors were capable of generating high-resolution 3D maps of biofilm pH with distinct local heterogeneities where pH values ranged from 5 to >7.

As a result of these chemical heterogeneities, biofilms typically consist of multiple layers of cells in different physiological states that might express distinct metabolic pathways (Stewart and Franklin, 2008). Accumulation of acidic waste products in the biofilm interior, for example, could lower the local pH and affect the physiological state of the bacteria (Hidalgo *et al.*, 2009). As the genome of certain bacterial species is composed of up to 15% of regulatory genes, bacteria are exquisitely adept at sensing and responding to changes in environmental conditions (Stewart and Franklin, 2008). Therefore, gene expression and physiological activities of the bacteria varies as they adapt to the local microniches within a biofilm, even when separated by as little as 10 μm (Xu *et al.*, 2000). These observations of physiological heterogeneities and adaptation lead to the increasingly acknowledged notion that biofilm microorganisms can undergo a process of cell specialization that is analogous to differentiation in multicellular organisms (Nadell *et al.*, 2009).

Developing biofilm-based catalysis one has to take complex transport mechanisms into account, as the active cell layers ('active' in the sense of catalytically active) will be defined by the depth of penetration of the necessary compounds including nutrients, biotransformation substrates, or oxygen in the case of aerobic reactions (Casey *et al.*, 1999). In parts, this problem can be solved by using intelligent reactor configurations like the membrane attached biofilm reactors (MABR), discussed previously. However, sophisticated reactor configurations are often hampered by excessive biomass growth leading to clogging of the system and abetting mass transfer limitations (Nicolella *et al.*, 2000a; Stewart, 2003). A combination of nutrient limitation and appropriate reactor design have been reported to successfully control biofilm growth (Förberg and Haystrom, 1985; Qureshi *et al.*, 2005). In the respective study a biofilm of *Clostridium acetobutylicum* adsorbed to beechwood shavings has been kept in a non-growing (resting) state to produce acetone and butanol in a fermentative process. As resting cells were shown to lose activity over time, a nutrient dosing technique was applied to achieve a constant productivity. The feeding technique comprised two different media, one with high glucose concentration and a nitrogen source and one without, containing only the typical buffer components. Every 8 hours glucose containing medium was pulsed to keep the activity on a relatively constant level.

Moreover, unproductive biomass, such as dead cells, dormant cells, and EPS has to be controlled to improve (specific and volumetric) productivity and lower carbon source consumption. EPS function as a 'glue' adhering cells onto the surface and sheathing them to develop a biofilm, therefore it is an essential part of the biofilm. EPS consist mainly of polysaccharides; the composition and amount of EPS formed is strain and nutrient dependent. For example, excess of a carbon source and limitation of other nutrients such as nitrogen or potassium may promote the production of EPS (Sutherland, 2001), which could make up to ~90% of the biofilm biomass (Wingender *et al.*, 1999). A recent study investigated the activity of d-amino acid oxidase in a recombinant *Acetobacter xylinum* biofilm (Setyawati *et al.*, 2009). *Acetobacter* sp. produce cellulose nanofibres, by which they may either attach to surfaces or form a kind of pellicle (thin protective membrane). In the above mentioned study, only 25% of the total biomass was actually composed of cells, while the other 75% consisted of cellulose. This was assumed to have a negative impact on the biotransformation activity, due to mass transfer limitations across the EPS matrix, as biofilm activities were about 30% lower than planktonic cultures (Setyawati *et al.*, 2009). It has also been suggested that accumulation of unproductive EPS material within bioreactors may reduce the available reactor space for active cells and therefore hamper the volumetric productivity of catalytic biofilm processes (Gross *et al.*, 2007). EPS are thus a 'double-edged sword' as they are on the one hand an integral part of biofilms needed for attachment, structure and protection, while on the other hand they may hamper mass transfer and thus the volumetric productivities of the biofilm. EPS synthesis should be controlled to the minimum necessary for biofilm formation. This is an as yet unsolved issue and so far may only be influenced by the choice of organism and medium composition. First engineering approaches aiming to manipulate the respective genes encoding for EPS production and/or protein domains have had a negative impact on EPS production (as earlier described). However, in all these studies a reduction in EPS production always concomitantly reduced the total biomass and gave way to weak biofilms or even repressed biofilm formation totally.

A further challenge is the physiological heterogeneity in the biofilm population, which makes it difficult, if not impossible to control the specific activities of the cells. The fraction of active over inactive cells is influenced by many factors which we are only slowly beginning to understand. Apart from the abovementioned various adaption processes to microniches inside the biofilm, which might also affect the target reaction of a biocatalytic process, local limitations in essential substrates such as carbon or oxygen might also play an important role, due to mass transfer limitations.

Nevertheless, in our opinion the main bottleneck for the establishment of biofilm based technologies is that biofilms adapt to an enormous variety of conditions under which they develop very specific structures. These specificities may well be unique to that single environment and microflora, which necessitates tailor-made approaches not only for every target product, but also for each reactor configuration.

Future trends and outlook

So far, biofilm based processes have mostly been developed for fermentations utilizing wild type, often anaerobic, organisms. This might be due to the fact, that most published examples were developed some decades ago in the 1980s, when strain engineering tools were still in their fledgling stages. This obviated a broader application towards the production of high value compounds. As biofilm research has been predominantly focused on the genetics of clinically relevant strains and methods for the destruction and prevention of biofilms rather than on their promotion, knowledge about the implications of biofilm formation on the productivity of industrially relevant strains with recombinant genes is still rare. However, in the recent years biofilm studies utilizing genetically engineered strains have been conducted with promising results (Gross *et al.*, 2007; Halan *et al.*, 2010). A study on plasmid maintenance and heterologous protein production in *E. coli* biofilms showed interesting results: even in the absence of antibiotic selective pressure a high copy number of plasmids could be maintained with high heterologous protein levels over several days (O'Connell *et al.*, 2007). Recently, an *E. coli* strain containing chromosomally integrated genes for ethanol production increased phenotypic stability, generating higher yields, longer production periods, and better ethanol tolerance as compared with the corresponding planktonic culture (Zhou *et al.*, 2008). Further investigations to better our molecular understanding of such phenomena are now required. Rosche *et al.* (2009) suggest the development of a few biofilm hosts as industrial workhorses, which would help to deepen our understanding of biofilm biology for industrial relevant strains. Such biofilm hosts could then be genetically modified to express enzymes required to catalyse the desired transformation of substrate to product.

More research has to be invested in biofilm reactor engineering. As many biocatalytic reactions performed with whole cells have high oxygen demands, reactor configurations with large surface to volume ratios are required. Recent developments in membrane aerated reactors with hollow fibre modules for wastewater treatment have to be evaluated for their potential to be implemented in productive biocatalysis. The use of structured packing materials in biofilm reactors has also been considered. These packing materials have been applied in the chemical industry for distillation and surface catalysis on a large scale, and observed to allow efficient mass transfer between gas and liquid phases (Rosche *et al.*, 1997).

Apart from the prevailing challenges in biofilm driven catalysis due to our still insufficient understanding of these highly dynamic microbial communities, studies conducted in recent years show great promise for establishing biofilm-based technology permanently in applied industrial biocatalysis. We are just beginning to explore the tremendous potential of this fascinating microbial lifestyle.

References

Abraham, W.R. (2006). Controlling biofilms of Gram-positive pathogenic bacteria. Curr. Med. Chem. *13*, 1509–1524.

Allegrucci, M., and Sauer, K. (2007). Characterization of colony morphology variants isolated from *Streptococcus pneumoniae* biofilms. J. Bacteriol. *189*, 2030–2038.

Anderson, G.G., and O'Toole, G.A. (2008). Innate and induced resistance mechanisms of bacterial biofilms. Curr. Top. Microbiol. Immunol. 322, 85–105.

Avnir, D., Coradin, T., Lev, O., and Livage, J. 2006. Recent bio-applications of sol–gel materials. J. Mater. Chem. *16*, 1013 - 1030.

Bagge, N., Schuster, M., Hentzer, M., Ciofu, O., Givskov, M., Greenberg, E.P., and Hoiby, N. (2004). *Pseudomonas aeruginosa* biofilms exposed to imipenem exhibit changes in global gene expression and beta-lactamase and alginate production. Antimicrob. Agents Chemother. *48*, 1175–1187.

de Beer, D., and Stoodley, P. (1995). Relation between the structure of an aerobic biofilm and transport phenomena. Water Sci. Technol. 32, 11–18.

de Beer, D., Stoodley, P., Roe, F., and Lewandowski, Z. (1994). Effects of biofilm structures on oxygen distribution and mass transport. Biotechnol. Bioeng. *43*, 1131–1138.

de Beer, D., Stoodley, P., and Lewandowski, Z. (1997). Measurement of local diffusion coefficients in biofilms by microinjection and confocal microscopy. Biotechnol. Bioeng. 53,151–158.

Beyena, H., and Lewandowski, Z. (2001). Mass-transport dynamics, activity, and structure of sulfate-reducing biofilms. Aiche J. 47, 1689–1697.

Bickerstaff, G.F. (1997). Immobilization of Enzymes and Cells. (Springer, Berlin).

Boles, B.R., Thoendel, M., and Singh, P.K. (20050. Self-generated diversity produces 'insurance effects' in biofilm communities. Proceed. Natl.

Acad. Sci. *102*, 955–955.

de Bont, J.A.M. (1998). Solvent-tolerant bacteria in biocatalysis. Trends Biotechnol. *16*, 493–499.

Bommarius, A., and Riebel, B. (2004). Biocatalysis Fundamentals and Applications. (Wiley-VCH, Weinheim).

Branda, S.S., Vik, A., Friedman, L., and Kolter, R. (2005). Biofilms: the matrix revisited. Trends Microbiol. *13*, 20–26.

Brindle, K., and Stephenson, T. (1996). The application of membrane biological reactors for the treatment of wastewaters. Biotechnol. Bioeng. *49*, 601–610.

Bryers, J.D. (2008). Medical biofilms. Biotechnol. Bioeng. *100*, 1–18.

Buchholz, K., and Seibe,lJ. (2008). Industrial carbohydrate biotransformations. Carbohydr. Res. *343*, 1966–1979.

Buchholz, K., Kasche, V., and Bornscheuer, U.T. (2005). Biocatalysts and Enzyme Technology. (Wiley-VCH, Weinheim).

Buckland, B.C. (1999). Microbial conversion of indene to indandiol: a key intermediate in synthesis of Crixivan. Metab. Eng. *1*, 63–74.

Cao, N., Du, J., Gong, C.S., and Tsao, G.T. (1996). Simultaneous production and recovery of fumaric acid from immobilized *Rhizopus oryzae* with a rotary biofilm contactor and an adsorption column. Appl. Environ. Microbiol. *62*, 2926–2931.

Casey, E., Glennon, B., and Hamer, G. (1999). Review of membrane aerated biofilm reactors. Resour. Conservat. Recycl. 27, 203–215.

Characklis, W.G., and Marshall, K.C. (1990). Biofilms. (John Wiley, New York).

Cho, H.Y., Yousef, A.E., and Yang, S.T. (1996). Continuous production of pediocin by immobilized *Pediococcus acidilactici* P02 in a packed-bed bioreactor. Appl. Environ. Microbiol. *45*, 589–594.

Ciesarova, Z., Domeny, Z., Smogrovicova, D., Patkova, J., and Sturdik, E. (1998). Comparison of ethanol tolerance of free and immobilized *Saccharomyces uvarum* yeasts. Folia Microbiol. *43*, 55–58.

Costerton, J.W., and Stewart, P.S. (2001). Battling biofilms. Sci. Am. *285*, 74–81.

Costerton, J.W., Cheng, K.J., Geesey, G.G., Ladd, T.I., Nickel, J.C., Dasgupta, M., and Marrie, T.J. (1987). Bacterial biofilms in nature and disease. Ann. Rev. Microbiol. *41*, 435–464.

Costerton, J.W., Lewandowski, Z., Caldwell, D.E., Korber, D.R., and Lappin-Scott, H.M. (1995). Microbial biofilms. Ann. Rev. Microbiol. *49*, 711–745.

Crueger, W., and Crueger, A. (1990). Acetic acids. In Biotechnology: A Textbook of Industrial Microbiology, Brock, T.D., editor. (Science Tech Publishers, Sunderland, Ma, USA). pp. 134–147.

Davies, J.A., Harrison, J.J., Marques, L.L.R., Foglia, G.R., Stremick, C.A., Storey, D.G., Turner, R.J., Olson, M.E., and Ceri, H. (2007). The GacS sensor kinase controls phenotypic reversion of small colony variants isolated from biofilms of *Pseudomonas aeruginosa* PA14. FEMS Microbiol. Ecol. *59*, 32–46.

De Kievit, T.R., Parkins, M.D., Gillis, R.J., Srikumar, R., Ceri, H., Poole, K., Iglewski, B.H., and Storey, D.G. (2001). Multidrug efflux pumps: Expression patterns and contribution to antibiotic resistance in *Pseudomonas aeruginosa* biofilms. Antimicrob. Agents Chemother. *45*, 1761–1770.

Desimone, M.F., Alvarez, G.S., Foglia, M.L., and Diaz, L.E. (2009). Development of sol–gel hybrid materials for whole cell immobilization. Recent Patent. Biotechnol. *3*, 55–60.

Doig, S.D., Avenell, P.J., Bird, P.A., Gallati, P., Lander, K.S., Lye, G.J., Wohlgemuth, R., and Woodley, J.M. (2002). Reactor operation and scale-up of whole cell Baeyer-Villiger catalyzed lactone synthesis. Biotechnol. Progr. *18*, 1039–1046.

Donlan, R.M. (2002). Biofilms: microbial life on surfaces. Emerg. Infect. Diseases *8*, 881–890.

Drenkard, E., and Ausubel, F.M. (2002). *Pseudomonas* biofilm formation and antibiotic resistance are linked to phenotypic variation. Nature *416*, 740–743.

Faber, K. (2004). Biotransformations in Organic Chemistry: a Textbook. (Springer, Berlin).

Flemming, H.C. (1993). Biofilms and environmental protection. Water Sci. Technol. *27*, 1–10.

Flemming, H.C. (2002). Biofouling in water systems: cases, causes and countermeasures. Appl. Microbiol. Biotechnol. *59*, 629–640.

Förberg, C., and Haggstrom, L. (1985). Control of cell-adhesion and activity during continuous production of acetone and butanol with adsorbed cells. Enzyme Microb. Technol. *7*, 230–234.

Fujita, M., Ike, M., and Hashimoto, S. (1991). Feasibility of waste-water treatment using genetically engineered microorganisms. Water Res. *25*, 979–984.

Fux, C.A., Costerton, J.W., Stewart, P.S., and Stoodley, P. (2005). Survival strategies of infectious biofilms. Trends Microbiol. *13*, 34–40.

Gbewonyo, K., Buckland, B.C., and Lilly, M.D. (1991). Development of a large-scale continuous substrate feed process for the biotransformation of simvastatin by *Nocardia sp*. Biotechnol. Bioeng. *37*, 1101–1107.

Gjermansen, M., Ragas, P., and Tolker-Nielsen, T. (2006). Proteins with GGDEF and EAL domains regulate *Pseudomonas putida* biofilm formation and dispersal. FEMS Microbiol. Lett. *265*, 215–224.

Godia, F., and Sola, C. (1995). Fluidized-bed bioreactors. Biotechnol. Prog. *11*, 479–497.

Govender, S., Pillay, V., and Odhav, B. (2010). Nutrient manipulation as a basis for enzyme production in a gradostat bioreactor. Enzyme Microb. Technol. *46*, 603–609.

van Groenestijn, J.W., Geelhoed, J.S., Goorissen, H.P., Meesters, K.P.M., Stams, A.J.M., and Claassen, P.A.M. (2009). Performance and population analysis of a non-sterile trickle bed reactor inoculated with *Caldicellulosiruptor saccharolyticus*, a thermophilic hydrogen producer. Biotech. Bioeng. *102*, 1361–1367.

Gross, R., Hauer, B., Otto, K., and Schmid, A. (2007). Microbial biofilms: New catalysts for maximizing productivity of long-term biotransformations. Biotechnol Bioeng. *98*, 1123–1134.

Gross, R., Lang, K., Buehler, K., and Schmid, A. (2010). Characterization of a biofilm membrane reactor and its prospects for fine chemical synthesis. Biotechnol. Bioeng. *105*, 705–717.

Halan, B., Schmid, A., and Buehler, K. (2010). Maximizing the productivity of catalytic biofilms on solid supports in membrane reactors. Biotechnol. Bioeng. *106*, 516–527.

Hall-Stoodley, L., and Stoodley. P. (2002). Developmental regulation of microbial biofilms. Curr. Opin. Biotechnol. *13*, 228–233.

Hall-Stoodley, L., Costerton, J.W., and Stoodley, P. (2004). Bacterial biofilms: From the natural environment to infectious diseases. Nat. Rev. Microbiol. *2*, 95–108.

Hammer, B.K., and Bassler, B.L. (2003). Quorum sensing controls biofilm formation in *Vibrio cholerae*. Mol. Microbiol. *50*, 101–114.

Hansen, S.K., Haagensen, J.A.J., Gjermansen, M., Jorgensen, T.M., Tolker-Nielsen, T., and Molin, S. (2007a). Characterization of a *Pseudomonas putida* rough variant evolved in a mixed-species biofilm with *Acinetobacter sp* strain C6. J. Bacteriol *189*, 4932–4943.

Hansen, S.K., Rainey, P.B., Haagensen, J.A., and Molin, S. (2007b). Evolution of species interactions in a biofilm community. Nature *445*, 533–536.

Harrison, J.J., Ceri, H., and Turner, R.J. (2007). Multimetal resistance and tolerance in microbial biofilms. Nat. Rev. Microbiol. *5*, 928–938.

Hassett, D.J., Ma, J.F., Elkins, J.G., McDermott, T.R., Ochsner, U.A., West, S.E., Huang, C.T., Fredericks, J., Burnett, S., Stewart, P.S., *et al.* (1999). Quorum sensing in *Pseudomonas aeruginosa* controls expression of catalase and superoxide dismutase genes and mediates biofilm susceptibility to hydrogen peroxide. Mol. Microbiol. *34*, 1082–1093.

Hausner, M., and Wuertz, S. (1999). High rates of conjugation in bacterial biofilms as determined by quantitative *in situ* analysis. Appl. Environ. Microbiol. *65*, 3710–3713.

Haussler, S. (2004). Biofilm formation by the small colony variant phenotype of *Pseudomonas aeruginosa*. Environ. Microbiol. *6*, 546–551.

Heipieper, H.J., Weber, F.J., Sikkema, J., Keweloh, H., and Debont, J.A.M. (1994). Mechanisms of resistance of whole cells to toxic organic-solvents. Trends Biotechnol. *12*, 409–415.

Heipieper, H.J., Neumann, G., Cornelissen, S., and Meinhardt, F. (2007). Solvent-tolerant bacteria for biotransformations in two-phase fermentation systems. Appl. Microbiol. Biotechnol. *74*, 961–973.

Hekmat, D., Bauer, R., and Fricke, J. (2003). Optimization of the microbial synthesis of dihydroxyacetone from glycerol with *Gluconobacter oxydans*. Bioproc. Biosyst. Eng. *26*, 109–116.

Hengge, R. (2009). Principles of c-di-GMP signalling in bacteria. Nat. Rev. Microbiol. 7, 263–273.

Hickey, R., Datta, R., Tsai, S.P., and Basu, R (2008). Membrane supported bioreactor for conversion of syngas components to liquid products. Patent WO2008154301.

Hidalgo, G., Burns, A., Herz, E., Hay, A.G., Houston, P.L., Wiesner, U., and Lion, L.W. (2009). Functional tomographic fluorescence imaging of pH-microenvironments in microbial biofilms by use of silica nanoparticle sensors. Appl. Environ. Microbiol. *75*, 7426–7435.

Ho, K-L.G., Pometto, A.L., and Hinz, P.N. (1997). Optimization of L-(+)-lactic acid production by ring and disc plastic composite supports through repeated-batch biofilm fermentation. Appl. Environ. Microbiol. *63*, 2533–2542.

Horn, H., and Hempel, D.C. (1997). Growth and decay in an auto-/heterotrophic biofilm. Water Res. *31*, 2243–2252.

Irie, Y., and Parsek, M.R. (2008). Quorum sensing and microbial biofilms. Curr. Top. Microbiol. Immunol. *332*, 67–84.

Ishige, T., Honda, K., and Shimizu, S. (2005). Whole organism biocatalysis. Curr. Opin. Chem. Biol. *9*, 174–180.

Ishihara, Y., Hyodo, M., Hayakawa, Y., Kamegaya, T., Yamada, K., Okamoto, A., Hasegawa, T., and Ohta, M. (2009). Effect of cyclic bis(3′–5′) diguanylic acid and its analogs on bacterial biofilm formation. FEMS Microbiol. Lett. *301*, 193–200.

Jesus, A., Romano-Baez, F., Leyva-Amezcua, L., Juarez-Ramirez, C., Ruiz-Ordaz, N., and Galindez-Mayer, J. (2009). Biodegradation of 2,4,6-trichlorophenol in a packed-bed biofilm reactor equipped with an internal net draft tube riser for aeration and liquid circulation. J. Hazard. Mater. *161*, 1140–1149.

Jirku, V. (1999). Whole cell immobilization as a means of enhancing ethanol tolerance. J. Ind. Microbiol. Biotechnol. 22, 147–151.

Ju, D.H., Shin, J.H., Lee H.K., Kong, S.H., Kim, J.I., and Sang, B.I. (2008). Effects of pH conditions on the biological conversion of carbon dioxide to methane in a hollow-fiber membrane biofilm reactor (Hf-MBfR). Desalination *234*, 409–415.

Kirisits, M.J., and Parsek, M.R. (2006). Does *Pseudomonas aeruginosa* use intercellular signalling to build biofilm communities? Cell. Microbiol. *8*, 1841–1849.

Kirisits, M.J., Prost, L., Starkey, M., and Parsek, M.R. (2005). Characterization of colony morphology variants isolated from *Pseudomonas aeruginosa* biofilms. Appl. Environ. Microbiol. *71*, 4809–4821.

Klausen, M., Aaes-Jorgensen, A., Molin, S., and Tolker-Nielsen, T. (2003a). Involvement of bacterial migration in the development of complex multicellular structures in *Pseudomonas aeruginosa* biofilms. Mol. Microbiol. *50*, 61–68.

Klausen, M., Heydorn, A., Ragas, P., Lambertsen, L., Aaes-Jorgensen, A., Molin, S., and Tolker-Nielsen, T. (2003b). Biofilm formation by *Pseudomonas aeruginosa* wild type, flagella and type IV pili mutants. Mol. Microbiol. *48*, 1511–1524.

Knapp, F. (1847). Lehrbuch der chemischen Technologie. (F. Vieweg und Sohn, Braunschweig).

Koh, K.S., Lam, K.W., Alhede, M., Queck, S.Y., Labbate, M., Kjelleberg, S., and Rice, S.A. (2007). Phenotypic diversification and adaptation of *Serratia marcescens* MG1 biofilm-derived morphotypes. J. Bacteriol. *189*, 119–130.

Kornaros, M., and Lyberatos, G. (2006). Biological treatment of wastewaters from a dye manufacturing company using a trickling filter. J. Hazard. Mater. *136*, 95–102.

Krahe, M. (2005). Biochemical Engineering. Ullmann's Encyclopedia of Industrial Chemistry. (Wiley, Weinheim, Germany).

Krug, T.A., and Daugulis, A.J. (1983). Ethanol production using *Zymomonas mobilis* immobilized on an ion exchange resin. Biotechnol. Lett. 5, 159–164.

Kunduru, M.R., and Pometto, A.L. (1996). Continuous ethanol production by *Zymomonas mobilis* and *Saccharomyces cerevisiae* in biofilm reactors. J. Indust. Microbiol. *16*, 249–256.

Kussell, E., Kishony, R., Balaban, N.Q., and Leibler, S. (2005). Bacterial persistence: A model of survival in changing environments. Genetics *169*, 1807–1814.

Laane, C., Boeren, S., Vos, K., and Veeger, C. (1987). Rules for optimization of biocatalysis in organic-solvents. Biotechnol. Bioeng. *30*, 81–87.

Leid, J.G., Shirtliff, M.E., Costerton, J.W., and Stoodley, P. (2002). Human leukocytes adhere to, penetrate, and respond to *Staphylococcus aureus* biofilms. Infect. Immun. *70*, 6339–6345.

Leon, R., Fernandes, P., Pinheiro, H.M., and Cabral, J.M.S. (1998). Whole-cell biocatalysis in organic media. Enzyme Microb. Tech. 23, 483–500.

Lewis K. (2007). Persister cells, dormancy and infectious disease. Nat. Rev. Microbiol. 5, 48–56.

Lewis, V.P., and Yang, S.T. (1992). Continuous propionic acid fermentation by immobilized *Propionibacterium acidipropionici* in a novel packed-bed bioreactor. Biotechnol. Bioeng. *40*, 465–474.

Li, X.Z., Webb, J.S., Kjelleberg, S., and Rosche, B. (2006). Enhanced benzaldehyde tolerance in *Zymomonas mobilis* biofilms and the potential of biofilms in fine-chemical production. Appl. Environ. Microbiol. *72*, 1639–1644.

Liese, A., Seelbach, K., and Wandrey, C. (2006). Industrial Biotransformations. (Wiley-VCH, Weinheim).

Maddox, I.S., Qureshi, N., and McQueen, J.M. (1988). Continuous production of 2,3-butanediol from whey permeate using cells of *Klebsiella pneumoniae* immobilized on to bonechar. New Zeal. J. Dairy Sci. Tech. 23, 127–132.

Molin, S., and Tolker-Nielsen, T. (2003). Gene transfer occurs with enhanced efficiency in biofilms and induces enhanced stabilisation of the biofilm structure. Curr. Opin. Biotech. *14*, 255–261.

Monds, R.D., and O'Toole, G.A. (2009). The developmental model of microbial biofilms: Ten years of a paradigm up for review. Trends Microbiol. *17*, 73–87.

Moons, P., Michiels, C.W., and Aertsen, A. (2009). Bacterial interactions in biofilms. Crit. Rev. Microbiol. *35*, 157–168.

Murdock, D., Ensley, B.D., Serdar, C., and Thalen, M. (1993). Construction of metabolic operons catalyzing the *de novo* biosynthesis of indigo in *Escherichia coli*. Nat. Biotechnol. *11*, 381–386.

Nadell, C.D., Xavier, J.B., and Foster, K.R. (2009). The sociobiology of biofilms. FEMS Microbiol. Rev. *33*, 206–224.

Nehring, D., Czermak, P., Vorlop, J., and Lubben, H. (2004). Experimental study of a ceramic microsparging aeration system in a pilot-scale animal cell culture. Biotechnol. Prog. *20*, 1710–1717.

Nickel, J.C., Ruseska, I., Wright, J.B., and Costerton, J.W. (1985). Tobramycin resistance of *Pseudomonas aeruginosa* cells growing as a biofilm on urinary catheter material. Antimicrob. Agents Chemother. *27*, 619–624.

Nicolella, C., Loosdrecht, M.C.M., and Heijnen, S.J. (2000a). Particle-based biofilm reactor technology. Tibtech *18*, 312–320.

Nicolella, C., van Loosdrecht, M.C.M., and Heijnen, J.J. (2000b). Wastewater treatment with particulate biofilm reactors. J. Biotechnol. *80*, 1–33.

Ntwampe, S., Sheldon, M., and Volschenk, H. 2007. The membrane gradostat reactor: Secondary metabolite production, bioremediation and commercial potential. Afr. J. Biotechnol. *6*,1164–1170.

O'Connell, H.A., Niu, C., and Gilbert, E.S. (2007). Enhanced high copy number plasmid maintenance and heterologous protein production in an *Escherichia coli* biofilm. Biotechnol. Bioeng. *97*, 439–446.

O'Toole, G.A., and Kolter, R. (1998). Flagellar and twitching motility are necessary for *Pseudomonas aeruginosa* biofilm development. Mol. Microbiol. *30*, 295–304.

O'Toole, G.A., and Stewart, P.S. (2005). Biofilms strike back. Nat. Biotechnol. *23*, 1378–1379.

O'Toole, G., Kaplan, H., and Kolter, R. (2000). Biofilm formation as microbial development. Ann. Rev. Microbiol. *54*, 49–79.

Park, J.B., Bühler, B., Panke, S., Witholt, B., and Schmid, A. (2007). Carbon metabolism and product inhibition determine the epoxidation efficiency of solvent-tolerant *Pseudomonas* sp. Strain VLB120 Delta C. Biotechnol. Bioeng. *98*, 1219–1229.

Parsek, M.R., and Greenberg, E.P. (2000). Acyl-homoserine lactone quorum sensing in Gram-negative bacteria: A signaling mechanism involved in associations with higher organisms. Proc. Natl. Acad. Sci. *97*, 8789–8793.

Parsek, M.R., and Greenberg, E.P. (2005). Sociomicrobiology: The connections between quorum sensing and biofilms. Trends Microbiol. *13*, 27–33.

Patel, R. (2006). Biocatalysis: Synthesis of chiral intermediates for drugs. Curr. Opin. Drug Discov. Devel. *9*, 741–764.

Patrauchan, M.A., Sarkisova, S.A., and Franklin, M.J. (2007). Strain-specific proteome responses of *Pseudomonas aeruginosa* to biofilm-associated growth and to calcium. Microbiol. *153*, 3838–3851.

Paulus, W. (2005). Directory of Microbicides for the Protection of Materials: A Handbook. (Springer, Dordrecht, The Netherlands).

Peacock, A.D., Chang, Y.J., Istok, J.D., Krumholz, L., Geyer, R., Kinsall, B., Watson, D., Sublette, K.L., and White, D.C. (2004). Utilization of microbial biofilms as monitors of bioremediation. Microb. Ecol. *47*, 284–292.

Pollard, D., and Woodley, J. (2007). Biocatalysis for pharmaceutical intermediates: the future is now. Trends Biotechnol. *25*, 66–73.

Qureshi, N., and Maddox, I.S. (1988). Reactor design for the ABE fermentation using cells of *Clostridium acetobutylicum* immobilized by adsorption onto bonechar. Bioproc. Eng. *3*, 69–72.

Qureshi, N., Karcher, P., Cotta, M., and Blaschek, H.P. (2004). High-productivity continuous biofilm reactor for butanol production. Appl. Biochem. Biotechnol. 713–721.

Qureshi, N., Annous, B.A., Ezeji, T.C., Karcher, P., and Maddox, I.S. (2005). Biofilm reactors for industrial bioconversion processes: employing potential of enhanced reaction rates. Microb. Cell Fact. *4*, doi:10.1186/1475-2859-4-24.

Ren, D., Bedzyk, L.A., Thomas, S.M., Ye, R.W., and Wood, T.K. (2004). Gene expression in *Escherichia coli* biofilms. Appl. Microbiol. Biotech. *64*, 515–524.

Richards, J.J., and Melander, C. (2009). Controlling bacterial biofilms. Chembiochem *10*, 2287–2294.

Rosche, B., Tshisuaka, B., Hauer, B., Lingens, F., and Fetzner, S. (1997). 2-oxo-1,2-dihydroquinoline 8-monooxygenase: Phylogenetic relationship to other multicomponent nonheme iron oxygenases. J. Bacteriol. *179*, 3549–3554.

Rosche, B., Li, X.Z., Hauer, B., Schmid, A., and Buehler, K. (2009). Microbial biofilms: a concept for industrial catalysis? Trends Biotech. *27*, 636–643.

Sakuragi, Y., and Kolter, R. (2007). Quorum-sensing regulation of the biofilm matrix genes (pel) of *Pseudomonas aeruginosa*. J. Bacteriol. *189*, 5383–5386.

Salter, G.J., and Kell, D.B. (1991). New materials and technology for cell immobilization. Curr. Opin. Biotech. *2*, 385–389.

dos Santos, L.M.F., and Livingston, A.G. (1995). Membrane-attached biofilms for voc waste-water treatment.1. Novel *in situ* biofilm thickness measurement technique. Biotechnol. Bioeng. *47*, 82–89.

dos Santos, L.M.F., Pavasant, P., Strachan, L.F., Pistikopoulos, E.N., and Livingston, A.G. (1997). Membrane attached biofilms for waste treatment – fundamentals and applications. Pure Appl. Chem. *69*, 2459–2469.

Schembri, M.A., Kjaergaard, K., and Klemm, P. (2003). Global gene expression in *Escherichia coli* biofilms. Mol. Microbiol. *48*, 253–267.

Schmid, A., Dordick, J.S., Hauer, B., Kiener, A., Wubbolts, M.G., and Witholt, B. (2001). Industrial biocatalysis today and tomorrow. Nature *409*, 258–268.

Schmidt, A.J., Ryjenkov, D.A., and Gomelsky, M. (2005). The ubiquitous protein domain EAL is a cyclic diguanylate-specific phosphodiesterase: Enzymatically active and inactive EAL domains. J. Bacteriol. *187*, 4774–4781.

Scott, A.I. (2003). Discovering nature's diverse pathways to vitamin B12: a 35-year odyssey. J.Organ. Chem. *68*, 2529–2539.

Setyawati, M.I., Chien, L.J., and Lee, C.K. (2009). Self-immobilized recombinant *Acetobacter xylium* for biotransformation. Biochem. Eng. J. *43*, 78–84.

Shah, D., Zhang, Z., Khodursky, A., Kaldalu, N., Kurg, K., and Lewis, K. (2006). Persisters: a distinct physiological state of *E. coli*. BMC Microbiol. *6*, doi:10.1186/1471-2180-6-53.

Sheldon, M., and Small, H. (2005). Immobilisation and biofilm development of *Phanerochaete chrysosporium* on polysulfone and ceramic membranes. J Memb. Sci. *263*, 30–37.

Sikkema, J., Debont, J.A.M., and Poolman, B. (1995). Mechanisms of membrane toxicity of hydrocarbons. Microbiol. Rev. *59*, 201–222.

Simm, R., Morr, M., Kader, A., Nimtz, M., and Romling, U. (2004). GGDEF and EAL domains inversely regulate cyclic di-GMP levels and transition from sessility to motility. Mol. Microbiol. 53, 1123–1134.

Simm, R., Fetherston, J.D., Kader, A., Romling, U., and Perry, R.D. (2005). Phenotypic convergence mediated by GGDEF-domain-containing proteins. J. Bacteriol. *187*, 6816–6823.

Singh, R., Paul, D., and Jain, R.K. (2006). Biofilms: Implications in bioremediation. Trends Microbiol. *14*, 389–397.

Smidsrød, O., and Skjak-Brk, G. (1990). Alginate as immobilization matrix for cells. Trends Biotech. *8*, 71–78.

Stanley, N., and Lazazzera, B. (2004). Environmental signals and regulatory pathways that influence biofilm formation. Mol. Microbiol. 52, 917–924.

Stephanopoulos, G., Aristidou, A., and Nielsen, J. (1998). Metabolic Engineering. (Academic Press, San Diego, Ca).

Stewart, P.S. (2003). Diffusion in biofilms. J. Bacteriol *185*, 1485–1491.

Stewart, P.S., and Franklin, M.J. (2008). Physiological heterogeneity in biofilms. Nat. Rev. Microbiol. *6*, 199–210.

Stoodley, P., Sauer, K., Davies, D.G., and Costerton, J.W. (2002). Biofilms as complex differentiated communities. Annual Rev. Microbiol. 56, 187–209.

Straathof, A.J.J., Panke, S., and Schmid, A. (2002). The production of fine chemicals by biotransformations. Curr. Opin. Biotech. *13*, 548–556.

Sutherland, I.W. (2001). Biofilm exopolysaccharides: A strong and sticky framework. Microbiology *147*, 3–9.

Syron, E., and Casey, E. (2008). Membrane-aerated biofilms for high rate biotreatment: Performance appraisal, engineering principles, scale-up, and development requirements. Environ. Sci. Technol. *42*, 1833–1844.

Vansonsbeek, H.M., Beeftink, H.H., and Tramper, J. (1993). 2-Liquid-phase bioreactors. Enzyme Microb. Technol. *15*, 722–729.

Videla, H.A., and Characklis, W,G. (1992). Biofouling and microbially influenced corrosion. Int. Biodeter. Biodeg. *29*, 195–212.

Vlamakis, H., Aguilar, C., Losick, R., and Kolter, R. (2008). Control of cell fate by the formation of an architecturally complex bacterial community. Genes Dev. *22*, 945–953.

Walton,A.Z., and Stewart, J.D. (2004). Understanding and improving NADPH-dependent reactions by nongrowing *Escherichia coli* cells. Biotech. Prog. *20*, 403–411.

Wang, Z-W., and Chen, S. (2009). Potential of biofilm-based biofuel production. Appl. Microbiol. Biotechnol. *83*, 1–18.

Wäsche, S., Horn, H., and Hempel, D.C. (2000). Mass transfer phenomena in biofilm systems. Water Sci. Tech. *41*, 357–360.

Wäsche, S., Horn, H., and Hempel, D.C. (2002). Influence of growth conditions on biofilm development and mass transfer at the bulk/biofilm interface. Water Res. *36*, 4775–4784.

Waters, C.M., and Bassler, B.L. (2005). Quorum sensing: Cell-to-cell communication in bacteria. Ann. Rev. Cell Dev. Biol. *21*, 319–346.

Webb, C. and Dervakos, G.A. (1996). Studies in viable cell immobilization. (Academic Press, San Diego, California, US).

Weuster-Botz, D., Aivasidis, A., and Wandrey, C. (1993). Continuous ethanol production by *Zymomonas mobilis* in a fluidized bed reactor. Part II: Process development for the fermentation of hydrolysed B-starch without sterilization. Appl. Microbiol. Biotech. *39*, 685–690.

Whiteley, M., Bangera, M.G., Bumgarner, R.E., Parsek, M.R., Teitzel, G.M., Lory, S., and Greenberg, E.P. (2001). Gene expression in *Pseudomonas aeruginosa* biofilms. Nature *413*, 860–864.

Willaert, R., and Nedovic, V.A. (2006). Primary beer fermentation by immobilised yeast – a review on flavour formation and control strategies. J. Chem. Tech. Biotechnology *81*, 1353–1367.

Wingender, J., Neu, R.T., and Flemming, H.C. (1999). What are bacterial extracellular polymeric substances? In Microbial Extracellular Polymeric Substances: Characterization, Structure, and Function, Wingender, J., Neu, R.T., and Flemming, H.C., editors. (Springer-Verlag, Berlin). pp. 1–19.

Wohlgemuth, R. (2009). The locks and keys to industrial biotechnology. Nat. Biotech. 25, 204–213.

Wöltinger, J., Karau, A., Leuchtenberger, W., and Drauz, K. (2005). Membrane reactors at Degussa. Adv. Biochem. Eng. Biotech. *92*, 289–316.

Woodley, J. (2006a). Choice of biocatalyst form for scalable processes. Biochem. Soc. Trans *34*, 301–303.

Woodley, J. (2006b). Microbial biocatalytic processes and their development. Adv. Appl. Microbiol. *60,* 1–15.

Woodley, J. (2008). New opportunities for biocatalysis: Making pharmaceutical processes greener. Trends Biotech. *26,* 321–327.

Woodley, J.M., and Lilly, M.D. (1990). Extractive biocatalysis – the use of 2-liquid phase biocatalytic reactors to assist product recovery. Chem. Eng. Sci. *45,* 2391–2396.

Xia, L., Chung, Y.K., Yang, S.T., and Yousef, A.E. (2005). Continuous Nisin production in laboratory media and whey permeate by immobilized *Lactococcus lactis.* Proc. Biochem. *40,* 13–24.

Xu, K.D., McFeters, G.A., and Stewart, P.S. (2000). Biofilm resistance to antimicrobial agents. Microbiol. *146,* 547–549.

Yildiz, F.H., and Schoolnik, G.K. (1999). *Vibrio cholerae* O1 El Tor: Identification of a gene cluster required for the rugose colony type, exopolysaccharide production, chlorine resistance, and biofilm formation. Proc. Natl. Adac. Sci. *96,* 4028–4033.

Zhang, S., Splendiani, A., dos Santos, L.M., and Livingston, A.G. (1998). Determination of pollutant diffusion coefficients in naturally formed biofilms using a single tube extractive membrane bioreactor. Biotechnol Bioeng. *59,* 80–89.

Zhang, S., Norrlow, O., Wawrzynczyk, J., and Dey, E.S. (2004). Poly(3-hydroxybutyrate) biosynthesis in the biofilm of *Alcaligenes eutrophus,* using glucose enzymatically released from pulp fiber sludge. Appl. Environ. Microbiol. *70,* 6776–6782.

Zhao, H.M., and van der Donk, W.A. (2003). Regeneration of cofactors for use in biocatalysis. Curr. Opin. Biotech. *14,* 583–589.

Zhou, B., Martin, G.J.O., and Pamment, N.B. (2008.) Increased phenotypic stability and ethanol tolerance of recombinant *Escherichia coli* KO11 when immobilized in continuous fluidized bed culture. Biotech. Bioeng *100,* 627–633.

Index

I

K

L

M

N

O

www.ingramcontent.com/pod-product-compliance
Lightning Source LLC
LaVergne TN
LVHW062123200926
848484LV00007B/1746
* 9 7 8 1 9 0 4 4 5 5 9 6 7 *